INTERNATIONAL CENTRE FOR MECHANICAL SCIENCES

COURSES AND LECTURES - No. 304

UNILATERAL PROBLEMS IN STRUCTURAL ANALYSIS - 2

PROCEEDINGS OF THE SECOND MEETING
ON UNILATERAL PROBLEMS IN
STRUCTURAL ANALYSIS
PRESCUDIN, JUNE 17-20, 1985

EDITED BY

G. DEL PIERO
UNIVERSITA' DI UDINE

F. MACERI
II UNIVERSITA' DI ROMA

Springer-Verlag Wien GmbH

Le spese di stampa di questo volume sono in parte coperte da contributi
del Consiglio Nazionale delle Ricerche.

This volume contains 86 illustrations.

Originally published by CISM, Udine in 1987.

ISBN 978-3-211-82036-0 ISBN 978-3-7091-2967-8 (eBook)
DOI 10.1007/978-3-7091-2967-8

PREFACE

This Volume collects the texts of the contributions to the third Meeting on Unilateral Problems in Structural Analysis, held at Villa Emma in the Prescudin Valley, near Udine, on June 17 to 20, 1985.

The first two meetings took place at Villa Emma, on May 1982, and at Ravello, on the Amalfi Coast in Southern Italy, on September 1983. The first one was very informal in character, but resulted in an important occasion of contact between researchers coming from different schools and pursuing different approaches to unilateral problems in Mechanics and Engineering.

The success of the first meeting induced us to organize a second one, with a more formal structure and with a larger participation. In that occasion, the discussion was very lively and fruitful, and the idea came of transforming the presented contributions into written communications, and of collecting them in a volume of Proceedings. The volume appeared in 1985, and became soon a reference book for researchers working in the area. The same formula was adopted for the third meeting. Once again, the publication of the proceedings was decided at the end of the meeting, and this explains in part the delay in the issue of the present book.

The communications collected here deal, mainly with developments of subjects already treated in the first volume, such as unilateral contact with friction, adhesion, and dynamical effects, or unilateral internal constraints and non-convex constitutive equations. In part, they deal with new topics, such as unilateral analytical mechanics and probabilistic approach to structural analysis in the presence of unilateral constraints.

A look at the contents of this Volume puts in evidence that for a good part of the topics a firm theoretical ground has been reached. For them, the interest shifts to the development of appropriate solution techniques. On the other hand, there are subjects for which the theoretical formulation is not yet satisfactory. In both cases, there is still much work to be done, and the whole area seems to be in a fast evolution. This is confirmed by the success of other recent meetings, as the ISIMM minisymposium on Unilateral Problems in Mechanics, held at the Second University of Rome on April 1987, and the CISM Course "Non-Smooth Mechanics and Applications", held at Udine on September 1987.

Returning to our Meeting, it is our wish to thank the CISM for the support in the organization, and personally Mrs. Elsa Venir for her active and efficient presence at Villa Emma. Also acknowledged are the sponsorships of the University of Udine and of the Second University of Rome, and of the Azienda delle Foreste della Regione Autonoma Friuli-Venezia Giulia, whose hospitality in the severe and beautiful Park of the Prescudin Valley gave us the opportunity of developing a peaceful and constructive work.

Gianpietro Del Piero, Franco Maceri

Contents

Page

AN ANALYSIS OF THE UNILATERAL CONTACT PROBLEM WITH FRICTION OF BEAMS AND PLATES ON AN ELASTIC HALF-SPACE

L. Ascione, D. Bruno
Dipartimento di Strutture, University of Calabria

Sommario. In questo lavoro si sviluppa una analisi del problema di contatto con attrito di alcuni modelli monodimensionali di travi e piastre circolari vincolati monolateralmente su un semispazio elastico. Vengono innanzitutto richiamati gli aspetti fondamentali del modello continuo, sviluppato in un precedente lavoro. Si fornisce quindi un modello agli elementi finiti e si presentano alcuni risultati numerici che mostrano l'influenza del fenomeno di attrito sulla natura dell'interazione suolo-struttura.

Summary. In this paper we analyze the unilateral contact problem with friction of some one-dimensional models, like beams and circular plates, constrained on an elastic half-space. Firstly, the main features of the continuous model, which has been developed in a previous work, are discussed. Then a finite element model is presented and some numerical results, which show the influence of the friction on the nature of the soil-foundation interaction, are given.

1. INTRODUCTION

In recent years the contact problems with friction have been an active subject of research.

From a theoretical point of view, the equilibrium problem of an elastic body, unilaterally constrained with Coulomb's friction against a rigid obstacle, is still an open problem [1]. In fact, the question of existence of solutions of the friction problem has an answer only for some very special cases [2]. As pointed out by Duvaut [3], the source of difficulties in establishing an existence theory for Signorini's problem with Coulomb friction is the lack of smoothness of the normal contact pressure.

A way to overcome this obstacle is to introduce some special friction laws, like that put forth by Oden and Pires [4 - 5].

In the engineering field the unilateral contact problems with friction between structures and soils have been less investigated and only a few numerical results are available in literature [6].

The main objective of this paper is to develop a numerical investigation on this subject. The friction law, that we take into consideration, can be seen as a regularization of the classical Coulomb law. Such friction law allows us to develop a variational formulation of the problem in terms of the plate displacement parameters and of the interaction forces between plate and half-space. This formulation is very useful in view of numerical applications, since in this case it is possible to express the half-space displacements in terms of the surface forces. In fact, as well-known, the inverse relations cannot be put in an explicit form.

The same type of formulation, applied to the study of frictionless contact problems between plates and half-space, is utilized in [7 - 8].

The main mathematical aspects relative to this approach (existence, uniqueness and approximations) have been discussed in a previous paper [9].

In the present one we utilize those results to analyze from a numerical point of view some one-dimensional examples, like beams and axisymmetric circular plates unilaterally constrained with friction against a homogeneous elastic half-space.

2. FORMULATION OF THE PROBLEM

With reference to Fig. 1, let the symbols that follow denote:

Ω : the open region corresponding to the middle plane of the plate,

$\underset{\sim}{u} = (u_x, u_y, u_z)$: the plate displacement,

$\underset{\sim}{q} = (q_x, q_y, q_z)$: the load acting on the upper surface Ω^- of the plate,

$\underset{\sim}{\sigma} = (\sigma_x, \sigma_y, \sigma_z)$: the interaction between plate and foundation,

$\underset{\sim}{\tau} = (\sigma_x, \sigma_y, 0)$: the tangential component of $\underset{\sim}{\sigma}$,

$\underset{\sim}{u}_f = (u_f, v_f, w_f)$: the displacement of the surface of the foundation,

$\underset{\sim}{\delta} = (\delta_x, \delta_y, \delta_z)$: the relative displacement between plate and foundation ($\underset{\sim}{\delta} = \underset{\sim}{u}_f - \underset{\sim}{u}$),

$\bar{\underset{\sim}{\delta}} = (\delta_x, \delta_y, 0)$: the tangential component of $\underset{\sim}{\delta}$.

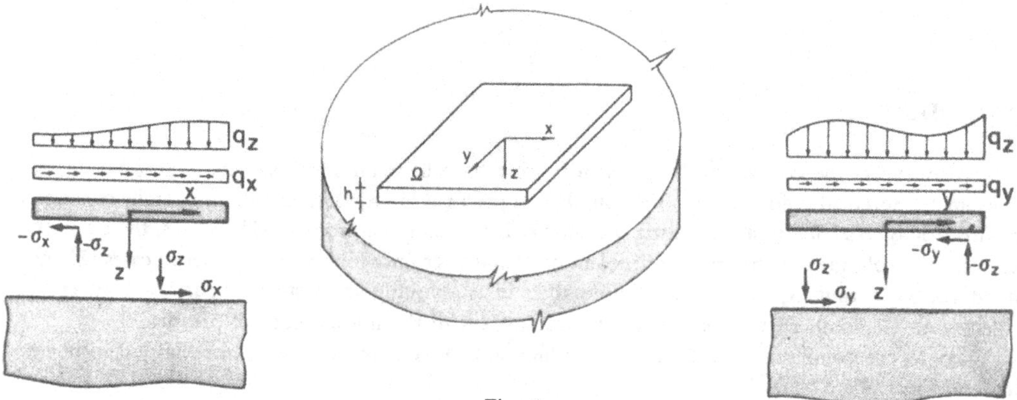

Fig. 1

According to Mindlin's theory [10], the plate displacement components can be put in the form:

$$u_x = u_s(x,y) - z\,\psi_x(x,y), \tag{2.1a}$$

$$u_y = v_s(x,y) - z\,\psi_y(x,y), \tag{2.1b}$$

$$u_z = w(x,y) \tag{2.1c}$$

where u_s and v_s correspond to the plate stretching and ψ_x, ψ_y represent the bending slopes along the x and y axes.

Denoting by $\underset{\sim}{V} = \{\underset{\sim}{v} = (u_s, v_s, w, \psi_x, \psi_y),\ \underset{\sim}{v} \in (H^1(\Omega))^5\}$ the space of the plate admissible displacements, the equilibrium equations of the plate can be written in the following variational form (virtual work equation):

$$< \underset{\sim}{A}\ \underset{\sim}{v}, \delta\underset{\sim}{v} > - < \underset{\sim}{R}^-\ \underset{\sim}{q}, \delta\underset{\sim}{v} > + < \underset{\sim}{R}^+\ \underset{\sim}{\sigma}, \delta\underset{\sim}{v} > = 0, \tag{2.2}$$

$$\forall\ \delta\underset{\sim}{v} \in \underset{\sim}{V},$$

where $<\bullet\ ,\ \bullet>$ is the duality pairing .

In eq. (2.2) the linear operator $\underset{\sim}{A} : \underset{\sim}{V} \rightarrow \underset{\sim}{V}'$, mapping the space $\underset{\sim}{V}$ into its dual $\underset{\sim}{V}'$ (the load space), represents the elastic equilibrium operator of the plate. As well known, operator $\underset{\sim}{A}$ is continuous; moreover, because the space $\underset{\sim}{V}$ includes plate rigid body displacements it is positive semidefinite:

$$\|\underset{\sim}{A}\ \underset{\sim}{v}\|_{\underset{\sim}{V}'} \le c\ \|\underset{\sim}{v}\|_{\underset{\sim}{V}}, \tag{2.3a}$$

$$< \underset{\sim}{A}\ \underset{\sim}{v}, \underset{\sim}{v} > \ge 0\ , \quad c \ge 0 \quad , \forall\ \underset{\sim}{v} \in \underset{\sim}{V}. \tag{2.3b}$$

The operators $\underset{\sim}{R}^-$ and $\underset{\sim}{R}^+$, defined as:

$$\underset{\sim}{R}^-: \underset{\sim}{q} = (q_x, q_y, q_z) \in (H^1(\Omega)')^3 \rightarrow (q_x, q_y, q_z, \frac{h}{2} q_x, \frac{h}{2} q_y) \in \underset{\sim}{V}', \tag{2.4a}$$

$$\underset{\sim}{R}^+: \underset{\sim}{\sigma} = (\sigma_x, \sigma_y, \sigma_z) \in (H^1(\Omega)')^3 \rightarrow (\sigma_x, \sigma_y, \sigma_z, -\frac{h}{2} \sigma_x, -\frac{h}{2} \sigma_y) \in \underset{\sim}{V}', \tag{2.4b}$$

map the surface forces $\underset{\sim}{q}$ and $\underset{\sim}{\sigma}$ into the generalized loads corresponding to the kinematical variables $\underset{\sim}{v} \in \underset{\sim}{V}$.

Further we define the linear operator $\underset{\sim}{T}$ which gives the displacements $\underset{\sim}{u}_f$ in the part Ω of the half-space surface, as a function of the loads $\underset{\sim}{\sigma}$:

$$\underset{\sim}{u}_f = \underset{\sim}{T}\ \underset{\sim}{\sigma}. \tag{2.5}$$

The displacement $\underset{\sim}{u}_f$ are boundary values (traces) of the elastic half-space displacement and therefore can be defined as elements of the space $\underset{\sim}{V}_f = V_f^3 = (H^{1/2}(\Omega))^3$. Consequently, the interactions $\underset{\sim}{\sigma}$ can be taken as elements of the dual space $\underset{\sim}{V}_f'$. The operator $\underset{\sim}{T}$ is continuous and coercive [11], i.e.:

$$\|\underset{\sim}{T}\ \underset{\sim}{\sigma}\|_{\underset{\sim}{V}_f} \le c_f \|\underset{\sim}{\sigma}\|_{\underset{\sim}{V}_f'} \tag{2.6a}$$

$$< \underset{\sim}{\sigma}, \underset{\sim}{T}\ \underset{\sim}{\sigma} > \ \ge c_f' \ \|\underset{\sim}{\sigma}\|^2_{\underset{\sim}{V}_f'}\ , \quad c_f, c_f' \ge 0\ , \quad \forall\ \underset{\sim}{\sigma} \in \underset{\sim}{V}_f'\ . \tag{2.6b}$$

In the case of a homogeneous elastic half-space, if the pressures $\underset{\sim}{\sigma}$ are sufficiently regular, the operator $\underset{\sim}{T}$ can be represented in the well-known Cerruti-Boussinesq's integral form [12], which we will use later in numerical applications.

Firstly, the unilateral contact hypotesis implies that the pressure σ_z be positive, that is the admissible interactions $\underset{\sim}{\sigma}$ between plate and half-space are elements of the cone $\underset{\sim}{\mathbb{K}}'$ in $\underset{\sim}{V}_f'$:

$$\underset{\sim}{\mathbb{K}}' = \{ \underset{\sim}{\sigma} = (\sigma_x, \sigma_y, \sigma_z) \in \underset{\sim}{V}_f' : < \sigma_z, w_f > \ \geq 0 \ , \ \forall\, w_f \geq 0 \}. \tag{2.7}$$

Moreover, the following two conditions:

$$\delta_z = w_f - w \geq 0 \tag{2.8a}$$

$$< \sigma_z, \delta_z > \ = 0 \tag{2.8b}$$

must be satisfied.

Now we assume that a relative tangential displacement $\underset{\sim}{\bar{\delta}} = (\delta_x, \delta_y, 0)$ occurs at the interface Ω^+ when a nonzero interaction $\underset{\sim}{\tau}$ is acting:

$$\underset{\sim}{\bar{\delta}} = \underset{\sim}{f}_\epsilon(\underset{\sim}{\sigma}) = \epsilon \underset{\sim}{J}\underset{\sim}{\tau} + \frac{1-\epsilon^2}{\epsilon} (|\underset{\sim}{J}\,\underset{\sim}{\tau}| - f J\sigma_z)^+ \frac{\underset{\sim}{J}\,\underset{\sim}{\tau}}{|\underset{\sim}{J}\,\underset{\sim}{\tau}|}, \tag{2.9}$$

where we denote by

J : the canonical isometry from V_f' onto V_f ,

$\underset{\sim}{J}$: the canonical isometry from $\underset{\sim}{V}_f'$ onto $\underset{\sim}{V}_f$,

$|\underset{\sim}{u}| = (u_x^2 + u_y^2 + u_z^2)^{1/2}$, $\forall\, \underset{\sim}{u} \in \underset{\sim}{V}_f$,

$(x)^+ = \max\{0,x\}$ for each real number x,

ϵ, f : two assigned positive constants.

Relationship (2.9) is depicted in Fig. 2 for a onedimensional case:

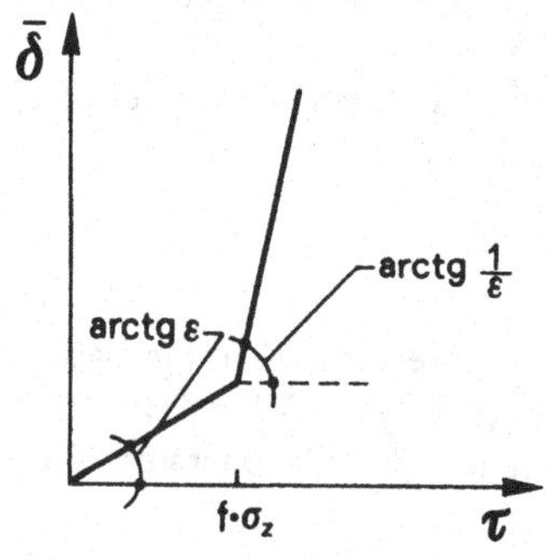

Fig. 2

It is easy to recognize that, for $\epsilon \to 0$, eq. (2.9) represents an approximation of the well-known Coulomb's friction law with coefficient of friction equal to f. We observe that the presence of a nonzero relative displacement $\bar{\delta}$, given by eq. (2.9) for any value of the tangential stress τ, has a realistic physical meaning as pointed out in [5].

After these preliminaries, our problem can be formulated as follows:

"Find $(v, \sigma) \in V \times \mathbb{K}'$ such that:

$$< A\, v - R^- g + R^+ \sigma, \delta v > = 0, \tag{2.10a}$$

$$< \delta\sigma - \sigma, R^{+\prime} v - T\sigma - f_\epsilon(\sigma) > \leq 0, \tag{2.10b}$$

$\forall\, (\delta v, \delta\sigma) \in V \times \mathbb{K}'$",

$R^{+\prime}$ being the dual operator of R^+.

We recall that eq. (2.10a) represents the virtual work equation; further it is easy to show that eq. (2.10b) is equivalent to the contact conditions (2.8) and (2.9).

3. EXISTENCE AND UNIQUENESS OF THE SOLUTION

Of central importance in the study of problem (2.10) is the following auxiliary problem corresponding to the hypothesis that the friction law is independent of the unknown normal stress σ_ν, i.e.:

$$\bar{\delta} = f_{o\epsilon}(\sigma) = \epsilon J \tau + \frac{1-\epsilon^2}{\epsilon} (|J\tau| - \tau_o)^+ \frac{J\tau}{|J\tau|}, \tag{3.1}$$

where $\tau_o \geq 0$ is an assigned element of the positive cone $\mathbb{K} \subseteq V_f$.

It is not difficult to recognize that the nonlinear operator $f_{o\epsilon}$ is the weak differential of the quadratic functional:

$$\phi_{o\epsilon}(\sigma) = \frac{1}{2} \epsilon \| J\tau \|_o^2 + \frac{1-\epsilon^2}{2\epsilon} \| (|J\tau| - \tau_o)^+ \|_o^2, \tag{3.2}$$

$\|\cdot\|_o$ being the L^2-norm.

The auxiliary problem can be formulated by means of the same equations (2.10), substituting to f_ϵ the operator $f_{o\epsilon}$ defined by eq. (3.1).

We observe now that the auxiliary problem is equivalent to the stationarity condition of the saddle functional [9]:

$$F_o(v, \sigma) = \frac{1}{2} < A\, v, v > - \frac{1}{2} < \sigma, T\sigma > - < R^- g, v > + < R^+ \sigma, v > + \\ - \phi_{o\epsilon}(\sigma), \tag{3.3}$$

defined on the convex set $V \times \mathbb{K}'$.

The following theorem holds:

Theorem 1

Let $\underset{\sim}{\mathcal{R}} \subseteq \underset{\sim}{V}$ be the subspace of the plate rigid displacement fields. Assume that the external loads $\underset{\sim}{q}$ are such that:

$$\exists\, \underset{\sim}{\sigma} \in \underset{\sim}{K}' : < \underset{\sim}{R}^{+} \underset{\sim}{\sigma} - \underset{\sim}{R}^{-} \underset{\sim}{q},\ \underset{\sim}{v}^{R} > \ = 0, \quad \forall\, \underset{\sim}{v}^{R} \in \underset{\sim}{\mathcal{R}}. \tag{3.4}$$

Then there exists a unique element $(\underset{\sim}{v}_{o}, \underset{\sim}{\sigma}_{o}) \in \underset{\sim}{V} \times \mathbb{K}'$ at which the functional F_o is stationary.

We can now define the operator:

$$\underset{\sim}{\mathcal{B}}: \tau_o \in \mathbb{K} \rightarrow f\, J\, \sigma_z^{(0)} \in \mathbb{K}\,, \tag{3.5}$$

where $\sigma_z^{(0)}$ is the vertical component of $\underset{\sim}{\sigma}_o$.

The following properties hold:

Theorem 2

i) The operator $\underset{\sim}{\mathcal{B}}$ is weakly continuous,
ii) there exists a bounded and closed ball B_r with radius r in the space V_f such that the operator $\underset{\sim}{\mathcal{B}}$ maps the set $B_r \cap \mathbb{K}$ into itself.

Two statements i) and ii) of Theorem 2 allow us to recognize the existence of at least one solution of Problem (2.10), which is assured by the classical Shauder-Tychonoff fixed point theorem [13]. In fact, it is easy to observe that the solution of the auxiliary problem, corresponding to a fixed point τ_o of $\underset{\sim}{\mathcal{B}}$, is also a solution of our problem (2.10).

Moreover, if f is sufficiently small, it is possible to show that operator $\underset{\sim}{\mathcal{B}}$ is a contraction mapping and then the uniqueness of the solution is assured.

The proofs of previous results are given in [9].

3.1. Approximations

We observe now that the main difficulty related with the previous variational formulation (3.3) consists in the constraint conditions which define the convex set $\underset{\sim}{V} \times \mathbb{K}'$.

In order to overcome this difficulty we will use a penalty approach. More precisely we introduce the exterior penalty functional:

$$\frac{1}{\eta} P(\underset{\sim}{\sigma}) = \frac{1}{2\eta} \|\sigma_z^{-}\|^2_{V_f'}, \tag{3.6}$$

where σ_z^{-} is the negative part of σ_z , defined as the projection of σ_z on the negative cone of V_f'.

Then, the penalized functional is:

$$F_{o\eta}(\underset{\sim}{v}, \underset{\sim}{\sigma}) = F_o(\underset{\sim}{v}, \underset{\sim}{\sigma}) - \frac{1}{\eta} P(\underset{\sim}{\sigma}) \tag{3.7}$$

and the approximate solution is obtained by finding the stationary point of the saddle functional $F_{o\eta}$, i.e. by solving the problem:

"Find $(\underset{\sim}{v}_\eta, \underset{\sim}{\sigma}_\eta) \in \underset{\sim}{V} \times \underset{\sim}{V}'_f$ such that:

$$\delta F_{o\eta}(\underset{\sim}{v}_\eta, \underset{\sim}{\sigma}_\eta; \delta \underset{\sim}{v}, \delta \underset{\sim}{\sigma}) = 0 \tag{3.8}$$

$\forall (\delta \underset{\sim}{v}, \delta \underset{\sim}{\sigma}) \in \underset{\sim}{V} \times \underset{\sim}{V}'_f$"

$\delta F_{o\eta}$ being the weak differential of $F_{o\eta}$ at $(\underset{\sim}{v}, \underset{\sim}{\sigma})$ with increment $(\delta \underset{\sim}{v}, \delta \underset{\sim}{\sigma})$.

As far as uniqueness and convergence of the penalized solutions are concerned, we refer [14] to the following theorem.

Theorem 3

If the assumption of Theorem 1 is satisfied, the penalized solutions $(\underset{\sim}{v}_{o\eta}, \underset{\sim}{\sigma}_{o\eta})$ converge strongly to the solution $(\underset{\sim}{v}_o, \underset{\sim}{\sigma}_o)$ of the auxiliary problem, i.e.:

$$\lim_{\eta \to 0} \| \underset{\sim}{v}_{o\eta} - \underset{\sim}{v}_o \|_{\underset{\sim}{V}} = \lim_{\eta \to 0} \| \underset{\sim}{\sigma}_{o\eta} - \underset{\sim}{\sigma}_o \|_{\underset{\sim}{V}'_f} = 0. \tag{3.9}$$

In [14] the same penalty formulation was utilized to study a frictionless unilateral contact problem. It is interesting to observe that convergence is assured without any stability condition contrary to the case of a rigid obstacle [15,16].

The last result that we give in this section concerns the finite element approximation. The following theorem holds:

Theorem 4

Assume that the family of subspaces $\{\underset{\sim}{V}_h \times \underset{\sim}{V}'_{fh}\}_{0 \le h \le 1}$ is complete in $\underset{\sim}{V} \times \underset{\sim}{V}'_f$, and that the external loads can be equilibrated by interactions $\underset{\sim}{\sigma}_h \in \mathbb{K}' \cap \underset{\sim}{V}'_{fh}$. Then there exists a unique solution $(\underset{\sim}{v}^h_{o\eta}, \underset{\sim}{\sigma}^h_{o\eta})$ of the problem (3.8) in the subspace $\underset{\sim}{V}_h \times \underset{\sim}{V}'_{fh}$, and:

$$\lim_{h \to 0} \| \underset{\sim}{v}^h_{o\eta} - \underset{\sim}{v}_{o\eta} \|_{\underset{\sim}{V}'} = \lim_{h \to 0} \| \underset{\sim}{\sigma}^h_{o\eta} - \underset{\sim}{\sigma}_{o\eta} \|_{\underset{\sim}{V}'_f} = 0. \tag{3.10}$$

The proof of this theorem is given in [9].

4. APPLICATIONS

Here we utilize the results referred to in the previous section to study the two simple structural examples shown in Fig. 3:

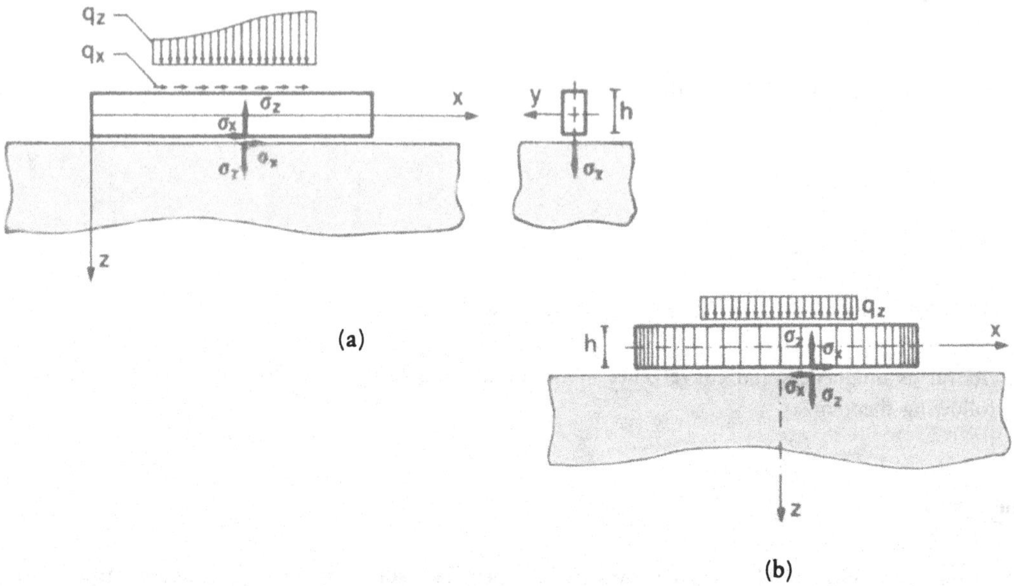

Fig. 3

The first corresponds to a Timoshenko beam, the second one to a Mindlin axisymmetric circular plate.

In order to get finite element approximations of the above examples, we give the expression of all terms appairing in the functional $F_{\sigma\eta}$. Putting:

C : extensional stiffness

D : flexural stiffness

k : shear stiffness

ν : Poisson's ratio of the plate (beam),

ν_f : Poisson's ratio of the elastic half-space,

G : shear modulus of the plate (beam),

G_f : shear modulus of the elastic half-space,

we have:

BEAM

$$\frac{1}{2} < \underset{\sim}{A}\, \underset{\sim}{v},\ \underset{\sim}{v} > = \frac{1}{2}\int_0^\ell C\, u_s'^2\, dx + \frac{1}{2}\int_0^\ell D\psi'^2\, dx + \frac{1}{2}\int_0^\ell k\,(w' - \psi)^2\, dx\ ;$$

$$\frac{1}{2} < \underset{\sim}{\sigma},\ \underset{\sim}{T}\, \underset{\sim}{\sigma} > = \frac{1}{2}\int_0^\ell (\sigma_x u_f + \sigma_z w_f)\, dx\ ,$$

where:

$$u_f(x) = \frac{1}{4\pi G_f}\left\{ \int_0^\ell \left[\frac{2(1-\nu_f)}{d} + \frac{2\nu_f (x-x')^2}{d^3}\right] \cdot \sigma_x(x')\, dx' - \int_0^\ell \frac{(1-2\nu_f)(x-x')}{d^2}\, \sigma_z(x')\, dx' \right\},$$

$$w_f(x) = \frac{1}{4\pi G_f}\left\{ \int_0^\ell \frac{(1-2\nu_f)(x-x')}{d^2}\, \sigma_x(x')\, dx' + \int_0^\ell \frac{2(1-\nu_f)}{d}\, \sigma_z(x')\, dx' \right\},$$

and

$$d = |x - x'|\ ;$$

$$< \underset{\sim}{R}^-\, \underset{\sim}{q},\ \underset{\sim}{v} > = \int_0^\ell \left(q_x u_s + \frac{h}{2}\, q_x \psi + q_z w\right) dx\ ;$$

$$< \underset{\sim}{R}^+\, \underset{\sim}{\sigma},\ \underset{\sim}{v} > = \int_0^\ell \left(\sigma_x u_s - \frac{h}{2}\, \sigma_x \psi + \sigma_z w\right) dx\ ;$$

$$\phi_{o\epsilon}(\underset{\sim}{\sigma}) = \frac{1}{2}\int_0^\ell \left\{ \epsilon\sigma_x^2 + \frac{1-\epsilon^2}{\epsilon}\left[(|\sigma_x| - \tau_o)^+\right]^2 \right\} dx;$$

$$P(\underset{\sim}{\sigma}) = \frac{1}{2}\int_0^\ell [\sigma_z^-]^2\, dx.$$

CIRCULAR PLATE

$$\frac{1}{2} < A\underset{\sim}{v},\ \underset{\sim}{v} > = \pi \int_0^\ell \Big\{ C\Big(u_s'^2 + 2\,\frac{\nu}{x}\, u_s' u_s + \frac{1}{x^2}\, u_s^2\Big) + D\Big(\psi'^2 + 2\,\frac{\nu}{x}\,\psi\psi' + \frac{1}{x^2}\psi^2\Big) + k\,(w' - \psi)^2 \Big\}\, x dx\ ; \qquad (4\text{-}1)$$

$$\frac{1}{2} < \underset{\sim}{\sigma},\ \underset{\sim}{T}\, \underset{\sim}{\sigma} > = \pi \int_0^\ell (\sigma_x u_f + \sigma_z w_f)\, x dx\ , \qquad (4\text{-}2)$$

where:

$$u_f(x) = \frac{1}{4\pi G_f}\left\{ \int_0^\ell \int_0^{2\pi} \left[\frac{2(1-\nu_f)\cos\vartheta}{d} + \frac{2\nu_f}{d^3} (x^2\cos\vartheta + x'^2\cos\vartheta - 2xx'\cos^2\vartheta - xx'\sin^2\vartheta)\right] \cdot \sigma_x(x')\, x' dx' d\vartheta - \int_0^\ell \int_0^{2\pi} \frac{1-2\nu_f}{d^2}\, (x - x'\cos\vartheta) \cdot \sigma_z(x') x' dx' d\vartheta \right\}, \qquad (4\text{-}3)$$

$$w_f(x) = \frac{1}{4\pi G_f}\left\{ \int_0^\ell \int_0^{2\pi} \frac{1-2\nu_f}{d^2}\, (x\cos\vartheta - x')\, \sigma_x(x') x' dx' d\vartheta + \int_0^\ell \int_0^{2\pi} \frac{2(1-\nu_f)}{d}\, \sigma_z(x') x' dx' d\vartheta \right\}, \qquad (4\text{-}4)$$

and

$$d = (x^2 + x'^2 - 2xx'\cos\vartheta)^{1/2}; \qquad (4\text{-}5)$$

$$< \underset{\sim}{R}^-\, \underset{\sim}{q},\ \underset{\sim}{v} > = 2\pi \int_0^\ell \left(q_x u_s + \frac{h}{2}\, q_x \psi + q_z w\right) x dx\ ; \qquad (4\text{-}6)$$

$$< \underset{\sim}{R}^+\, \underset{\sim}{\sigma},\ \underset{\sim}{v} > = 2\pi \int_0^\ell \left(\sigma_x u_s - \frac{h}{2}\, \sigma_x \psi + \sigma_z w\right) x dx\ ; \qquad (4\text{-}7)$$

$$\phi_{o\epsilon}(\underset{\sim}{\sigma}) = \pi \int_0^\ell \left\{ \epsilon\sigma_x^2 + \frac{1-\epsilon^2}{\epsilon}\left[(|\sigma_x| - \tau_o)^+\right]^2 \right\} x dx\ ; \qquad (4\text{-}8)$$

$$P(\underset{\sim}{\sigma}) = \pi \int_0^\ell [\sigma_z^-]^2\, x dx. \qquad (4\text{-}9)$$

4.1. Finite element approximation

With reference to Fig. 4 we discretize the region $[0,\ell]$ into N_e finite elements of length ℓ_e.

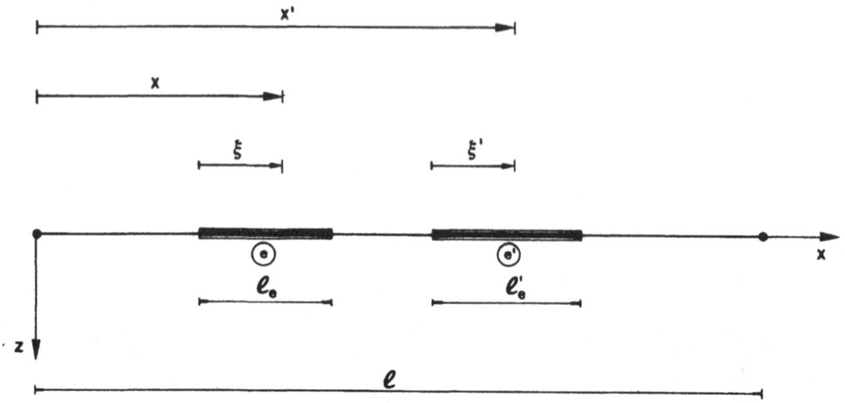

Fig. 4

Over each element the unknown functions u_s, w, ψ, σ_x, σ_z are interpolated as follows:

$$u_s^{(e)}(\xi) = \sum_\alpha u_\alpha^{(e)} f_{u\alpha}^{(e)}(\xi) = \underset{\sim}{F}_u^T \underset{\sim}{u}^{(e)}, \tag{4.10a}$$

$$w^{(e)}(\xi) = \sum_\beta w_\beta^{(e)} f_{w\beta}^{(e)}(\xi) = \underset{\sim}{F}_w^T \underset{\sim}{w}^{(e)}, \tag{4.10b}$$

$$\psi^{(e)}(\xi) = \sum_\gamma \psi_\gamma^{(e)} f_{\psi\gamma}^{(e)}(\xi) = \underset{\sim}{F}_\psi^T \underset{\sim}{\psi}^{(e)}, \tag{4.10c}$$

$$\sigma_x^{(e)}(\xi) = \sum_\lambda r_{x\lambda}^{(e)} f_{x\lambda}^{(e)}(\xi) = \underset{\sim}{F}_x^T \underset{\sim}{r}_x^{(e)}, \tag{4.10d}$$

$$\sigma_z^{(e)}(\xi) = \sum_\lambda r_{z\lambda}^{(e)} f_{z\lambda}^{(e)}(\xi) = \underset{\sim}{F}_z^T \underset{\sim}{r}_z^{(e)}, \tag{4.10e}$$

Consequently, the terms of functional $F_{\sigma\eta}$ are expressed in the following discrete form:

$$\frac{1}{2} < \underset{\sim}{A}\,\underset{\sim}{v}, \underset{\sim}{v} > = \frac{1}{2}\sum_e \underset{\sim}{v}^{(e)T} \underset{\sim}{K}^{(e)} \underset{\sim}{v}^{(e)} = \frac{1}{2} \underset{\sim}{v}^T \underset{\sim}{K}\,\underset{\sim}{v}, \tag{4.11}$$

where:

$$\underset{\sim}{v}^{(e)} = \begin{bmatrix} \underset{\sim}{u}^{(e)} \\ \underset{\sim}{\psi}^{(e)} \\ \underset{\sim}{w}^{(e)} \end{bmatrix} \qquad \underset{\sim}{K}^{(e)} = \begin{bmatrix} \underset{\sim}{K}_u^{(e)} & \underset{\sim}{0} & \underset{\sim}{0} \\ & \underset{\sim}{K}_\psi^{(e)} & \underset{\sim}{K}_{\psi w}^{(e)} \\ \text{sym} & & \underset{\sim}{K}_w^{(e)} \end{bmatrix} \tag{4.12}$$

being:

$$\underset{\sim}{K}_u^{(e)} = \int_0^{\ell_e} C\, \underset{\sim}{F}_u' \underset{\sim}{F}_u'^T d\xi ,$$

$$\underset{\sim}{K}_u^{(e)} = 2\pi C \Big\{ \int_0^{\ell_e} \underset{\sim}{F}_u' \underset{\sim}{F}_u'^T x d\xi + \nu \int_0^{\ell_e} (\underset{\sim}{F}_u \underset{\sim}{F}_u'^T + \underset{\sim}{F}_u' \underset{\sim}{F}_u^T) d\xi + \int_0^{\ell_e} \frac{1}{x} \underset{\sim}{F}_u \underset{\sim}{F}_u^T d\xi \Big\}, \tag{4.13}$$

$$\underset{\sim}{K}_\psi^{(e)} = \int_0^{\ell_e} D\, \underset{\sim}{F}_\psi' \underset{\sim}{F}_\psi'^T d\xi + \int_0^{\ell_e} k\, \underset{\sim}{F}_\psi \underset{\sim}{F}_\psi^T d\xi ,$$

$$\underset{\sim}{K}_\psi^{(e)} = 2\pi \Big\{ D \Big[\int_0^{\ell_e} x\, \underset{\sim}{F}_\psi' \underset{\sim}{F}_\psi'^T d\xi + \nu \int_0^{\ell_e} (\underset{\sim}{F}_\psi \underset{\sim}{F}_\psi'^T + \underset{\sim}{F}_\psi' \underset{\sim}{F}_\psi^T) d\xi + \int_0^{\ell_e} \frac{1}{x} \underset{\sim}{F}_\psi \underset{\sim}{F}_\psi^T d\xi \Big] + k \int_0^{\ell_e} x\, \underset{\sim}{F}_\psi \underset{\sim}{F}_\psi^T d\xi \Big\} , \tag{4.14}$$

$$\underset{\sim}{K}_w^{(e)} = \int_0^{\ell_e} k\, \underset{\sim}{F}_w' \underset{\sim}{F}_w'^T d\xi ,$$

$$\underset{\sim}{K}_w^{(e)} = 2\pi k \int_0^{\ell_e} x\, \underset{\sim}{F}_w' \underset{\sim}{F}_w'^T d\xi , \tag{4.15}$$

$$\underset{\sim}{K}_{\psi w}^{(e)} = - \int_0^{\ell_e} k\, \underset{\sim}{F}_\psi \underset{\sim}{F}_w'^T d\xi ,$$

$$\underset{\sim}{K}_{\psi w}^{(e)} = - 2\pi k \int_0^{\ell_e} x\, \underset{\sim}{F}_\psi \underset{\sim}{F}_w'^T d\xi , \tag{4.16}$$

and, $(\ )' = d(\ )/d\xi$;

$$\frac{1}{2} < \underset{\sim}{\sigma}, \underset{\sim}{T}\, \underset{\sim}{\sigma} > = \frac{1}{2} \sum_{e,e'} \underset{\sim}{r}^{(e)T} \underset{\sim}{T}^{(e,e')} \underset{\sim}{r}^{(e')} = \frac{1}{2} \underset{\sim}{r}^T \underset{\sim}{T}\, \underset{\sim}{r} , \tag{4.17}$$

where:

$$\underset{\sim}{r}^{(e)} = \begin{bmatrix} \underset{\sim}{r}_x^{(e)} \\ \underset{\sim}{r}_z^{(e)} \end{bmatrix} , \quad \underset{\sim}{T}^{(e,e')} = \left[\begin{array}{c:c} \underset{\sim}{T}_{xx}^{(e,e')} & \underset{\sim}{T}_{xz}^{(e,e')} \\ \hdashline \underset{\sim}{T}_{zx}^{(e,e')} & \underset{\sim}{T}_{zz}^{(e,e')} \end{array} \right] \tag{4.18}$$

being

$$\underset{\sim}{T}_{xx}^{(e,e')} = \frac{1}{4\pi G_f} \int_0^{\ell_e} \int_0^{\ell'_e} c_{xx}(\xi, \xi') \underset{\sim}{F}_x \underset{\sim}{F}_x^T d\xi d\xi' ,$$

$$\underset{\sim}{T}_{xx}^{(e,e')} = \frac{1}{4\pi G_f} \int_0^{\ell_e} \int_0^{\ell'_e} \int_0^{2\pi} c_{xx}(\xi, \xi', \vartheta) \cdot \underset{\sim}{F}_x \underset{\sim}{F}_x^T d\xi d\xi' d\vartheta , \tag{4.19a}$$

$$\underset{\sim}{T}_{xz}^{(e,e')} = \frac{1}{4\pi G_f} \int_0^{\ell_e} \int_0^{\ell'_e} c_{xz}(\xi,\xi') \underset{\sim}{F}_x \underset{\sim}{F}_z^T d\xi d\xi' ,$$

$$\underset{\sim}{T}_{xz}^{(e,e')} = \frac{1}{4\pi G_f} \int_0^{\ell_e} \int_0^{\ell'_e} \int_0^{2\pi} c_{xz}(\xi, \xi', \vartheta) \cdot \underset{\sim}{F}_x \underset{\sim}{F}_z^T d\xi d\xi' d\vartheta , \tag{4.19b}$$

$$\underset{\sim}{T}_{zz}^{(e,e')} = \frac{1}{4\pi G_f} \int_0^{\ell_e} \int_0^{\ell'_e} c_{zz}(\xi,\xi') \underset{\sim}{F}_z \underset{\sim}{F}_z^T d\xi d\xi' ,$$

$$\underset{\sim}{T}_{zz}^{(e,e')} = \frac{1}{4\pi G_f} \int_0^{\ell_e} \int_0^{\ell'_e} \int_0^{2\pi} c_{zz}(\xi, \xi', \vartheta) \cdot \underset{\sim}{F}_z \underset{\sim}{F}_z^T d\xi d\xi' d\vartheta , \tag{4.19c}$$

$$\underset{\sim}{T}_{xx}^{(e,e')} = \frac{1}{4\pi G_f}\int_0^{\ell_e}\int_0^{\ell'_e} c_{xx}(\xi,\, \xi') \cdot \underset{\sim}{F}_x \underset{\sim}{F}_x^T \, d\xi d\xi' \,, \qquad \underset{\sim}{T}_{xx}^{(e,e')} = \frac{1}{4\pi G_f}\int_0^{\ell_e}\int_0^{\ell'_e}\int_0^{2\pi} c_{xx}(\xi,\, \xi',\, \vartheta) \cdot \underset{\sim}{F}_x \underset{\sim}{F}_x^T \, d\xi d\xi' d\vartheta \,, \tag{4.19d}$$

and:

$$c_{xx}(\xi,\, \xi') = \frac{2(1-\nu_f)}{d} + \frac{2\nu_f}{d} \,, \qquad c_{xx}(\xi,\xi',\vartheta) = xx'\left[\frac{2(1-\nu_f)}{d}\cos\vartheta + \frac{2\nu_f}{d^3} \cdot (x^2\cos\vartheta + x'^2\cos\vartheta - xx'\cos^2\vartheta - 1)\right] , \tag{4.20a}$$

$$c_{zz}(\xi,\, \xi') = \frac{2(1-\nu_f)}{d} \,, \qquad c_{zz}(\xi,\, \xi',\, \vartheta) = xx'\,\frac{2(1-\nu_f)}{d} \,, \tag{4.20b}$$

$$c_{xz}(\xi,\, \xi') = c_{zx}(\xi,\xi') = \frac{1-2\nu_f}{d^2}(x - x') ; \qquad c_{xz}(\xi,\, \xi',\vartheta) = c_{zx}(\xi,\xi',\vartheta) = \frac{1-2\nu_f}{d^2} \cdot (x\cos\vartheta - x')xx' \,; \tag{4.20c}$$

$$< \underset{\sim}{R}^{-}\underset{\sim}{q},\, \underset{\sim}{v}> \; = \sum_e \underset{\sim}{v}^{(e)T} \underset{\sim}{q}^{(e)} = \underset{\sim}{v}^T \underset{\sim}{q} \,, \tag{4.21}$$

where:

$$\underset{\sim}{q}^{(e)} = \begin{bmatrix} \underset{\sim}{q}_u^{(e)} \\ \underset{\sim}{q}_v^{(e)} \\ \underset{\sim}{q}_w^{(e)} \end{bmatrix} \tag{4.22}$$

being:

$$\underset{\sim}{q}_u^{(e)} = \int_0^{\ell_e} q_x \underset{\sim}{F}_u \, d\xi \,, \qquad \underset{\sim}{q}_u^{(e)} = 2\pi \int_0^{\ell_e} x q_x \underset{\sim}{F}_u \, d\xi \,, \tag{4.23a}$$

$$\underset{\sim}{q}_{x\psi}^{(e)} = \int_0^{\ell_e} \frac{h}{2}\, q_x\, \underset{\sim}{F}_\psi\, d\xi\,, \qquad \underset{\sim}{q}_{x\psi}^{(e)} = 2\pi \int_0^{\ell_e} x\, \frac{h}{2}\, q_x\, \underset{\sim}{F}_\psi\, d\xi\,, \tag{4.23b}$$

$$\underset{\sim}{q}_{xw}^{(e)} = \int_0^{\ell_e} q_x\, \underset{\sim}{F}_w\, d\xi\,, \qquad \underset{\sim}{q}_{xw}^{(e)} = 2\pi \int_0^{\ell_e} x\, q_x\, \underset{\sim}{F}_w\, d\xi\,, \tag{4.23c}$$

$$< \underset{\sim}{R}^-\, \underset{\sim}{\sigma},\, \underset{\sim}{v} > \; = \sum_e \underset{\sim}{r}^{(e)T}\, \underset{\sim}{M}^{(e)}\, \underset{\sim}{v}^{(e)} = \underset{\sim}{r}^T \underset{\sim}{M}\, \underset{\sim}{v}, \tag{4.24}$$

where:

$$\underset{\sim}{M}^{(e)} = \left[\begin{array}{c:c:c} \underset{\sim}{M}_{xu}^{(e)} & \underset{\sim}{M}_{x\psi}^{(e)} & \underset{\sim}{0} \\ \hdashline \underset{\sim}{0} & \underset{\sim}{0} & \underset{\sim}{M}_{xw}^{(e)} \end{array}\right], \tag{4.25}$$

being:

$$\underset{\sim}{M}_{xu}^{(e)} = \int_0^{\ell_e} \underset{\sim}{F}_x\, \underset{\sim}{F}_u^T\, d\xi\,, \qquad \underset{\sim}{M}_{xu}^{(e)} = 2\pi \int_0^{\ell_e} x\; \underset{\sim}{F}_x\, \underset{\sim}{F}_u^T\, d\xi\,, \tag{4.26a}$$

$$\underset{\sim}{M}_{x\psi}^{(e)} = -\int_0^{\ell_e} \underset{\sim}{F}_x\, \underset{\sim}{F}_\psi^T\, d\xi\,, \qquad \underset{\sim}{M}_{x\psi}^{(e)} = -2\pi \int_0^{\ell_e} x\, \frac{h}{2}\, \underset{\sim}{F}_x\, \underset{\sim}{F}_\psi^T\, d\xi\,, \tag{4.26b}$$

$$\underset{\sim}{M}_{xw}^{(e)} = \int_0^{\ell_e} \underset{\sim}{F}_x\, \underset{\sim}{F}_w^T\, d\xi\,; \qquad \underset{\sim}{M}_{xw}^{(e)} = 2\pi \int_0^{\ell_e} x\; \underset{\sim}{F}_x\, \underset{\sim}{F}_w^T\, d\xi\,; \tag{4.26c}$$

$$\phi_{oe}(\underset{\sim}{\sigma}) = \frac{1}{2} \sum_e \{\, \underset{\sim}{r}_x^{(e)T}\, \underset{\sim}{S}_o^{(e)}\, \underset{\sim}{r}_x^{(e)} + (\,|\, \underset{\sim}{r}_x^{(e)} | - \underset{\sim}{r}_o^{(e)})^T \cdot$$

$$\cdot\, \underset{\sim}{\bar{G}}^{(e)}(\, \underset{\sim}{r}_x^{(e)})(\,|\, \underset{\sim}{r}_x^{(e)} | - \underset{\sim}{r}_o^{(e)})\} = \frac{1}{2}\, \underset{\sim}{r}^T\, \underset{\sim}{S}_o \underset{\sim}{r} + \frac{1}{2} (\,|\underset{\sim}{r}| - \underset{\sim}{r}_o)^T\, \underset{\sim}{G}\,(\, \underset{\sim}{r}\,)\,(\,|\, \underset{\sim}{r}| - \underset{\sim}{r}_o) \tag{4.27}$$

where

$$|\, \underset{\sim}{r}_x^{(e)} | = \begin{bmatrix} |r_{x1}^{(e)}| \\ \vdots \\ |r_{xl}^{(e)}| \\ \vdots \\ \; \end{bmatrix}, \tag{4.28}$$

and $\underset{\sim}{r}_o$ is the global vector of the nodal values corresponding to the datum τ_o; moreover:

$$\underset{\sim}{S}_o^{(e)} = \epsilon \int_0^{\ell_e} \underset{\sim}{F}_x\, \underset{\sim}{F}_x^T\, d\xi\,, \qquad \underset{\sim}{S}_o^{(e)} = 2\pi\epsilon \int_0^{\ell_e} x\, \underset{\sim}{F}_x\, \underset{\sim}{F}_x^T\, d\xi\,, \tag{4.29}$$

$$\underset{\sim}{\bar{G}}^{(e)}(\underset{\sim}{r}_x^{(e)}) = \frac{1-\epsilon^2}{\epsilon} \sum_{\gamma_e} P_{\gamma e}\quad \underset{\sim}{F}_x(\xi_{\gamma e})\, \underset{\sim}{F}_x^T(\xi_e)\,, \qquad \underset{\sim}{\bar{G}}^{(e)}(\underset{\sim}{r}_x^{(e)}) = 2\pi\, \frac{1-\epsilon^2}{\epsilon} \sum_{\gamma_e} x(\xi_{\gamma e})\, P_{\gamma e}\quad \underset{\sim}{F}_x(\xi_{\gamma e})\, \underset{\sim}{F}_x^T(\xi_e), \tag{4.30}$$

where:

$$P_{\gamma e} = \begin{cases} 0 & \text{if } |\sigma_x^{(e)}(\xi_{\gamma e})| \leq \tau_o(\xi_{\gamma e}) \\ \text{Gaussian weight at the point } \xi_{\gamma e} & \text{if } |\sigma_x^{(e)}(\xi_{\gamma e})| > \tau_o(\xi_{\gamma e})\,; \end{cases} \tag{4.31}$$

$$P(\underset{\sim}{\sigma}) = \frac{1}{2}\sum_e \underset{\sim}{r}_z^{(e)T}\,\underset{\sim}{G}^{(e)}(\underset{\sim}{r}_z^{(e)})\,\underset{\sim}{r}_z^{(e)} = \frac{1}{2}\underset{\sim}{r}^T\underset{\sim}{G}(\underset{\sim}{r})\,\underset{\sim}{r}, \tag{4.32}$$

where:

$$\underset{\sim}{G}^{(e)}(\underset{\sim}{r}_z^{(e)}) = \sum_{\gamma_e} P_{\gamma e}\,\underset{\sim}{F}_z(\xi_{\gamma e})\,\underset{\sim}{F}_z^T(\xi_e)\,, \tag{4.33}$$

and:

$$P_{\gamma e} = \begin{cases} 0 & \text{if } \sigma_z^{(e)}(\xi_{\gamma e}) > 0 \\ \text{Gaussian weight at the point } \xi_{\gamma e} & \text{if } \sigma_z^{(e)}(\xi_{\gamma e}) \leq 0\,. \end{cases} \tag{4.34}$$

4.2. Numerical algorithm

Previous results allow us to express functional $F_{o\eta}$ in the following discrete form:

$$\begin{aligned} F_{o\eta}(\underset{\sim}{v}, \underset{\sim}{r}) = {} & \frac{1}{2}\,\underset{\sim}{v}^T\underset{\sim}{K}\,\underset{\sim}{v} - \frac{1}{2}\,\underset{\sim}{r}^T\underset{\sim}{T}\,\underset{\sim}{r} - \underset{\sim}{g}^T\underset{\sim}{v} + \underset{\sim}{r}^T\underset{\sim}{M}\,\underset{\sim}{v} + {} \\ & -\frac{1}{2}\,\underset{\sim}{r}^T\underset{\sim}{S}_o\,\underset{\sim}{r} - \frac{1}{2}(|\underset{\sim}{r}| - \underset{\sim}{r}_o)^T\bar{\underset{\sim}{G}}(\underset{\sim}{r})(|\underset{\sim}{r}| - \underset{\sim}{r}_o) - \frac{1}{2\eta}\,\underset{\sim}{r}^T\underset{\sim}{G}(\underset{\sim}{r})\,\underset{\sim}{r}\,; \end{aligned} \tag{4.35}$$

consequently, the stationary condition (3.8) gives:

$$\underset{\sim}{K}\,\underset{\sim}{v} - \underset{\sim}{g} + \underset{\sim}{M}^T\underset{\sim}{r} = \underset{\sim}{0}\,, \tag{4.36a}$$

$$-\underset{\sim}{T}\,\underset{\sim}{r} + \underset{\sim}{M}\,\underset{\sim}{v} - \underset{\sim}{S}_o\,\underset{\sim}{r} - \bar{\underset{\sim}{G}}(\underset{\sim}{r})\,\underset{\sim}{r} + \bar{\underset{\sim}{G}}(\underset{\sim}{r})\,\underset{\sim}{r}_o^* - \frac{1}{\eta}\underset{\sim}{G}(\underset{\sim}{r})\,\underset{\sim}{r} = \underset{\sim}{0}\,, \tag{4.36b}$$

$\underset{\sim}{r}_o^*$ being the vector whose nonzero components are:

$$r_o^{*(i)} = r_o^{(i)}\,\frac{r_x^{(i)}}{|r_x^{(i)}|}\,. \tag{4.37}$$

We observe now that relations (4.36) represent the finite element approximation of problem (3.8) relative to the datum τ_o. As already pointed out in Sec. 3, the solution of our primary problem can be reached by successive rearrangements of $\mathbf{r}_o$.
Hence, the iterative scheme that we use in numerical applications is the following:

$$\underset{\sim}{\mathbf{v}}_{h+1} = \underset{\sim}{\mathbf{K}}^{*-1}_{h-1}\, \underset{\sim}{\mathbf{q}}^*_h ,$$

$$\underset{\sim}{\mathbf{r}}_{h+1} = \overline{\underset{\sim}{\mathbf{K}}}_h\, \underset{\sim}{\mathbf{v}}_{h+1} + \underset{\sim}{\mathbf{B}}_h \overline{\underset{\sim}{\mathbf{G}}}(\underset{\sim}{\mathbf{r}}_h)\, \underset{\sim}{\mathbf{r}}^*_{om} ,$$

where:

$$\underset{\sim}{\mathbf{g}}^*_h = \underset{\sim}{\mathbf{g}} - \underset{\sim}{\mathbf{M}}^T \underset{\sim}{\mathbf{B}}_h\, \overline{\underset{\sim}{\mathbf{G}}}(\underset{\sim}{\mathbf{r}}_h)\, \underset{\sim}{\mathbf{r}}^*_{om} ,$$

$$\underset{\sim}{\mathbf{K}}^*_h = \underset{\sim}{\mathbf{K}} + \underset{\sim}{\mathbf{M}}^T \overline{\underset{\sim}{\mathbf{K}}}_h ,$$

$$\overline{\underset{\sim}{\mathbf{K}}}_h = [\underset{\sim}{\mathbf{T}} + \underset{\sim}{\mathbf{S}}_o + \overline{\underset{\sim}{\mathbf{G}}}(\underset{\sim}{\mathbf{r}}_h) + \frac{1}{\eta}\, \underset{\sim}{\mathbf{G}}(\underset{\sim}{\mathbf{r}}_h)]^{-1}\, \underset{\sim}{\mathbf{M}} = \underset{\sim}{\mathbf{B}}_h \underset{\sim}{\mathbf{M}}.$$

The datum $\underset{\sim}{\mathbf{r}}^*_{om}$ relative to the *m-th* auxiliary problem is evaluated on the basis of the solution corresponding to the *(m-1)-th* one. More precisely the new limit tangential stress $\tau_{o(m)}$, relative to the *m-th* auxiliary problem, is assumed equal to $\mathrm{f} \cdot \sigma^{(o)}_{s(m-1)}$.

4.3) Numerical results and concluding remarks

In this last section we present some numerical results relative to the load conditions shown in Fig. 5.

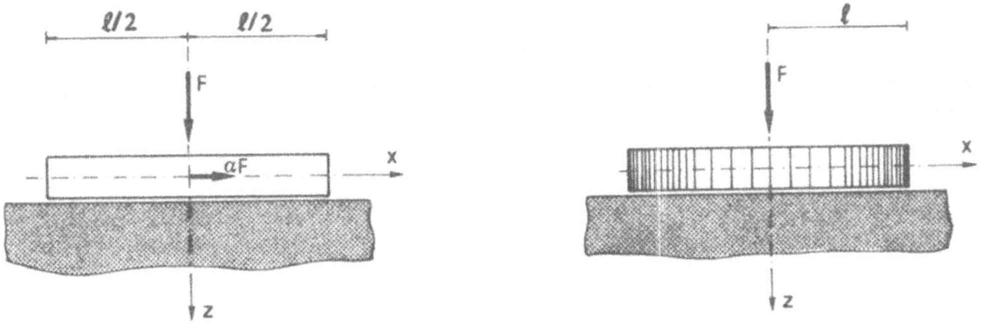

Fig. 5

In order to deal with nondimensional quantities, we put:

BEAM	CIRCULAR PLATE	
$t = \dfrac{x}{\ell} \qquad t \in [-\dfrac{1}{2}, \dfrac{1}{2}],$	$t = \dfrac{x}{\ell} \qquad t \in [0,1],$	(4.40a)
$U = \dfrac{D}{F\ell^3} u,$	$U = \dfrac{D}{F\ell^2} u,$	(4.40b)
$W = \dfrac{D}{F\ell^3} w,$	$W = \dfrac{D}{F\ell^2} w,$	(4.40c)
$\Psi = \dfrac{D}{F\ell^2} \psi,$	$\Psi = \dfrac{D}{F\ell} \psi,$	(4.40d)
$\Sigma_x = \dfrac{\ell}{F} \sigma_x,$	$\Sigma_x = \dfrac{\pi\ell^2}{F} \sigma_x,$	(4.40e)
$\Sigma_z = \dfrac{\ell}{F} \sigma_z,$	$\Sigma_z = \dfrac{\pi\ell^2}{F} \sigma_z,$	(4.40f)

It is easy to show that the solution of the problem depends on the following dimensionless parameters:

$\zeta = \dfrac{h}{\ell},$	$\zeta = \dfrac{h}{\ell},$	(4.41a)
$\varrho = \dfrac{32\, D\, (1-\nu_f^2)}{\pi\, E_f\, \ell^4}$	$\varrho = \dfrac{2\, D\, (1-\nu_f^2)}{E_f\, \ell^3}$	(4.41b)
$\nu,\ \nu_f$	$\nu,\ \nu_f,$	(4.41c)
f,	f,	(4.41d)
$\bar{\epsilon} = 4\pi\, G_f\, \epsilon,$	$\bar{\epsilon} = \dfrac{4\pi G_f}{\ell} \epsilon,$	(4.41e)

The numerical results that we give in the following correspond to a uniform mesh of N_e elements, over which the unknown functions are approximated by linear polynomials. From here we assume $\nu = 0.3$ and $\nu_f = 0.1$.

Firstly, we investigate on the convergence of our numerical model by refering to the beam example. In Tab. 1 the values of the parameters W and Σ_z at the center of the beam are given for some values of N_e.

Tab. 1

N_e	W	Σ_z
10	0.3350×10^{-3}	4.9438
20	0.3804×10^{-3}	4.8717
30	0.4029×10^{-3}	4.8759
40	0.4180×10^{-3}	4.7927

($\varrho = 0.00025$; $f = 0.6$; $\alpha = 0$; $\bar{\epsilon} = 10^{-6}$)

Tab. 2

$\bar{\epsilon}$	W	Σ_x	Σ_z
10^{-1}	0.21138×10^{-2}	0.13397	2.3934
10^{-4}	0.21235×10^{-2}	0.14348	2.4081
10^{-7}	0.21235×10^{-2}	0.14348	2.4082

($\varrho = 0.0025$; $f = 0.6$; $\alpha = 0.1$; $N_e = 20$)

Moreover, the influence of the parameter $\bar{\epsilon}$ is examined in Tab. 2 for the some beam example. We observe that for values of $\bar{\epsilon} < 10^{-4}$ the solution is practically independent of this parameter. Consequently, in this range of values our numerical results can be seen as approximations of the solution of a contact problem with Coulomb's friction law.

We remark that a similar behaviour with respect to $\bar{\epsilon}$ has been found also in the case of the plate example. Therefore, in the next numerical applications we will assume: $\bar{\epsilon} = 10^{-6}$.

The influence of the relative stiffness parameter ϱ is examined in Figs. 6, where the plots of W, Σ_x and Σ_z along the beam axis are given for $\varrho = 0.025, 0.0025, 0.00025$.

We observe that for $\varrho = 0.025$ the beam does not get up from the foundation, i.e. the contact is bilateral.

Obviously, as shown in Fig. 6d, near the boundary of the contact regions, where $\Sigma_z \to 0$, the tangential stress Σ_x reaches the limit value $f \cdot \Sigma_z$.

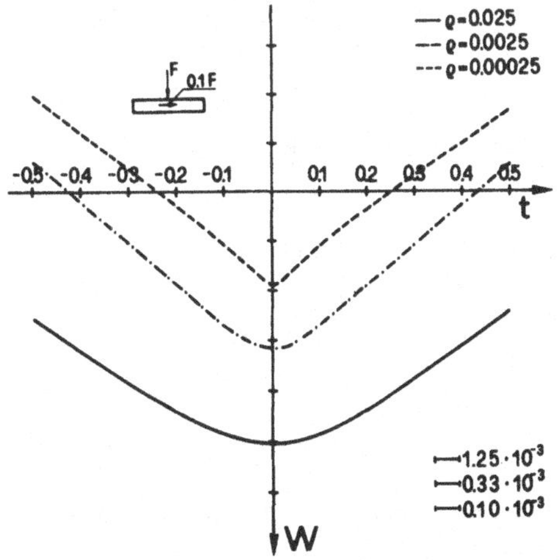

Fig. 6 a

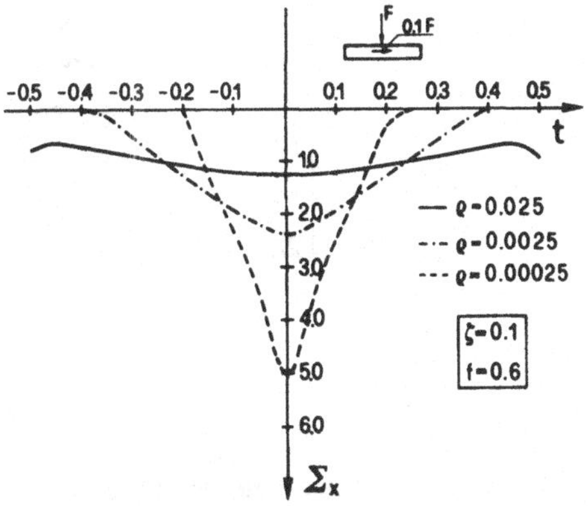

Fig. 6 b

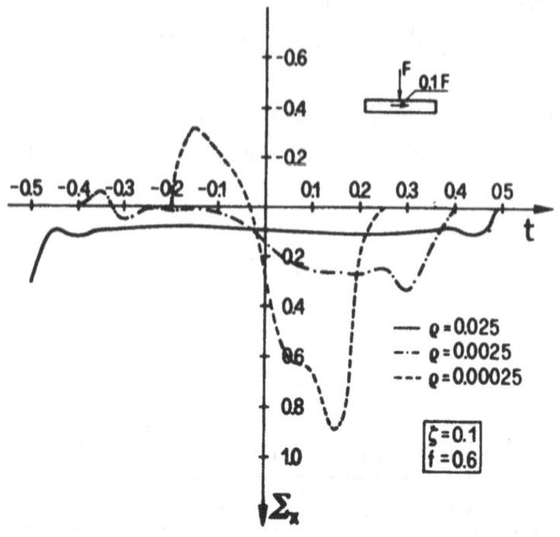

Fig. 6 c

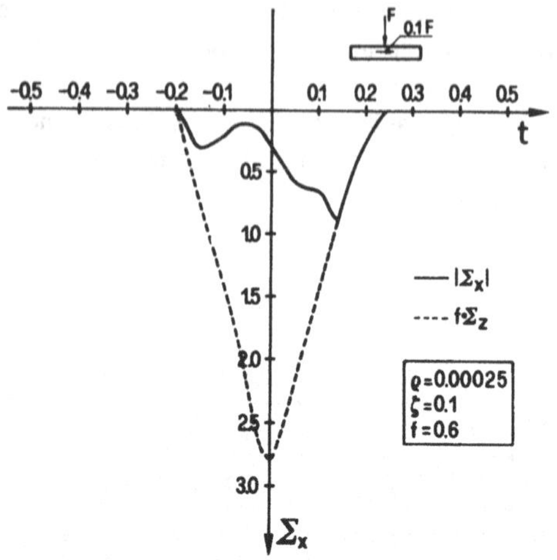

Fig. 6 d

Finally, the influence of the ratio ζ between the thickness and the length of the beam is analyzed in Fig. 7.

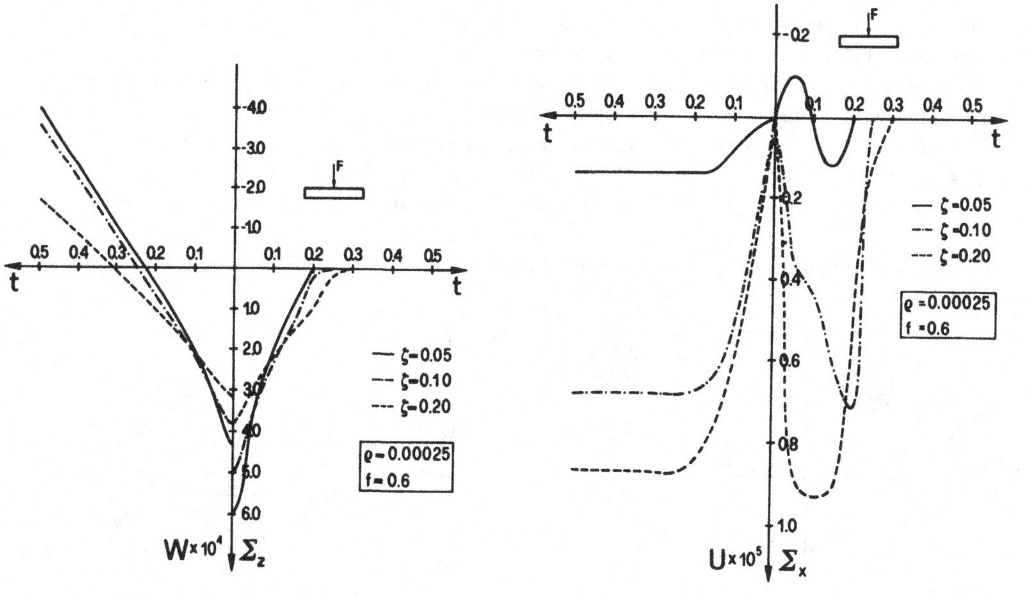

Fig. 7

We observe a remarkable influence of this parameter on the numerical results. In particular we observe that the contact region is also a function of ζ.

For what concerns the plate example, we investigate the influence of the friction coefficient f , which is particularly significant in this case. Figs. 8 show the plot of the variables, Σ_z, W and Σ_x, U along the plate radius for assigned values of ϱ and ζ.

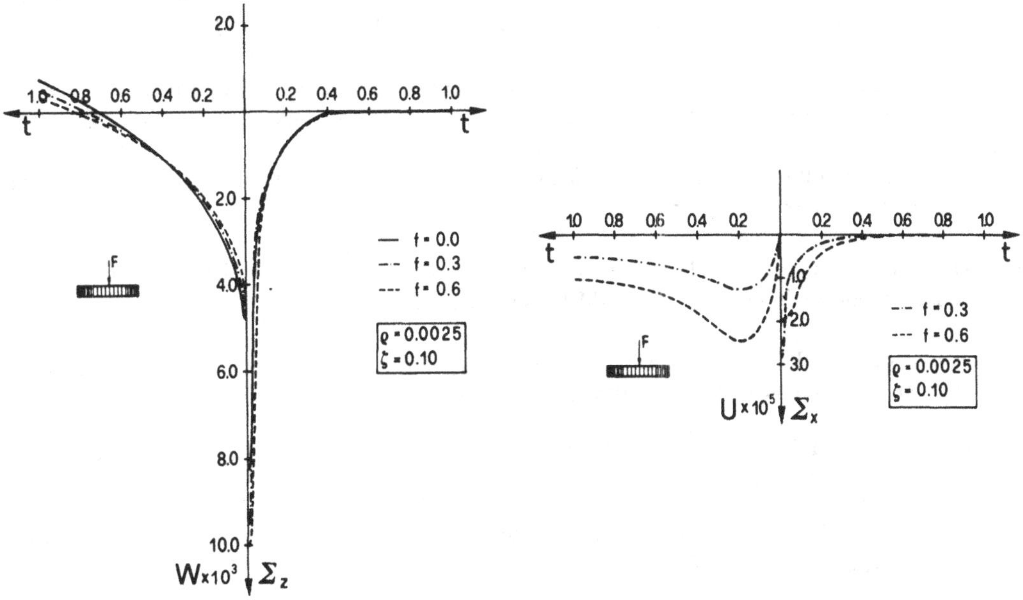

Fig. 8

A comparison with the frictionless case (f = 0) is also given. We remark that for the examined examples the influence of the friction can modify the values of displacements and stresses until a percentage of 20%.

REFERENCES

1. Duvaut, G. and Lions, J.L.: Les Inéquations en Mécanique et en Phisique, Dunod, Paris, 1972.

2. Nečas, J., Jarušec, J. and Haslinger, J.: On the solution of the variational inequality to the Signorini problem with small friction, Boll. U.M.I. (5), 17-B, 1980, 736-811.

3. Duvaut, G.: Problèmes mathématiques de la Mécanique-Équilibre d'un solide élastique avec contact unilateral et frottement de Coulomb, C.R. Acad. Sc., Paris, t. 290, Série A, 1980, 263, 265.

4. Oden, J.T. and Pires, E.: Contact problems in elastostatics with non-local friction laws, TICOM Report 81-12, November, 1981.

5. Oden, J.T. and Pires, E.: Nonlocal and nonlinear friction laws and variational principles for contact problems in elasticity, TICOM Report 82-3, April, 1982.

6. Panagiotopoulos, P.D.: A nonlinear programming approach to the unilateral contact and friction-boundary value problem in the theory of elasticity, Ingenieur-Archiv 44, 1975, 421-432.

7. Ascione, L. and Grimaldi, A.: Unilateral contact between a plate and an elastic foundation, Meccanica, 1984, 19,223-233.

8. Ascione, L. and Olivito, R.S.: Unbonded contact of a Mindlin plate on an elastic half-space, Meccanica, 20, 1985, 49-58.

9. Ascione, L. and Bruno, D.: The unilateral contact problem with friction of a plate resting on an elastic half-space, Rep. no. 71, Dept. of Structures, University of Calabria, 1984.

10. Mindlin, R.D.: Influence of rotatory inertia on shear and flexural motions of isotropic elastic plates, J. Appl. Mech., 18, 1951, 31-88.

11. Ascione, L. and Olivito, R.S.: Some topics about the force-displacement relationship of an elastic obstacle in view of contact problem solution, Rep. no. 31, Dept. of Structures, University of Calabria, 1981.

12. Gladwell, G.M.L.: Contact Problems in the Theory of Elasticity, Sijthoff & Noordhoff, 1980.

13. Dunford, N. and Schwartz, J.: Linear Operators, Part I, Interscience Public., 1958.

14. Ascione, L. and Grimaldi, A.: Penalty Formulations of the Unilateral Contact Problem Between Plates and an Elastic Half-space, in Penalty - Finite Element Methods in Mechanics (ed. J.N. Reddy), American Society of Mechanical Engineers, 1982.

15. Brezzi, F., Hager, W.W., Raviart, P.A.: Error estimates of the finite element solution of variational inequalities, Numer. Math. 28, 1977, 431-443.

16. Oden J.T. Mixed finite element approximations via interior and exterior penalties for contact problems in elasticity, Int. Sym. on Hibrid and Mixed Finite Element Methods, Georgia Institute of Technology, Atlanta, Georgia, April, 8-10, 1981.

PROBABILISTIC TREATMENT OF STRUCTURAL PROBLEMS WITH UNILATERAL CONSTRAINTS

A. Baratta
Facoltà di Architettura, University of Naples

ABSTRACT

In the paper, the problem to find statistics of the response of structures with unilateral constraints when the structural pattern is affected by uncertainty, is approached. In particular, the case of the elastic beam supported by an underlying rigid soil is investigated, under the assumption that the surface of the soil is not perfectly known, but is described by a suitable random process. It is shown that if the beam is discretized into a number of finite beam-elements, and only the ordinates of the soil profile under the nodal points are significant, the problem can be solved by the evaluation of a multidimensional integral. In most cases a Monte-Carlo procedure with "importance sampling" proves to be very efficient for numerical calculation of the above integral.

Acknowledgement
Paper supported by grants of the Italian Ministry for Public Education (M.P.I.)

1. INTRODUCTION

Structural problems are often affected by uncertainty. When the mathematical model is linear, it is rather simple in principle, although possibly very cumbersome in pratice,to treat such uncertainty in a probabilistic context |1|.

When the behaviour of the mechanical object is more complex, non-linear models must be introduced, capable to keep into proper account geometrical non-linearities, material anelasticity, and so on. In such cases, the probabilistic treatment is much less straightforward, unless one makes recourse to some Monte-Carlo simulation in order to get informations on the statistics of structural response, given the statistics of the input data.

With the exception of the case when the system is affected only by slight nonlinearities |2|, direct Monte-Carlo techniques are not practical in structural analysis, since they require that a (possibly) large number of similar cumbersome structural problems are set up, by randomly varying the input data, and solved, in order to build up the statistics of the structure response |3|.

In most cases, it should be recognized that the treatment of physical problems can be approached in different ways. Some of these approaches produce results only if one explicitly introduces uncertainty in setting the structural pattern, and are often disregarded in a fully deterministic context.

Unilateral constraints yield an example of nonlinearities that are frequently encountered in structural problems, although in many cases engineers try to ignore this feature by introducing some artificious devices in the calculations, or by making recourse to some trial-error procedure. In the last years, unilateral problems have been extensively investigated (see e.g. |4|) finding a rather general (at least for well-established problems) settlement, in conjunction with the progress of variational analysis of non-linear differential equations, and so on.

From the computational point of view, the solution of convex unilateral problems can be pursued through the search of the constrained optimal point of some suitable functional. Operationally, this search is accomplished by previously discretizing the problem in the time/space domain, and then by applying any programming search technique suited for the problem under examination.

The above brief discussion points out both the mathematical and operational complexity of unilateral problems, enhancing the circumstance that a very poor efficiency can be expected from pure simulation procedures.

In this paper, a first approach to the problem of structures with unilateral constraints under uncertainty is attempted, with reference to a particular problem, i.e. the elastic beam unilaterally supported by a rigid soil with stochastic profile.

2. BEAMS ON UNILATERAL RIGID SOIL

Consider the beam in Fig. 1, simply supported at the ends. Let the beam be discretized in n+1 beam elements, connected at n interior nodes. Let **F** and **u** be the n-dimensional vectors

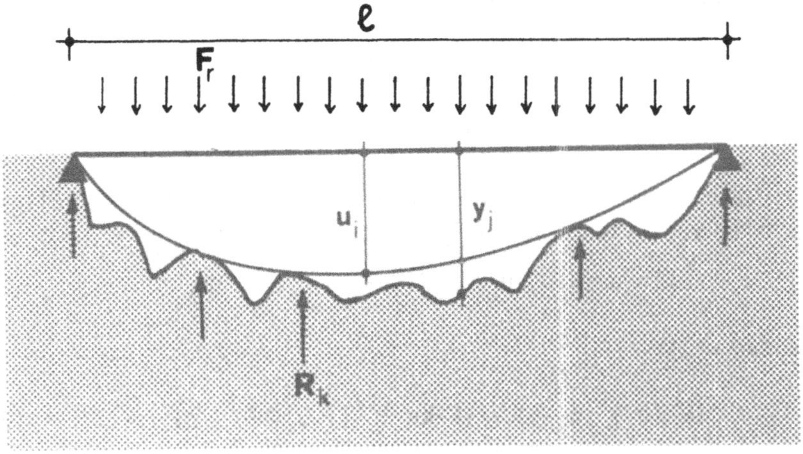

fig. 1 - The structural pattern

of active forces and nodal (vertical) displacements respectively, both assumed to be positive downwards, and **K** be the nxn stiffness matrix. If the beam is also unilaterally supported on a rigid soil at intermediate points, let **y** be the n-dimensional vector of the ordinates of the soil profile underlying the nodal points, and **R** be the reactive forces (positive upwards) at the same points. Linear interpolation of the profile is assumed between adjacent nodal points, so that contact cannot take place at points interior to the beam elements.

It is easy to prove that if one considers the Total Potential Energy function

$$\mathcal{E} = \tfrac{1}{2}\,\mathbf{u}^t\,\mathbf{K}\,\mathbf{u} - \mathbf{F}^t\,\mathbf{u} \qquad (2.1)$$

one finds that such function takes a minimum in solution, on the set of admissible displacement vectors, defined by the inequality $\mathbf{u} \leq \mathbf{y}$.

This theorem yields a powerful tool to get the equilibrated

solution $\mathbf{u}_o$, in that it makes possible to reduce the problem to the search of the constrained minimum of a quadratic function, a well-established and deeply investigated algorithm.

Unfortunately, when the profile of the underlying soil is not known, but must be modeled as a stochastic function, the above point of view is not practical any more, in that the problem of drawing statistics of the extremal point of a stochastic function, or of a deterministic function with random constraints as well, has not yet been investigated at present. Let therefore look for a different approach, perhaps more suited for the stochastic problem at hand.

To this aim, consider the following basic properties:

$$\mathbf{F} - \mathbf{K}\,\mathbf{u}_o \geq 0, \qquad \mathbf{u}_o \leqslant \mathbf{y}, \tag{2.2}$$

and let us express the first variation of the energy function following any admissibile variation $\delta\,\mathbf{u} = \mathbf{u} - \mathbf{u}_o$ starting from $\mathbf{u}_o$:

$$\delta\mathcal{E} = \mathbf{u}_o^{\,t}\mathbf{K}(\mathbf{u} - \mathbf{u}_o) - \mathbf{F}^t(\mathbf{u} - \mathbf{u}_o) \geqq 0 \quad \forall\mathbf{u}, \tag{2.3}$$

whence one can write

$$(\mathbf{F}-\mathbf{K}\mathbf{u}_o)^t(\mathbf{u}_o-\mathbf{y}) + (\mathbf{F}-\mathbf{K}\mathbf{u}_o)^t\,(\mathbf{y}-\mathbf{u}) \geqslant 0 \quad \forall\mathbf{u}. \tag{2.4}$$

The first member at the left side of ineq. (2.4) is non-positive:

$$(\mathbf{F}-\mathbf{K}\mathbf{u}_o)^t\,(\mathbf{u}_o-\mathbf{y}) \leqslant 0\,.$$

On the other side, if one takes $\mathbf{u} = \mathbf{y}$, ineq. (2.4) yields:

$$(\mathbf{F}-\mathbf{K}\mathbf{u}_o)^t\,(\mathbf{u}_o-\mathbf{y}) \geqslant 0,$$

whence the following complementarity relation is inferred:

$$(\mathbf{F}-\mathbf{K}\mathbf{u}_o)^t\,(\mathbf{u}_o-\mathbf{y}) = 0, \tag{2.5}$$

fully equivalent to the condition that the energy attains its minimum at u_o.

3. STOCHASTIC SOIL PROFILE

Let the soil profile be described by a random function $\tilde{y}(x)$

and let $m_y(x)$ and $\Sigma_y(x', x'')$ be the average value and the covariance function of the process respectively.

Let $\tilde{y}_i$ be the random variable that measures the depth of the ground under the i-th node (Fig. 1), and let

$$p_y (y_1, \ldots, y_n) \tag{3.1}$$

be the joint PMF (probability mass function) of the $\tilde{y}_i$'s. If the instantaneous PMF of $\tilde{y}(x)$ is Gaussian, or is however dependent only on the expected values and on the correlation matrix, the JPMF (3.1) can be simply obtained by the functions $m_y(x)$ and $\Sigma_y(x', x'')$.

Provided that the JPMF (3.1) has been assumed, consider the set of all possible configurations of the beam

$$\Omega_u = \{ \mathbf{u} \in R^n : \mathbf{F}\text{-}\mathbf{Ku} \geqslant \mathbf{0} \}. \tag{3.2}$$

For any $\mathbf{u} \in \Omega_u$, put

$$R_i = F_i - k_{ij} u_j , \tag{3.3}$$

and denote by I_0 the set of indices corresponding to the outcome $R_i = 0$, and by I the set of indices corresponding to supported points, where $R_i > 0$.

By eq. (2.5), combined with the second condition for admissibility ($\mathbf{y} \geqslant \mathbf{u}$), it is possible to calculate the PMF of the displacement vector $\mathbf{u}_0$ in solution. In fact, for any $\mathbf{u}$

$$\text{Prob}\{\tilde{\mathbf{u}}_0 = \mathbf{u}\} = \text{Prob} \left\{ [\bigcap_{i \in I_0} \{ \tilde{y}_i \geqslant u_i \}] \cap [\bigcap_{i \in I} \{ \tilde{y}_i = u_i \}] \right\} = f(\mathbf{u}) d\,\Omega_u, \tag{3.4}$$

with

$$d\,\Omega_u = \prod_{i \in I} du_i . \tag{3.4'}$$

Any statistics of the response $\mathbf{u}$, say $S(\mathbf{u})$, can thus be calculated by multivariate integration

$$E\,|\tilde{S}| = \int_{\Omega_u} S(\mathbf{u})\, f(\mathbf{u})\, d\,\Omega_u . \tag{3.5}$$

The situation is sketched in fig. 2 for the two-dimensional case. Here the level curves of the JPMF of the ordinates of the profile are drawn, together with the equilibrium straight lines of both nodes. The contoured area is the domain of possible deflections Ω_u (3.2), while the intersection of the two lines $R_1 = 0$ and $R_2 = 0$ represents the free solution $\mathbf{u}_f$ when contact between the beam and the soil does not take place and the probability that $\mathbf{u}_f$ is the solution coincides with the probability that the profile vector $\tilde{y}$ falls in the domain. Furtherly, for any line starting from the origin (such as r_1 or r_2 in fig. 2), let A_1 (resp. A_2) be the intersection point with the boundary of Ω_u. Consider the segment OA_1 (resp. OA_2). All points of this segment describe deflections that are everywhere supported by the soil (here, hence, $I_o = \emptyset$), except point A_1 (resp. A_2) where the beam is supported at the node 1 (resp. 2) but is free at node 2 (resp. 1).

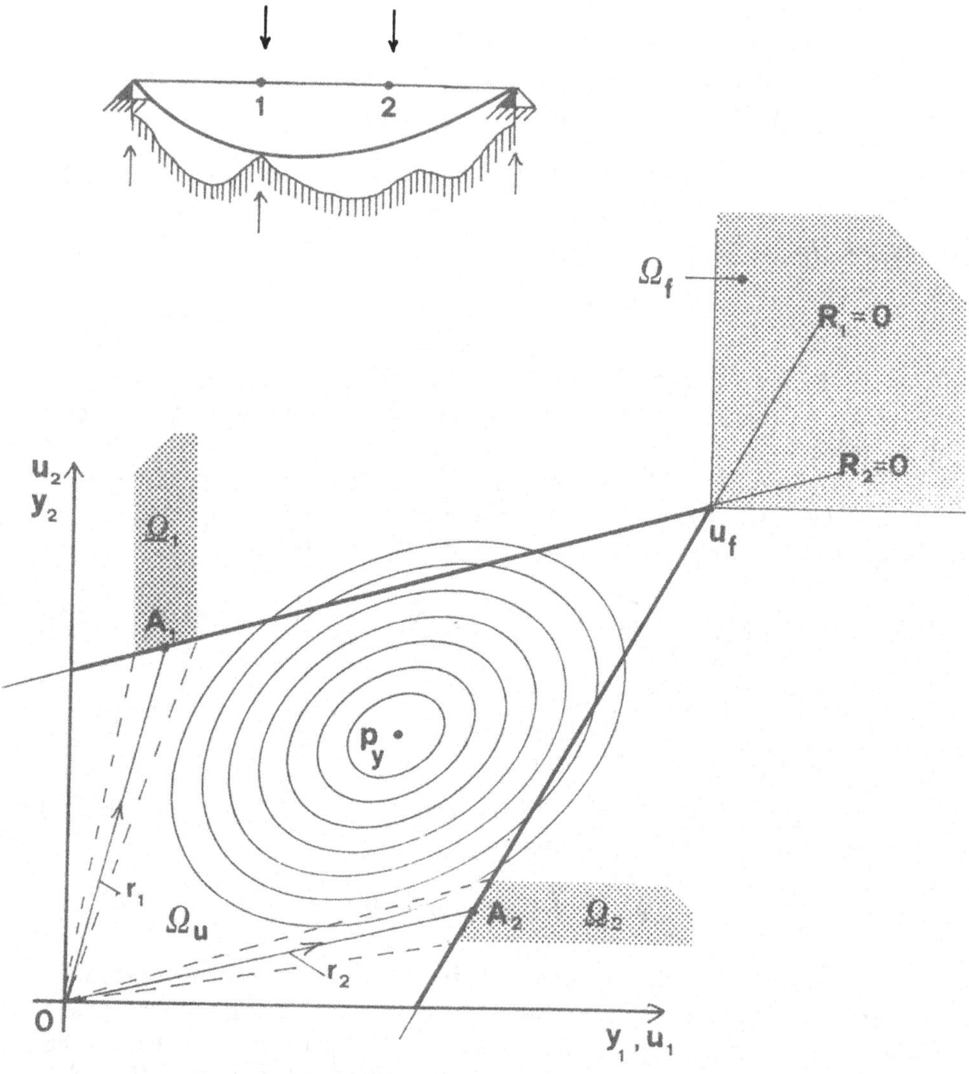

fig. 2 - Graphic explanation for the solution procedure

Thus in numerical calculation of the integral (3.5), the integral gets contributions from the points interior to the small triangle in the neighborhood of the line r_1 (resp. r_2) for deflections varying on the segment OA_1 (resp. OA_2), while any contribution deriving from the grey strip Ω_1 (resp. Ω_2) starting from A_1 (resp. A_2) is associated to the deflection A_1 (resp. A_2). It can thus be argued that the boundary of Ω_u yields a very significant contribution to the integral (3.5).

Being concerned with a multidimensional integral, possibly with large n, the evaluation of the integral may be very expensive, mainly because of the skewness of the domain of integration Ω_u, that does not allow to bound directly the variables. The calculations become easier if one introduces a new set of basic variables (the <u>reactions</u> R_i in place of the <u>displacements</u> u_i), by the transformation:

$$\mathbf{u} = \mathbf{C}\ (\mathbf{F}\text{-}\mathbf{R}), \tag{3.6}$$

with $\mathbf{C}$ the matrix inverse of $\mathbf{K}$. So, eq. (3.5) can be written:

$$E|\tilde{S}| = \int_{\Omega_R} S|\mathbf{u}(\mathbf{R})|\ f|\mathbf{u}(\mathbf{R})|\ J|\mathbf{u}(\mathbf{R})|\ d\Omega_R, \tag{3.7}$$

with

$$\Omega_R = \{\mathbf{R} \in R^n : \mathbf{R} \geqslant \mathbf{0}\}, \tag{3.8}$$

and J the Jacobian of the relation that expresses the constrained displacements u_i ($i \in I$) in function of the reactions R_i. The transformation (3.6) allows to simplify the domain of integration, at the cost of a more intricate evaluation of the integrand function.

The evaluation of the integral (3.7), simple in principle, may be rather cumbersome in practice, if the beam is discretized in a large number of elements. Here a Monte-Carlo technique proves useful to drastically reduce the time necessary for the calculations. In order to apply the procedure efficiently, it is convenient to introduce importance sampling, as follows.

Let $g(\mathbf{R})$ be the integrand function, and $p(\mathbf{R})$ be any JPMF of the basic variables $R_1, \ldots, R_n$. The integral (3.7) can then be written as follows:

$$E|\tilde{S}| = \int_{\Omega_\mathbf{R}} p(\mathbf{R})\ \frac{g(\mathbf{R})}{p(\mathbf{R})}\ d\,\Omega_R, \tag{3.9}$$

i.e., it can be viewed at as the expected value of the function $G(\mathbf{R})$ = g/p, when the basic variables are distributed according to $p(\mathbf{R})$. Hence, it can be calculated by statistical simulation, by generating N random samples of the vector **R**, distributed according to p(**R**), and then by calculating the average value of the N values $G(\mathbf{R}_k) = G_k$:

$$E|\tilde{S}| = \frac{\sum_k G_k}{N} \tag{3.10}$$

The problem is then to choose the basic distribution p(R) in an efficient way. The uniform distribution is obviously the easiest to be handled,so it is natural that one is inclined to use an uniform p(R). The problem is only to estimate upper and lower bounds for any R_i. This can be made by first executing a preliminary calculation of the integral (3.9), then by taking an uniform p(R) assumed different from zero in a very large part $\bar{\Omega}_R \subseteq \Omega_R$ such that the points R not lying in $\bar{\Omega}_R$ almost certainly yield negligible contribution to the integral. During this preliminary simulation it is possible to recognize the points **R** where the function G(**R**) is different from zero, so one can record maximum and minimum values for each R_i associated with significant contribution. After this preliminary search, one can restrict the domain $\bar{\Omega}_R$ and repeat the calculation for final results.

4. NUMERICAL RESULTS AND CONCLUSIONS

Let the profile of the underlying soil be described by a random process whose autocorrelation function is practically zero when evaluated at any couple of points x_j, x_k at which nodes of the beam are located :

$$\Sigma_y (x_j, x_k) \doteqdot 0 . \tag{4.1}$$

This implies that the elements by which the beam has been discretized cannot be smaller than a given length (the "autocorrelation extinction length" of the soil profile), otherwise some non-zero coefficients of correlation should be taken into account in building up the JPMF of the ordinates y_i. Even in the latter case, the procedure does not present difficulties, exeept that, in the integrand function, f(u) should be calculated by the multivariate integral:

$$f(\mathbf{u}) = \int_{u_1}^{\infty} .. \int_{u_m}^{\infty} p_y(y_1, \dots, y_m)\, dy_1 \dots dy_m, \tag{4.2}$$

where it is assumed that $I_U^m = \{1, \dots, m\}$ and $I = \{m+1, \dots, n\}$, with $m \leqslant n$.

For simplicity's sake, assume that eq. (4.1) holds and the

soil ordinates $\tilde{y}_i$ are all statistically independent, with the same expected value m_y and standard deviation s_y. With these assumptions, the range (y_1, y_2) for the soil ordinates at every nodal point x_i, is given by :

$$y_1 = m_y - \sqrt{3}\, s_y ,\qquad y_2 = m_y + \sqrt{3}\, s_y . \tag{4.3}$$

Put:

$$z_i = \min(u_i, y_2);\; w_i = \min(u_i, y_1), \tag{4.4}$$

the individual PMF and PDF of the soil ordinate at the point x_i are given by :

$$p_1(u_i) = \begin{cases} \dfrac{1}{y_2 - y_1} & \text{if } y_1 \leqslant u_i \leqslant y_2, \\ 0 & \text{otherwise,} \end{cases} \tag{4.5'}$$

$$\text{Prob}\{\tilde{y}_i \geqslant u_i\} = \frac{y_2 - z_i}{y_2 - w_i} = P_i(u_i). \tag{4.5''}$$

With these positions, in the integrand function the factor $f(\mathbf{u})$ can be expressed as:

$$f(\mathbf{u}) = \left[\prod_{i\in I_0} P_i(u_i)\right]\left[\prod_{i\in I} p_1(u_i)\right]. \tag{4.6}$$

Let u_f and M_f be the free deflection and the bending moment at the midspan of the beam, in the absence of contact. Assume that the average depth of the soil under the beam is a fraction of the free deflection, and that the range of variation is equal to the average depth:

$$m_y = t\, u_f \quad t \in (0,1), \tag{4.7}$$

$$y_1 = 0.5\ m_y\ ;\quad y_2 = 1.5\ m_y\ . \tag{4.8}$$

The expected value and variance of the abscissa z where the maximum value of the bending moment is attained, the expected value and variance of the maximum moment M, and the covariance between M and z have been evaluated.

Consider the case t = 0.1. This is a case where the supporting action of the soil heavily influences the beam response. The results are as follows.

Consider the location $\tilde{z}$ of the maximum moment. One gets:

$$\begin{aligned} m_z &= 0.5\ l, \\ s_z &= 0.274\ l. \end{aligned} \tag{4.9}$$

If one applies relations similar to eqs. (4.3) to estimate the range of variation of $\tilde{z}$ (although there is no reason to believe that $\tilde{z}$ is uniformly distributed), one finds

$$z_1 = 0.02\ l;\ z_2 = 0.98\ l\ . \tag{4.10}$$

This means that the point where the maximum moment is attained is extremely uncertain. Consider the statistics of the maximum moment itself; one gets :

$$\begin{aligned} m_M &= 0.14\ M_f, \\ s_M &= 0.01\ M_f\ , \end{aligned} \tag{4.11}$$

with a coefficient of variation

$$V_M = \frac{s_M}{m_M} = 0.07. \tag{4.12}$$

Consider moreover the covariance of M and z. One gets

$$\mathrm{Cov}\ |M,\ z| = 1.6 \times 10^{-4}\ M_f l, \tag{4.13}$$

whence the coefficient of correlation:

$$_{Mz} = \frac{\mathrm{Cov}|M,\ z)|}{s_M s_z} = -\ 0.06, \tag{4.14}$$

The results (4.13) and (4.15) suggest that the maximum moment is independent on the location where it is attained and can be evaluated with large confidence, despite the fact that its localization is highly uncertain. A rational choice to design such a beam is therefore to make it with uniform cross-section proportionated to resist a suitable fractile of the maximum moment.

The Monte-Carlo integration required to perform 10000 simulations in order to get an accuracy of a few percents in the evaluated integrals, corresponding to almost 1 hr of calculation time working in BASIC on the Victor 1/B microcomputer, with a waste of time much smaller than executing direct simulation of the structural problem, and almost 20 times smaller than if the integral was calculated by ordinary quadratures.

In order to check the credibility of the results, a sample soil profile has been generated (Fig. 3) and the unilateral problem has been solved by the energy approach as outlined in sec. 2. The elastic deflection of the beam is plotted in Fig. 3.a, while the diagram of the bending moment is plotted in Fig. 3.b.
The maximum moment calculated is :

$$M_{max} = 0.12\ M_f ,$$

with a percentage deviation of 14% from the expected value (4.12), in good agreement with the calculated coefficient of variation (4.12).

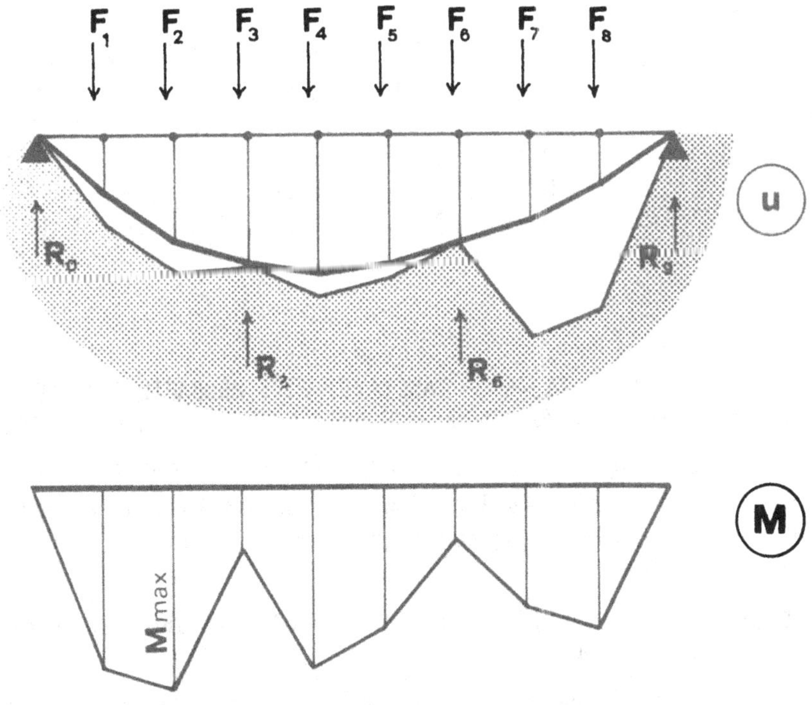

fig. 3 - Solution of the sample structure

Response statistics have been calculated for other values of the parameter t, always assuming that the range D_y of all variables y_i is equal to the expected value, in way that uncertainty increases with the depth of the underlying soil. The results are drawn in table I. It can be observed that the standard deviation of the abscissa z tends to decrease as the depth increases, as one would expect, although this effect is attenuated by the circumstance that -in this example- the standard deviation in the soil profile increases proportionally to the average ordinates. On the other side, the coefficient of variation of the maximum moment increases at first with increasing m_y, but henceforth tends to remain constant. This can be explained by considering that the increase in the standard deviation probably tends to compensate the trend of the beam to behave as simply supported at the ends, as the average ordinates of the profile become larger and larger.

TABLE I

m_y/u_f	$D_y m_y$	m_z/l	s_z/l	D_z/l	m_M/M_f	s_M/M_f	V_M	ϱ_{Mz}
0.10	1	0.5	0.274	0.96	1.4 E-1	1.0 E-2	0.07	-0.06
0.20	1	0.5	0.235	0.81	2.6 E-1	4.7 E-2	0.18	-0.09
0.30	1	0.5	0.200	0.69	3.3 E-1	5.8 E-2	0.18	0.08
0.40	1	0.5	0.163	0.56	3.9 E-1	7.1 E-2	0.18	0.02
0.50	1	0.5	0.157	0.54	4.5 E-1	7.9 E-2	0.17	0.00
1.00	1	0.5	0.082	0.28	7.5 E-1	1.4 E-1	0.18	-0.06

REFERENCES

|1| AUGUSTI, G. - BARATTA, A. - CASCIATI, F. : **PROBABILISTIC METHODS IN STRUCTURAL ENGINEERING**, Chapman & Hall Editors, London, 1984

|2| BARATTA, A. : NON-LINEAR TRUSS RELIABILITY BY MONTECARLO SAMPLING, **Proc. III Int. Conf. "Applications of Statistics and Probability in Soil and Structural Engineering"**, vol. I, pp. 136-148, Sydney, Australia, 1979

|3| SHREIDER, Yu. A. : **METHOD OF STATISTICAL TESTING**, Elsevier Publ. Co., New York, 1964

|4| DEL PIERO, G. - MACERI, F. (editors) : **UNILATERAL PROBLEMS IN STRUCTURAL ANALYSIS**, Springer-Verlag, Udine (Italy), 1985

A NUMERICAL ALGORITHM FOR SIGNORINI'S PROBLEM WITH COULOMB FRICTION

H.J.C. Barbosa, R.A. Feijóo
LNCC/CNPq, Rio de Janeiro

ABSTRACT

The problem of contact between a deformable solid and a rigid surface where friction forces following Coulomb's law can arise is studied in this work. The problem is approximated by a sequence of two simpler problems: (i) contact without friction and (ii) friction with prescribed normal stress. Within the context of linear elasticity steps (i) and (ii) are formulated as minimization problems. Lagrange multipliers are introduced and the finite element method is used for spatial discretization. Two quadratic programming problems arise and are solved by Gauss-Seidel algorithm with relaxation and projection. A numerical example is presented.

1. INTRODUCTION

The problem of the equilibrium of a deformable body which upon application of a load system may come into contact with a rigid surface (Signorini's problem) where friction forces can arise is an important and difficult problem in solid mechanics. Difficulties arise in finding a convenient formulation for the problem, in the issue of existence and uniqueness of solutions and in constructing efficient numerical algorithms for obtaining approximate solutions. It should be observed that even when friction is not included and within the context of linear elasticity, contact problems are non-linear due to the fact that the actual contact region is not known "a priori". As a result, most work related to numerical solutions for contact problems follows one of the two approaches: i) Iterative procedures (incremental or not) which may require the introduction of artificial interface elements in the contact region [1,2] or ii) Formulations based upon variational principles that lead to an optimization problem solved by mathematical programming techniques [3,4].

Following some previous work [5,6,7], the second approach is taken here as it allows for a more rigorous formulation.

2. FORMULATION

Consider a deformable body which occupies the open bounded region Ω in $\mathbb{R}^3$ with a smooth boundary $\Gamma=\overline{\Gamma}_u \cup \overline{\Gamma}_f \cup \overline{\Gamma}_c$. Along Γ_u (Γ_f) displacements (forces) are prescribed and Γ_c is the portion of Γ where contact between the body and a rigid foundation may take place upon application of a load system ℓ. It is assumed that $\Gamma_c \cap \Gamma_u = \Gamma_u \cap \Gamma_f = \emptyset$ and that meas (Γ_u) is positive.

Assuming infinitesimal strains (and displacements) and a Hookean material, the strain energy density function ϕ and the stress tensor σ are given by

$$\phi(E) = \frac{1}{2}\mathbb{D}E\cdot E \qquad \text{and} \qquad \sigma = \frac{\partial\phi}{\partial E} = \mathbb{D}E \quad ,$$

where E is the infinitesimal strain tensor, $\mathbb{D}$ is the fourth-order elasticity tensor, assumed symmetric and positive-definite, and the dot denotes scalar product. Within this context static problems can be formulated by the Principle of Virtual Work (PVW).

The first step is to define the space V of all possible displacements of the solid and the set Kin $\subset$ V of kinematically admissible displacements. Here, V will be the Hilbert space $(H^1(\Omega))^3$ of continuous functions whose derivatives are square-integrable in Ω. If the constraints are bilateral the PVW leads to a variational equation. However, for unilateral constraints the PVW leads to a variational inequality. Contact problems fall in this second category due to the fact that the admissible displacements $v \in$ Kin must satisfy non-interpenetration conditions and Kin becomes a convex set instead of a linear variety.

Applying the PVW, the solution of Signorini's problem (without friction) amounts to finding the displacement field $u \in Kin$ such that

$$-\int_{\Omega} \mathbb{D}E(u)\cdot E(v-u)d\Omega + \ell(v-u) \leq 0 \qquad \forall v \in Kin \ . \tag{1}$$

The first term is the virtual work associated to the internal stresses $\mathbb{D}E(u)$ and strains $E(v-u)$ and $\ell(v-u)$ denotes the virtual work of the loading system ℓ under the virtual displacements v-u. The set Kin is given by

$$Kin = \{v \in V\colon v\cdot n-s\leq 0 \text{ on } \Gamma_c \text{ and } v|_{\Gamma_c}=0\} \quad ,$$

where n is the unit vector outward and normal to the boundary Γ_c and s is the initial gap between the body and the rigid foundation. Rigid body motions have been excluded as the body is considered fixed in Γ_u.

However, due to the convexity and differentiability of ϕ, to solve inequality (1) is equivalent to find $u \in Kin$ which minimizes the total potential energy π over the convex set Kin

$$\pi(u) \leq \pi(v) \qquad \forall v \in Kin \quad , \tag{2}$$

where

$$\pi(v) = \frac{1}{2}\int_{\Omega} \mathbb{D}E(v)\cdot E(v)d\Omega - \ell(v) \ . \tag{3}$$

Existence and uniqueness of the solution are guaranteed [8].

If friction between the surfaces in contact must be taken into account, one can show that, assuming Coulomb's law, an extra term must be included in inequality (1). This term is

$$-\int_{\Gamma_c} \eta|\sigma_n(u)|(|v_t|-|u_t|)d\Gamma \tag{4}$$

where η is the coefficient of friction, $\sigma_n(u)$ is the normal stress on the boundary Γ_c and v_t and u_t are the tangential components of the displacements v and u on Γ_c.

Introducing the functional

$$j(u,v) = \int_{\Gamma_c} \eta|\sigma_n(u)||v_t|d\Gamma \tag{5}$$

the solution of Signorini's problem with Coulomb friction is given by $u \in Kin$ such that

$$-\int_{\Omega} \mathbb{D}E(u)\cdot E(v-u)d\Omega - j(u,v)+j(u,u)+\ell(v-u) \leq 0 \qquad \forall v \in Kin \ . \qquad (6)$$

It can be shown [3] that the solution of (6) satisfies (assuming sufficient regularity) equilibrium and constitutive equations in Ω, force (displacement) boundary conditions in $\Gamma_f(\Gamma_u)$ as well as the non-interpenetration and Coulomb-friction conditions in Γ_c:

- if $u\cdot n - s < 0$ then $\sigma_n = 0$, $\sigma_t = 0$.

- if $u\cdot n - s = 0$ then $\sigma_n \leq 0$

  if $|\sigma_t| < \eta|\sigma_n|$ then $u_t=0$

  if $|\sigma_t| = \eta|\sigma_n|$ then $u_t=-\lambda\sigma_t$ for some $\lambda\geq 0$

where σ_t is the tangential stress on Γ_c.

The issue of existence and uniqueness of solutions for (6) is still open. For a particular situation, Necas, Jarusek and Haslinger [9] showed the existence of solutions to (6) provided that η is sufficiently small. Duvaut [10] introduced the idea of a non-local friction law and established an existence result for "any" friction and also uniqueness for the case of "small" friction. Oden and Pires [11,12] proposed a class of non-local friction laws as well as numerical algorithms for obtaining approximate solutions for contact problems.

However, for the special case where the normal stress σ_n is prescribed along Γ_c, ($\sigma_n=\overline{F}_n$), Duvaut and Lions [3] established the existence and uniqueness of the solution. In this case the contact surface Γ_c is known in advance and u_n is not prescribed in Γ_c. The boundary conditions in Γ_c reduce to

$$|\sigma_t| < g \rightarrow u_t = 0$$

$$|\sigma_t| = g \rightarrow u_t = -\lambda\sigma_t \quad \text{for some } \lambda\geq 0$$

where $g=\eta|\overline{F}_n|$ is given and represents the maximum tangential stress that can be developed due to friction along Γ_c.

Introducing the functionals

$$j_g(v) = \int_{\Gamma_c} g|v_t|d\Gamma \qquad\qquad f_n(v) = \int_{\Gamma_c} \overline{F}_n v_n d\Gamma \qquad (7)$$

where g is a given positive function in Γ_c and $\overline{F}_n$ is a given normal stress distribution in Γ_C and defining the subspace $\mathbb{K}$

$$\mathbb{K} = \{v \in V: v|_{\Gamma_u} = 0\}$$

the special problem of friction with prescribed normal stress can be stated as:

find $u \in \mathbb{K}$ such that

$$-\int_\Omega \mathbb{D}E(u)\cdot E(v-u)d\Omega - j_g(v) + j_g(u) + \ell(v-u) + f_n(v-u) \leqq 0$$

$$\forall v \in \mathbb{K} . \qquad (8)$$

It can be shown [3] that to solve this inequality is equivalent to solving the following minimization problem

$$\inf_{v\in\mathbb{K}} [\pi(v)+j_g(v)-f_n(v)] \qquad (9)$$

For Signorini's problem with friction, inequality (6), the following iterative procedure has been suggested [4]:

i) Solve Signorini's problem without friction, inequality (1)

ii) With the normal stress in Γ_C found in (i) solve the special friction problem with prescribed normal stress, inequality (8)

iii) Tangential stress found in (ii) are then used as additional loads in Signorini's problem without friction (i) and the steps (i), (ii) and (iii) are repeated until convergence is (hopefully) achieved.

Introducing the functional

$$f_t(v) = \int_{\Gamma_c} \bar{F}_t \cdot v_t d\Gamma$$

where $\bar{F}_t$ is a given distribution of tangential forces along Γ_C, the procedure described above can be written as

1) Given $\bar{F}_t^{k-1}$ find u^k solution of the minimization problem

$$\inf_{v\in Kin} [\pi(v)-f_t(v)] \quad ;$$

2) Calculate $\bar{F}_n^k = \sigma_n(u^k)$ and $g = \eta|\sigma_n(u^k)|$;

3) Find u^* solution of the minimization problem

$$\inf_{v \in \mathbb{K}} [\pi(v)+j_g(v)-f_n(v)] \quad ;$$

4) Calculate $\bar{F}_t^k=\sigma_t(u^*)$ and repeat all steps for k=2,3... until convergence is achieved.

It is important to point out here that conditions under which the procedure just described converges to the solution of the original problem have not, to the authors knowledge, been given and unsuccessful applications have been reported in the literature [13].

The main difficulty in the first minimization problem is due to the constraint set Kin while in the second one the difficulty arises due to the non-differentiability of $j_g(v)$.

Panagiotopoulos [4] solves both minimization problems by non-linear programming techniques; Campos, Oden and Kikuchi [14] adopt a penalization technique in the first problem and a regularization technique in the second one. Some other possibilities are presented by Raous [15] and Haslinger and Panagiotopoulos [16]. The basic idea used here is duality [3]. The first minimization problem is substituted by the equivalent saddle-point problem

$$\inf_{v \in \mathbb{K}} \sup_{\lambda_n \geq 0} L_1(v,\lambda_n) \tag{10}$$

where the set Kin has been replaced by $\mathbb{K}$. The Lagrangean $L_1(v,\lambda_n)$ is given by

$$L_1(v,\lambda_n) = \pi(v) - f_t(v) + \int_{\Gamma_c} \lambda_n(v \cdot n - s)d\Gamma \quad .$$

In the second minimization problem the non-differentiable functional $j_g(v)$ is replaced by

$$\sup_{\lambda_t \in \Lambda} \int_{\Gamma_c} \lambda_t \cdot v_t d\Gamma \quad ,$$

where

$$\Lambda =\{\lambda =(\lambda_1,\lambda_2,\lambda_3): \sum_{i=1}^{3} \lambda_i^2(x) \leq g^2(x), \ x \in \Gamma_c\} \quad ,$$

and one is led to the equivalent saddle-point problem

$$\inf_{v \in \mathbb{K}} \sup_{\lambda_t \in \Lambda} L_2(v,\lambda_t) \quad , \tag{11}$$

with

$$L_2(v,\lambda_t) = \pi(v) - f_n(v) + \int_{\Gamma_c} \lambda_t \cdot v_t d\Gamma \quad .$$

The Lagrange multipliers λ_n and λ_t can be interpreted, by duality, respectively as the normal stress in Γ_c and the tangential stress, due to friction, in Γ_c.

3. APPROXIMATE SOLUTIONS

To obtain approximate solutions for the problems formulated in the preceding section the finite element method is used to construct finite-dimensional approximation spaces. For plane problems the following interpolation scheme can be adopted

$$v = \Phi q \quad , \quad \lambda_n = \psi p \qquad \text{and} \qquad \lambda_t = \psi t \quad ,$$

where Φ is the matrix of interpolation functions for the displacement field v in terms of the nodal unknowns q and ψ is a row-vector with interpolation functions for the Lagrange multipliers λ_n and λ_t in terms of the parameters p and t. The global interpolants Φ are constructed from local bilinear interpolants associated to a four-node quadrilateral isoparametric finite element. The interpolation of λ_t and λ_n is done by means of piecewise constant functions along the sides of the elements in Γ_c. In this way problems (10) and (11) are approximated respectively by

$$\min_{q} \max_{p \geq 0} \frac{1}{2} q'Kq - q'F - q'F_t + q'Mp + S'p \tag{12}$$

and

$$\min_{q} \max_{|\psi t| \leq g} \frac{1}{2} q'Kq - q'F - q'F_n + q'At \quad , \tag{13}$$

where K is the standard stiffness matrix, F, F_t and F_n are vectors of nodal loads which are equivalent, respectively, to the applied load system ℓ, tangential loads due to friction and normal reactions in the contact surface. The primes denote transposition and the matrices M and A and the vector S are given by

$$M = \int_{\Gamma_c} \Phi' \begin{Bmatrix} n_x \\ n_y \end{Bmatrix} \psi d\Gamma \quad , \quad A = \int_{\Gamma_c} \Phi' \begin{Bmatrix} t_x \\ t_y \end{Bmatrix} \psi d\Gamma \tag{14}$$

and

$$S = \int_{\Gamma_c} \psi' s \, d\Gamma \quad , \tag{15}$$

where (n_x, n_y) and (t_x, t_y) are the components of unit vectors respectively outward normal and tangent to the boundary Γ_c. It is clear that $q \in \mathbb{R}^n$, $p \in \mathbb{R}^m$ and $t \in \mathbb{R}^m$ where n is the number of degrees of freedom of the discrete model defined by the finite element mesh adopted and m is the number of elements along the boundary Γ_c.

As the minimization over q is unconstrained and K is assumed positive-definite this variable can be eliminated using the stationarity conditions

$$Kq - (F+F_t) + Mp = 0 \tag{16}$$

and

$$Kq - (F+F_n) + At = 0 \ . \tag{17}$$

Combining (12) with (16) and (13) with (17) one obtains

$$\min_{p>0} \frac{1}{2} p'Pp - p'd_1 \tag{18}$$

and

$$\min_{-\bar{g} \leq t \leq \bar{g}} \frac{1}{2} t'Tt - t'd_2 \tag{19}$$

where

$$P = M'K^{-1}M \ , \quad d_1 = M'K^{-1}(F+F_t) - S$$

$$T = A'K^{-1}A \ , \quad d_2 = A'K^{-1}(F+F_n) \ .$$

The vector $\bar{g} \in \mathbb{R}^m$ has its i-th entry equal to η times the absolute value of the (prescribed) normal stress along the side of the i-th element in Γ_c.

Due to the type of constraints that arise in problems (18) and (19) a very simple numerical algorithm can be used: Gauss-Seidel with relaxation and projection [17,5,6]. Finally, from (16) nodal unknowns q are obtained and element stresses can be computed.

Remark 1. Although a piecewise constant interpolation for the Lagrange multipliers λ_n and λ_t has been used to obtain the finite-dimensional approximations (12) and (13) it is important to note that matrices M and A in (12) and (13) result from the approximation of the integrals

$$\int_{\Gamma_c} \lambda_n (v \cdot n - s) d\Gamma \qquad \text{and} \qquad \int_{\Gamma_c} \lambda_t \cdot u_t d\Gamma$$

and, as such, can take different forms according to the numerical scheme adopted in the approximation of these integrals. In fact, in addition to the piecewise constant interpolation scheme already mentioned another scheme was tried which resulted in an improved performance of the numerical algorithm. The idea was to take concentrated Lagrange multipliers in the nodal points along Γ_C. In this case, M, A and S can still be given by (14) and (15) provided that we take the entries of the row-vector as Dirac's delta "functions" associated to the nodal points along Γ_C.

Remark 2. The problem of contact with Coulomb friction between two deformable bodies and the problem of indentation of a deformable body by a rigid one considering Coulomb friction can both be treated along the same basic lines.

4. A NUMERICAL EXAMPLE

This section describes the results of some numerical experiments performed with the algorithm discussed in the preceding section. The problem considered is that of a rectangular block pressed against a rigid horizontal foundation on which Coulomb's law of friction is assumed to hold. The block is also submitted to a horizontal uniformly distributed load as shown in Figure 1 and plane strain is assumed. The material of the block is homogeneous and isotropic with Young's modulus E=13000 daN/mm² and Poisson's coefficient ν=0.2. Due to the symmetry of the problem, only half of the block was discretized by means of 194 four-node isoparametric finite elements leading to a discrete model with 439 degrees of freedom.

The first load case considered here corresponds to F=15 daN/mm, f=5 daN/mm and a coefficient of friction η=1.0. Figure 2 shows the deformed mesh amplified by a factor of 100. Normal and tangential nodal

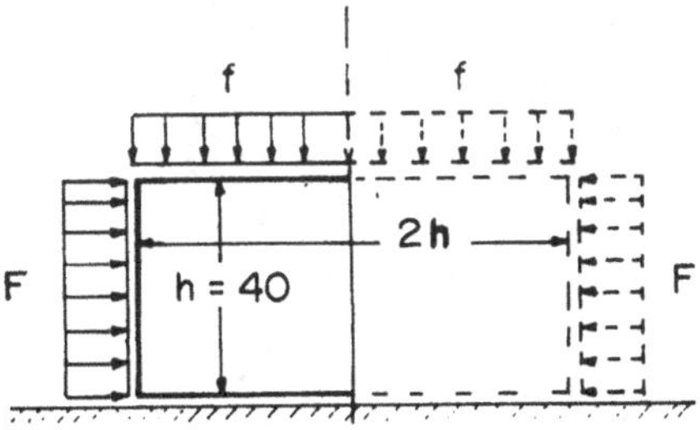

Fig. 1

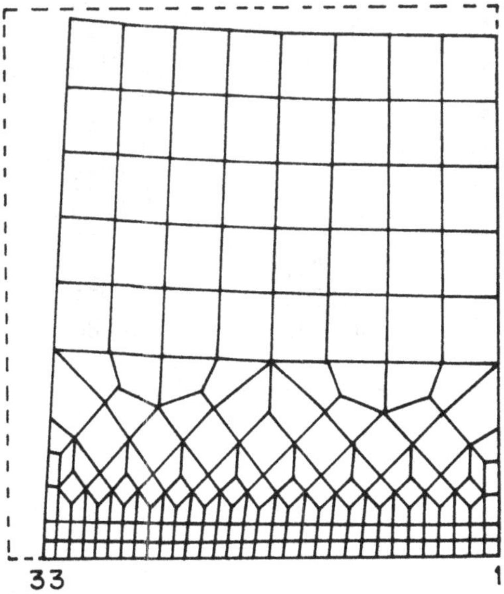

Fig. 2

displacements as well as normal and tangential nodal reactions along the

TABLE 1

NODE	u_n(mm)	u_t(mm)	F_n(dN)	F_t(dN)
1	0.	0.	5.23	0.
2	0.	0.	10.47	0.64
3	0.	0.	10.47	1.28
4	0.	0.	10.46	1.96
5	0.	0.	10.44	2.69
6	0.	0.	10.41	3.52
7	0.	0.	10.38	4.48
8	0.	0.	10.31	5.75
9	0.	0.	9.80	8.57
10	0.	0.0003	8.93	8.93
11	0.	0.0009	8.28	8.28
12	0.	0.0016	7.86	7.86
13	0.	0.0025	7.52	7.52
14	0.	0.0034	7.21	7.21
15	0.	0.0043	6.93	6.93
16	0.	0.0054	6.66	6.66
17	0.	0.0065	6.39	6.39
18	0.	0.0076	6.13	6.13
19	0.	0.0088	5.86	5.86
20	0.	0.0101	5.59	5.59
21	0.	0.0113	5.30	5.30
22	0.	0.0127	5.00	5.00
23	0.	0.0140	4.68	4.68
24	0.	0.0154	4.34	4.34
25	0.	0.0169	3.96	3.96
26	0.	0.0183	3.53	3.53
27	0.	0.0198	3.04	3.04
28	0.	0.0213	2.46	2.46
29	0.	0.0229	1.72	1.72
30	0.	0.0244	0.66	0.66
31	0.0001	0.0258	0.	0.
32	0.0003	0.0272	0.	0.
33	0.0006	0.0286	0.	0.

TABLE 2

NODE	u_t (mm)	F_n(dN)	F_t(dN)
1	0.	10.40	0.
2	0.	20.80	0.17
3	0.	20.80	0.34
4	0.	20.79	0.52
5	0.	20.77	0.70
6	0.	20.76	0.88
7	0.	20.74	1.08
8	0.	20.71	1.28
9	0.	20.68	1.50
10	0.	20.65	1.73
11	0.	20.61	1.99
12	0.	20.57	2.30
13	0.	20.52	2.66
14	0.	20.42	3.23
15	0.00003	20.17	4.03
16	0.00016	19.86	3.97
17	0.00036	19.63	3.93
18	0.00060	19.43	3.89
19	0.00088	19.24	3.85
20	0.00119	19.04	3.81
21	0.00153	18.84	3.77
22	0.00189	18.63	3.73
23	0.00228	18.40	3.68
24	0.00270	18.15	3.63
25	0.00314	17.86	3.57
26	0.00360	17.53	3.51
27	0.00409	17.14	3.43
28	0.00461	16.65	3.33
29	0.00515	16.03	3.21
30	0.00572	15.17	3.03
31	0.00633	13.88	2.78
32	0.00699	11.50	2.30
33	0.00776	3.64	0.73

contact surface are displayed in Table 1 where regions with different behaviour are easily identified: adhesion (from node 1 to node 9), sliding (from node 10 to node 30) and a region where contact was lost (from node 31 to node 33).

A second load case, corresponding to F=10 daN/mm and f=15 daN/mm, was analysed considering η=0.2 and the results are summarized in Table 2 where it can be seen that nodes 1 to 14 are in adhesion and nodes 15 to 33 are in a sliding condition.

Other load cases were also analysed and the results obtained agree with those found by Raous using a different algorithm [15].

5. CONCLUSIONS

A numerical algorithm for Signorini's problem with Coulomb friction was discussed in this work. Although the sequence of sub-problems proposed cannot be guaranteed to converge in all cases, the performance of the algorithm has been quite satisfactory in the numerical experiments performed so far. Extensions to contact between two deformable bodies and to the indentation of a deformable body by a rigid one are under way and will be reported soon.

ACKNOWLEDGEMENT

The authors would like to thank Prof. Michel Raous (Lab. de Mécanique et d'Acoustique, Marseille) for kindly providing the data for the numerical example and also for many interesting discussions during his stay in Rio de Janeiro.

REFERENCES

1. FRANCAVILLA, A.; ZIENKIEWICZ, O.C., "A Note on Numerical Computation of Elastic Contact Problems", *Int.J.Num.Meth.Engng.* 9, (1975), p. 913-924.
2. STADTER, J.T.; WEISS, R.D., "Analysis of Contact through Finite Element Gaps", *Computer and Structures* 10 (1979), p. 867-873.
3. DUVAUT, G.; LIONS, J.L., *Les Inéquations en Mécanique et en Physique*, Dunod, Paris (1972).
4. PANAGIOTOPOULOS, P.D., "A Nonlinear Programming Approach to the Unilateral Contact and Friction Boundary Value Problem in the Theory of Elasticity", *Ing.Arch.* 44 (1975), p. 421-432.
5. FEIJÓO, R.A.; BARBOSA, H.J.C., "Static Analysis of Piping Systems with Unilateral Supports", Proc. VII Brazilian Congress of Mechanical Engineering, Uberlândia, MG (1983), v. D, p. 269-279.
6. BARBOSA, H.J.C.; FEIJÓO, R.A., "Numerical Algorithms for Contact Problems in Linear Elastostatics", Proc.Conf.Struct.Anal. and Design of Nuclear Power Plants, Porto Alegre, RS (1984), v.1, p. 231-244.

7. BARBOSA, H.J.C.; FEIJÓO, R.A., "An Algorithm for the Rigid Indentation Problem in Elastostatics", Research and Development Report nº 025/84, LNCC/CNPq (in Portuguese).

8. CEA, J., *Optimization, Théorie et Algorithmes*, Dunod, Paris (1971).

9. NECAS, J.; JARUSEK, J. and HASLINGER, J., "On the Solution of the Variational Inequality to the Signorini Problem with Small Friction", *Boll.Un.Mat.Ital.*, 17B, p. 796-811 (1980).

10. DUVAUT, G., "Equilibre d'un Solide Elastique avec Contact Unilateral et Frottement de Coulomb", *C.R.Acad.Sc. Paris*, t 290, serie A, p. 263-265 (1980).

11. ODEN, J.T.; PIRES, E.B., "Nonlocal and Nonlinear Friction Laws and Variational Principles for Contact Problems in Elasticity", *J.Appl. Mech.* 50(1), p. 67-76 (1983).

12. ODEN, J.T.; PIRES, E.B., "Numerical Analysis of Certain Contact Problems with Non-Classical Friction Laws", *Computers and Structures* 16, p. 471-478 (1983).

13. KALKER, J.J., "On the Contact Problem in Elastostatics", in *Unilateral Problems in Structural Analysis*, Ed. G. Del Piero, F. Maceri, CISM Courses and Lectures nº 288.

14. CAMPOS, L.T.; ODEN, J.T. and KIKUCHI, N., "A Numerical Analysis of a Class of Contact Problems with Friction in Elastostatics", *Comp.Meth. Appl.Mech.Engng.* 34 (1982), p. 821-845.

15. RAOUS, M., "Contacts Unilatéraux avec Frottement en Visco-élasticité", in *Unilateral Problems in Structural Analysis*, Ed. G. Del Piero, F. Maceri, CISM Courses and Lectures nº 288.

16. HASLINGER, J.; PANAGIOTOPOULOS, P.D., "The Reciprocal Variational Approach to the Signorini Problem with Friction. Approximation Results", *Proc. Royal Soc. of Edinburgh*, 98A, p. 365-383 (1984).

17. GLOWINSKI, R.; LIONS, J.L.; TREMOLIERES, R., *Analyse Numérique des Inéquations Variationelles*, Dunod, Paris (1976).

GENERALIZATION OF NEWTON-TYPE METHODS OF INELASTIC CONTACT PROBLEMS

D. Bischoff
Institute of Structural Mechanics
and Numerical Mechanics, University of Hannover

ABSTRACT

Just as for elastic contact problems one can state for the elastoplastic one a minimization problem in terms of stress- and strain-rates. After a discretization in time by finite differences and in space by finite elements one has to solve either a sequence of nonlinear optimization problems with constraints or a sequence of Kuhn-Tucker inequalities. For this, sequential quadratic programming algorithms (as the natural extension of Newton-methods to problems with constraints) have proved best. So far, these methods were restricted to problems with small dimension and thus not suited for problems in the field of structural mechanics.

It is shown how to implement this method in general finite element codes by help of an algorithm given by Bertsekas. In the case that contact appears on the boundary of the body it is demonstrated how to accelerate the convergence substantially by a condensation of the linearized problems to the contact-variables.

1. INTRODUCTION

Many numerical methods for bilateral problems in nonlinear elasticity of solids use the minimum properties of the solutions. In this field Newton and quasi-Newton methods have been proved to be effective |1|. This strategy was also applied to elastoplastic problems [2], [3], [4] as far as the rate type plastic formulations can be reduced to minimum problems depending on the strain- and displacement-rates |5|, |6|, |7|.

In case of contact with an obstacle, inequality constraints are added to the free minimum problem. The numerical treatment rests either on a reduction to problems without these constraints or on a generalization of Newton or quasi-Newton methods to constrained problems |8|.

Here we consider the second case where the system of equalities and inequalities, given by the Kuhn-Tucker-conditions (which generalize the equilibrium conditions in the unconstrained case) are solved by a sequential quadratic programming method [9]. The quadratic optimization problem in every Newton step is solved by a generalized active set method, proposed by Bertsekas [10] which is not sensitive to the dimension of the problem. A considerable improvement of the performance of the algorithm was achieved by a condensation technique. Numerical tests showed that the CPU time was increased only by a factor of about 1.5 compared with the same problem with prescribed contact zone i.e. the contact problem was solved nearly in the same time as the equivalent problem without contact.

2. CHARACTERIZATION OF PLASTIFYING MATERIALS BY A MINIMUM PRINCIPLE

2.1 Continuous problem

2.1.1 Notations and basic assumptions

We consider a body of volume Ω and surface Γ in a Cartesian coordinate system $\{x_1, x_2, x_3\}$. The body is subjected to a body force $f_i(x_k)$ on Ω, surface tractions $\bar{t}_i(x)$ on Γ_t and prescribed displacements $\bar{u}_i(x)$ on the surface Γ_u. Displacements are denoted by u_i, strains by ε_{ij} and stresses by σ_{ij}. The displacements and their gradients are assumed to be small. Strains and displacements are connected by

$\varepsilon_{ij} := \frac{1}{2}\left(\frac{\partial u_i}{\partial x_j} + \frac{\partial u_j}{\partial x_i}\right)$ and loading is quasistatic.

To model plastic behavior we use a von Mises yield criterion with the yield function

$$F(\sigma_{ij}) := \frac{1}{2}\sigma_{ij}^D \sigma_{ij}^D - k_o^2$$

(σ_{ij}^D : deviatoric stress)

to define the set of admissible stresses: $R := \{\sigma \,|\, F(\sigma) \leqslant 0\}$.

The surface of the obstacle is given by

$$g(x) := \psi\ (x_1, x_2) - x_3 = 0$$

and thus the displacements are restricted by the non-penetration condition

$$\phi\ (u, x) := \psi\ (x_1+u_1(x), x_2+u_2(x)) - (x_3+u_3(x)) \leqslant 0$$

for all x on the potential contact surface Γ_c.

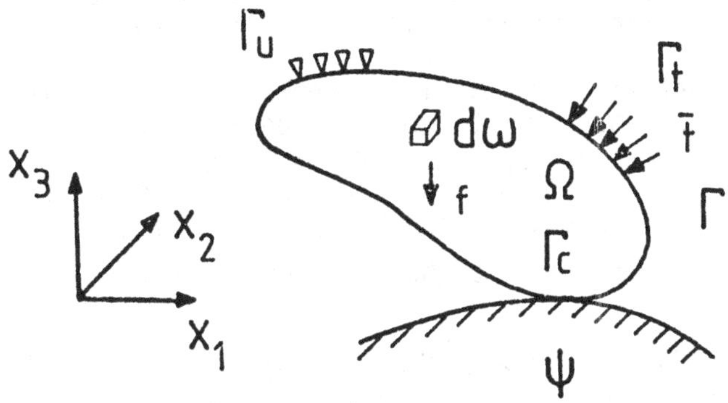

Fig. 2.1, Body Ω in contact with obstacle ψ

What we are looking for are the displacements u(t) in the equilibrium states, where t is the parameter describing the load history. The quantities which are given by a minimum property are however the displacement rates $\dot{u}$ of a body at equilibrium. One gets them as the solution of

$$\Pi(\dot{u}) = \min \{ \Pi(\dot{v}) | \dot{v} \in \mathbb{V}_c \}$$

$$\text{where} \quad \Pi(\dot{v}) := \int_\Omega (\tfrac{1}{2}\ \dot{\varepsilon}_{ij}\ P\ D_{ijk\ell}\ \dot{\varepsilon}_{k\ell} - \dot{f}_i \dot{u}_i) dw - \int_{\Gamma_t} \dot{\bar{t}}_i \dot{u}_i\ d\gamma \quad ,$$

P is the projection with respect to the metric D^{-1} on the tangential halfspace at the set Q:= R-σ in the origin (see e.g. [11], [12]) and $\mathbb{V}_c$ is the space of kinematically admissible displacement rates, that is

$$\mathbb{V}_c := \{ \dot{u} \in \mathbb{V} |\ \dot{u}_i = \dot{\bar{u}}_i \text{ on } \Gamma_u, \frac{\partial}{\partial u_i} \dot{u}_i \leqslant 0 \text{ if } \phi(u,x)=0 \text{ on } \Gamma_c \}$$

$\mathbb{V}$ is an appropriate function space that ensures the existence of solutions. In our case we should choose $\mathbb{V}$ as the space of bounded deformation, introduced by Temam and Strang [13].

The displacement u(t) itself is then evaluated by

$$u(t) := u_0 + \int_0^t \dot{u}(\tau)d\tau$$

2.2 Discretized problem

The discretization of the problem is performed in two steps: First, in regard to time the differential quotients are approximated by difference-quotients, that is by $\Delta_n t := t_n - t_{n-1}$ and analogous definitions for $\Delta_n \varepsilon_{ij}$, $\Delta_n f_i, \ldots$, and the projection $_nP$ on the halfspace by the projection on the yieldsurface itself (called P_n in the sequel).

Then the space is discretized by finite elements, that is $\mathbb{V}$ is replaced by a finite dimensional space $\mathbb{V}_h$ by choosing an appropriate basis of shape functions $\{N_h(x)\}$. Thus $\Delta_n u(x)$ is replaced by $\Delta_n u_h(x) := \Delta_n u_h N_h(x)$ with $\Delta_n u_h \in {_n}\mathbb{K}_h \subset \mathbb{R}^M$, $\Delta_n \varepsilon(x)$ by $B\Delta_n u_h$ and so on. The constraints $_n\mathbb{K}_h$ are determined by the contact condition $\phi({_n}u_h + \Delta_n v_h)|_{\Gamma_c} \leqslant 0$ and the boundary condition is equivalent to the M_c inequalities:

$${_n}\phi_h(\Delta_n v_h) := \phi(({_n}u_h + \Delta_n v_h)N_h(x_h), x_h) \leqslant 0 \quad \text{for all } x_h \in \Gamma_c$$

($M_c := |I_c|$, where I_c is the index set of the potential contact nodes and

$${_n}u_h := \sum_{\nu=1}^{n-1} \Delta_\nu u_h \text{).}$$

Thus $\quad {_n}\mathbb{K}_h := \{\Delta_n v_h \in \mathbb{R}^M | \; {_n}\phi_h(\Delta_n v_h) \leqslant 0, + \text{ boundary cond.}\}$

Details concerning the determination of $_n\mathbb{K}_h$ can be found in [14].

A neccessary condition for the minimization points of the fully discretized problem is the following system of equations and inequalities:

$$\nabla {_n}L(\Delta_n u_h, {_n}\ell) = \nabla {_n}\Pi_h(\Delta_n u_h) + {_n}\ell^T \nabla {_n}\phi_h(\Delta_n u_h) = 0 \qquad (2.8\text{ a})$$

$${_n}\ell^T \; {_n}\phi_h(\Delta_n u_h) = 0 \qquad (2.8\text{ b})$$

$${_n}\ell \geqslant 0 \qquad (2.8\text{ c})$$

$${_n}\phi_h(\Delta_n u_h) \leqslant 0 \qquad (2.8\text{ d})$$

with $\quad {_n}L(\Delta_n u_h, {_n}\ell) := {_n}\Pi_h(\Delta_n u_h) + {_n}\ell^T \, {_n}\phi_h(\Delta_n u_h) \qquad (2.9)$

and $\quad \nabla {_n}\Pi_h(\Delta_n v_h) := \int_\Omega (P_n DB)^T B dw \, \Delta_n v_h - \Delta_n f_h$

They generalize the equilibrium conditions for problems without constraints and are called "Kuhn-Tucker conditions" (see e.g. [10]) in the mathematical literature. The multipliers ${}_n\ell$ are the reaction forces in the contact nodes.

3. EXSTENSION OF NEWTON'S METHOD TO PROBLEMS WITH CONSTRAINTS

3.1 Sequential quadratic Programming

The linearization of the Kuhn-Tucker conditions (2.8) in the pair $(\Delta_n v_h^k, {}_n\ell)$ leads to the following system of equations and inequalities:

$$\nabla_n L(\Delta_n v_h^k, {}_n\ell^k) + \nabla_n^2 L(\Delta_n v_h^k, {}_n\ell) d^k + \nabla_n \phi_h(\Delta_n v_h^k) y^k = 0 \tag{3.1 a}$$

$$\nabla_n \phi_h(\Delta_n v_h^k)^T d^k + {}_n\phi_h(\Delta_n v_h^k) \leqslant 0 \tag{3.1 b}$$

$${}_n\ell^k + y^k \geqslant 0 \tag{3.1 c}$$

$$\{\nabla_n \phi_h(\Delta_n v_h^k)^T d^k + {}_n\phi_h(\Delta_n v_h^k)\}^T ({}_n\ell^k + y^k) = 0 \tag{3.1 d}$$

which is, after the substitution $\tilde{y}^k := {}_n\ell^k + y^k$ the Kuhn-Tucker condition of the minimum problem:

$$\frac{1}{2} d^{kT} \nabla_n L(\Delta_n v_h^k, {}_n\ell^k) d^k + \nabla_n \Pi_h(\Delta_n v_h^k) \rightarrow \min \tag{3.2 a}$$

$$\text{subject to } \nabla_n \phi_h(\Delta_n v_h^k)^T d^k + {}_n\phi_h(\Delta_n v_h^k) \leqslant 0. \tag{3.2 b}$$

In analogy to the classical Newton method we get as the new iterate for the displacements:

$${}_n v_h^{k+1} = {}_n v_h^k + d^k$$

and for the Lagrange multipliers

$${}_n\ell^{k+1} = {}_n\ell^k + y^k \quad \text{(i.e. the multipliers of the linearized problem)}$$

To ensure global convergence the augmented Lagrangian ${}_nL_r$ given by

$${}_nL_r(\Delta_n v_h^k, {}_n\ell^k) := {}_n\Pi_h(\Delta_n v_h^k) + {}_n\ell^{k^T} \max\{{}_n\phi_h(\Delta_n v_h^k), -\frac{{}_n\ell^k}{r}\} + \frac{1}{2} r|\max\{{}_n\phi_h(\Delta_n v_h^k), -\frac{{}_n\ell^k}{r}\}|^2 \quad , r > 0$$

can be used as a merit function in order to perform a line search in direction (d^k, y^k) [15]. Thus in every load increment $\Delta_n t$ we have to perform the following algorithm.

3.2 Algorithm for plastic contact problems

STEP 1: Given: ${}_{n-1}u_h$, ${}_{n-1}\sigma_h$, $\Delta_n v_h^{\,0}$, r, ${}_{n-1}\ell$,

constants eps1, eps2, eps3, eps4 > 0,

${}_n\ell^0 = {}_{n-1}\ell$, $\quad k = 0$.

STEP 2: Determine a new displacement increment $d_o^{\,k}$ and the corresponding multiplier $\tilde{y}_o^{\,k} (= {}_n\ell_o^{\,k})$ by solving the quadratic minimum problem (3.2),

$i = 0$.

STEP 3: Determine the stress increment by projection of a trial stress.

STEP 4: Determine ${}_nL_r(\Delta_n v_h^{\,k} + d_i^{\,k}, {}_n\ell_i^{\,k})$

IF $|\nabla {}_nL_r(\Delta_n v_h^{\,k} + d_i, {}_n\ell_i^{\,k})^T \Delta_n v_h^{\,k}| < \text{eps1}$

THEN ${}_n\ell^{k+1} = {}_n\ell_i^{\,k}$, $\quad \Delta_n v_h^{\,k+1} = \Delta_n v_h^{\,k} + d_i^{\,k}$, $\quad k = k + 1$,

ELSE $d_{i+1}^{\,k} = \alpha_i d_o^{\,k}$, $\quad {}_n\ell_{i+1}^{\,k} = {}_n\ell^k + \alpha_i(\tilde{y}_o^{\,k} - {}_n\ell^k)$

(α_i should be determined by a line search)

$i = i + 1$, GOTO STEP 1

STEP 5: IF $(\|\nabla {}_nL_r(\Delta_n v_h^{\,k}, {}_n\ell^k)\| < \text{eps2}$ AND ${}_n\ell^k > -\text{eps3}$

AND ${}_n\phi_h(\Delta_n v_h^{\,k}) < \text{eps4}$)

THEN ${}_nu_h = {}_{n-1}u_h + \Delta_n v_h^{\,k}$, $\quad {}_n\sigma_h = {}_{n-1}\sigma_h + \Delta_n \sigma_{h_i}^{\,k}$

STOP

The algorithm has the advantage that the solution procedure (STEP 1, STEP 2, STEP 4, STEP 5) is independent of the mechanical model and its discretization (STEP 3).

4. IMPLEMENTATION

4.1 Solution of a quadratic minimum problem

The crucial point in the algorithm for contact problems is the solution of the quadratic optimization problem (3.2) which has in an abstract setting the form

$$Q(d) := \frac{1}{2} d^T H d - b^T d \rightarrow \min \tag{4.1 a}$$

$$C(d) := Ad + c \leq 0 \tag{4.1 b}$$

where Q and A be nxn and mxn matrices, respectively. A has full rank and Q is positive definite and symmetric. This problem can be solved by an algorithm using an active set strategy. We will here outline only the basic ideas of this method - a more detailed description can be found in [16] or [17].

Basic ideas:

(i) Change the working set W (subset of the nodes currently being in contact) in a controlled way.

(ii) Descent type iteration.

Concerning (i):

Adding nodes to the working set: Every node coming into contact during the iterative solution of $Q(d) \rightarrow \min$ subject to $C(d)_i = 0$, $i \in I_W$ will be added to the working set.

Dropping nodes from the working set:
Let d_W be a feasible solution of $Q(d) \rightarrow \min$ subject to a working set W.

There arise two cases: In the first, all reaction forces in the contact surface are of compression type - then d_W is the solution of the original problem. In the second, there is a reaction force ℓ_i in the contact surface which is of tension type - then the node i can be removed from the working set and the iteration starts with the new set I_W.

Concerning (ii):

A descent method of the projected Newton or projected quasi-Newton type should be used, together with a line search technique to ensure convergence (see e.g. [18]).

By applying active set strategies to contact problems one must take care of the special structure of the matrices (sparse) in order to get an effective algorithm, (see e.g. [19]).

Since the sparsity of the matrices is being destroyed by most algorithms for general linear constraints it is essential to transform the contact problems to bound constraints by the use of local coordinate systems.

For such type of constraints we will give an algorithm in the next section which is well suited for sparse problems and has the advantage of adding and dropping active constraints in a simultane ousway.

4.2 Algorithm for bound constraints

Especially for bound constraints it is possible, as we will see, to construct algorithms which preserve the sparsity of the matrices. Thus we have first to transform the general constraints $Ad+c \leq 0$ to bound constraints. Since (4.1 b) decomposes in M_C independent linear inequalities with only three variables this can be done by local coordinate transformations (called K in the sequel). Graphically this means, that we have different coordinate systems in the contact nodes with axes parallel and perpendicular to the tangents of the contact surface. With the substitution $\hat{d} = K^{-1}d$ the problem (4.1) transforms to

$$\hat{Q}(\hat{d}) := \frac{1}{2} \hat{d}^T \hat{H} \hat{d} - \hat{b}^T d \qquad (4.2\text{ a})$$

$$\hat{C}(\hat{d}) := -\hat{d} \leq c_i \qquad i \in I_C \qquad (4.2\text{ b})$$

where $\hat{H} := K^T H K$ and $\hat{b} := b^T K$

In the following we consider problem (4.2) instead of (4.1).

With the following notations:

Projection of $z \in \mathbb{R}^M$ on the contact surface:

$$(z_C)_i := \begin{cases} \max\{-c_i, z_i\} & \text{if } i \in I_C \\ z_i & \text{otherwise} \end{cases}$$

Diagonalization of a matrix H with respect to I_W:

$$(H_W)_{ij} := \begin{cases} 0 & \text{if} \quad i \neq j \quad (i \in I_W \vee j \in I_W) \\ H_{ij} & \text{otherwise} \end{cases}$$

it holds [10]:

(i) a reduction of $\hat{Q}$ is possible along

$$(\hat{d} - \alpha(\hat{H}_W)^{-1} \nabla \hat{Q}(\hat{d}))_C =: \hat{d}(\alpha), \qquad \alpha \in \mathbb{R}_+$$

if $\hat{d}$ is not a solution of (4.2)

(ii) $\hat{d}$ is a solution of (4.2) $\Leftrightarrow$ $\hat{d} = \hat{d}(1)$

(iii) Algorithm for (4.2):

START: Given: constants $\varepsilon > 0$, β (0,1), γ (0, 1/2), $\hat{d}^0 \in \mathbb{R}^M$

$k = 0$

STEP 1: Determination of the working set I_{W^k} :

$$\delta_k := |\hat{d}^k - (\hat{d}^k - \nabla\hat{Q}(\hat{d}^k))_C|$$

$$\varepsilon_k := \min \{ \varepsilon, \delta_k \}$$

$$I_{W^k} := \{ i \in I_C | 0 \leq \hat{d}_i^k + c_i \leq \varepsilon_k \wedge (\nabla\hat{Q}(\hat{d}^k))_i > 0 \}$$

STEP 2: Reduction of $\hat{Q}$ in the plane given by I_{W^k}

2.1: Determination of a descent direction

$$D^k := (H_{W^k})^{-1}$$

$$p^k := D^k \nabla\hat{Q}(\hat{d}^k)$$

IF $\hat{d}^k(1) := (\hat{d}^k - p^k)_C = \hat{d}^k$: STOP

2.2: Line search (Armijo-like)

Determine m_k as the smallest positive integer for which it holds:

$$\hat{Q}(\hat{d}^k) - \hat{Q}(\hat{d}^k(\beta^m)) > \gamma\{\beta^m \sum_{i=1}^{M} (\nabla\hat{Q}(\hat{d}^k)_i (\hat{d}_i{}^k - (\hat{d}^k(\beta^m))_i)\}$$

Choose $\alpha_k := \beta^{m_k}$

$$\hat{d}^{k+1} := \hat{d}^k(\alpha_k)$$

$k = k + 1$

GOTO STEP 1

4.3 Convergence

Under suitable assumptions one can show [10] that

(i) $\hat{d}^k$ converges quadratically and

(ii) there exists a $\bar{k} \in \mathbb{N}$ for which the working set I_{W^k} corresponds to the active set I_A for all k greater or equal than $\bar{k}$.

In contrast with this nice theoretical result the algorithms works very poor, especially for problems with constraints only on the surface. The reason for this is the projection of penetrating nodes onto the contact surface. This results in very distorted elements storing large internal energies.

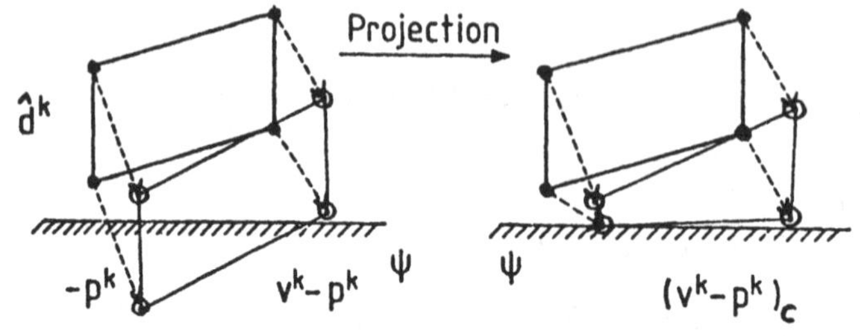

Fig. 4.1, Projection of penetrating nodes

Physically meaningful would be a simultaneous change of interior nodes lying in the neighbourhood of penetrating contact nodes.

This could be done by a preliminary condensation of the whole system to the potential contact variables ; that is to split the system into free variables $\hat{d}_f$ and constraint ones $\hat{d}_c$, accordingly the matrix $\hat{H}$ in $\hat{H}_{ff}$, $\hat{H}_{fc}$, $\hat{H}_{cf}$, $\hat{H}_{cc}$ and the load vector $\hat{b}$ in $\hat{b}_f$ and $\hat{b}_c$.

A necessary condition for $\hat{d}_f$ to be part of the solution of (4.2) would be:

$$\hat{H}_{ff}\hat{d}_f + \hat{H}_{fc}\hat{d}_c = \hat{b}_f.$$

Thus $\hat{d}_f$ can be eliminated and we are left with the condensed problem:

$$\bar{Q}(\hat{d}_c) = \frac{1}{2}\,\hat{d}_c^{\,T}\,\bar{H}\hat{d}_c - \bar{b}\hat{d}_c \quad , \tag{4.3 a}$$

$$\hat{C}(\hat{d}_c) = \hat{d}_{ci} \leqslant -\,c_i \quad , \qquad i \in I_c \tag{4.3 b}$$

where $\bar{H} := \hat{H}_{cc} - \hat{H}_{cf}\,\hat{H}_{ff}^{-1}\,\hat{H}_{fc}$,

and $\bar{b} := \hat{b}_c - \hat{H}_{cf}\,\hat{H}_{ff}^{-1}\,\hat{b}_f$.

With this condensed version of the Bertsekas-algorithm we got the same number of iteration steps for problems with and without contact and the CPU-times showed only an enlargement of a factor 1.5 in comparison with problems without contact.

It should be noted that it depends on the ratio of the number of free variables to constrained ones, whether it is more effective to use the condensed version explicitly or to change the original Bertsekas-algorithm in such a way, that $d^k(\alpha)$ is splitted into $d_f^k(\alpha)$ and $d_c^k(\alpha)$ where $d_c^k(\alpha) := (d^k - p^k)_c$ and $d_f^k(\alpha) := H_{ff}^{-1}(b_f - H_{fc}d_c^k(\alpha))$, that means, that internal nodes are adapted in an optimal way to the projected penetrating nodes.

5. EXAMPLE: CONTACT OF A PLASTIFYING BAR

For numerical tests we considered a parabolically curved symmetric bar with width 1 cm (see Fig. 5.1), whose material properties are given by the Young modulus E = 21000 kN/cm², the Poisson ratio υ = 0,3 and the yield limit k_0 = 40 kN/cm². The bar was discretized by 20 4-node, 20 8-node and 80 4-node plane elements, respectively, while the load of 50 kN was applied in 1, 2 and 4 increments.

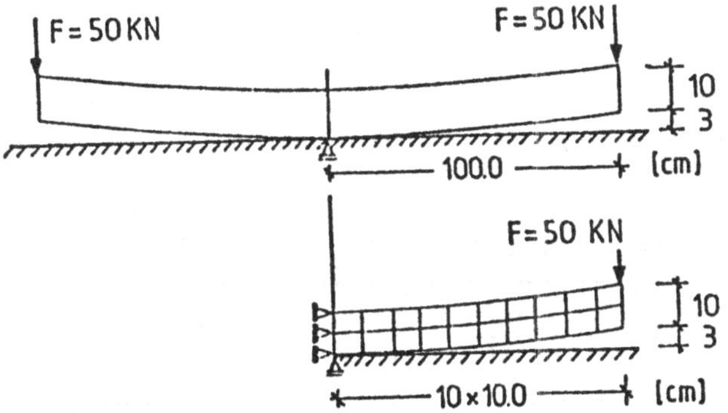

Fig. 5.1, Parabolically curved bar on a rigid plane with two point loads at the end

The results of calculation (for the discretization with 20 8-node elements and 4 load increments) are shown in Fig. 5.2 and a comparison with a purely elastic calculation shows a smaller contact zone and larger reaction forces for the plastic problem. An interesting effect is the strong changing of the plastic zone and plastic unloading in large parts of the beam in spite of increased applied force.

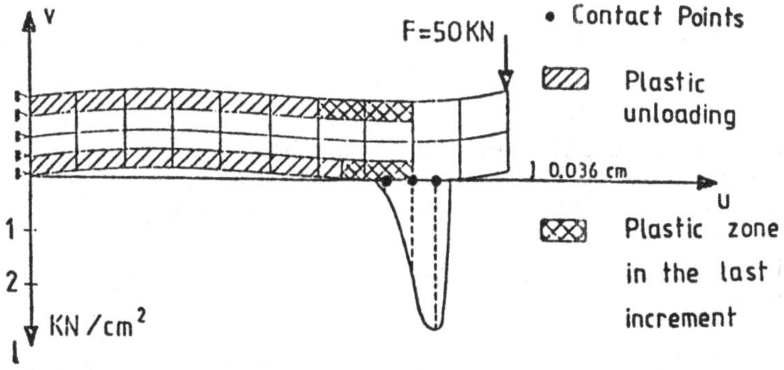

Fig. 5.2, Plastic contact (20 8-node elements, 4 increments)

The numerical behavior of the algorithm is given in table 5.1 which shows the number of Newton steps and in every Newton-step the number of active set steps times the number of Armijo-line search steps. As seen in the table, this line search is called only very rarely and thus doesn't consume almost any time - although it is essential for the convergence.

The determination of the exact contact zone is indicated by a dashed line. One can see that this zone is always reached within the first three iteration steps, i.e. before the plastic iteration has converged, which shows the excellent behavior of this algorithm. We have compared the 4-node and the 8-node element for the bar discretized with 20 elements and the load applied within 1 and 4 increments. (The results for other incrementations or discretizations show no significant differences).

Table 5.1: Numerical behavior of the algorithm

	20 4-node elements					20 8-node elements				
Newton steps	1 increm.	4 increments I	II	III	IV	1 increm.	4 increments I	II	III	IV
1	7x1	1x1	2x1	2x1	2x1	8x1	7x1	7x1	4x1	4x1
2	1x1	2x1	3x1	1x1	1x1	4x1	1x5+ 1x2	6x1	2x1	3x1
3	1x1	1x1	1x1	1x1	1x1	2x1	1x1	1x1	1x1	1x1
4	1x1	1x1	1x1	1x1	1x1	1x1	1x1	1x1	1x1	1x1
5	1x1	1x1	1x1	1x1	1x1	1x1	1x1	1x1	1x1	1x1
6			1x1	1x1		1x1	1x1	1x1		
7							1x1			

A comparison of the calculation with a prescribed contact zone shows that the algorithms need exactly the same number of Newton-steps as it does with a variable zone. The CPU time for the contact problem increases about the factor 1.5 due to the condensation and the optimization algorithm. This is just the amount which was recognized too for the purely elastic problem.

The support of the Stiftung Volkswagenwerk is gratefully acknowledged.

REFERENCES:

1. Dennis, J., Moré, J.: "Quasi Newton Methods, Motivation and Theory", SIAM Review 19, (1977), 46-89.

2. Strang, G., Matthies, H., Temam, R.: "Mathematical and Computational methods in plasticity", in: S. Nemat-Nasser (ed.): Variational Methods in the Mechanics of Solids, Pergamon Press, Oxford, (1980), 20-28.

3. Bathe, K.J., Snyder, M.D., Cimento, A.P., Rolph, W.D.: "On some current procedures and difficulties in finite element analysis of elastic-plastic response", Computers & Structures, 12 (1980), 607-624

4. Simo, J.C., Taylor, R.L.: "Consistent tangent operators for rate-independent elastoplasticity", Comp. Meths. in Appl. Mechs. and Eng., 48 (1985), 101-118.

5. Martin, J.B.: "On the kinematic minimum principle for the rate problem in classical plasticity", J. Mech. Phys. Solids, 23, (1975), 123-128.

6. Sewell, M.J.: "The governing equations and extremum principles of
7. elasticity and plasticity generated from a single functional", part I, II; J. Structural Mech., 2, (1972), 1-32, 135-158.

8. Bischoff, D., Plank, L., Stein, E.: "Solution strategies for elastic and inelastic contact problem of solids", in: E. Hinton,...(eds.): Numerical methods in transient and coupled problems, J. Wiley, L., to appear.

9. Schittkowski, K.: "The nonlinear programming method of Wilson, Han and Powell with an augmented Lagrangian type line search function", part I, II, Numerische Mathematik, 39, (1981), 83-127.

10. Bertsekas, D.P.: "Constrained optimization and Lagrange multiplier methods", Acad. Pr., N.Y., (1982).

11. Stein, E., Lambertz, K.H., Plank, L.: Ultimate load analysis of folded plate structures with large elastoplastic deformations - theoretical and practical comparisons of different FE-algorithms, Numeta 85, Swansea, U.K., (1985).

12. Gruttmann, F.: Konsistente Steifigkeitsmatrizen in der Elasto-Plastizitätstheorie, Workshop: Diskretisierungen in der Kontinuumsmechanik - Finite Elemente und Randelemente, Bad Honnef, Gr 1-Gr 14 (1985)

13. Temam, R., Strang, G.: "Functions of bounded deformation", Arch. Rat. Mech. Anal., 75 (1980), 7-21.

14. Wriggers, P., Nour-Omid, B.: "Solution methods for contact problems", Rep. UCB/SESM 84/09, Univ. of California, Berkeley, 1984.

15. Schittkowski, K.: "On the convergence of a sequential quadratic programming method with an augmented Lagrangian line search function", Mathematische Operationsforschung und Statistik,

Series Optimization, 14, (1983), 197-216.

16. Gill, P.E., Murray, W., Wright, M.H.: "Practical Optimization", Acad. Pr., London, (1981).

17. Bischoff, D., Mahnken, R.: "Zur Konvergenz von Kontaktalgorithmen, die auf Active Set Strategien beruhen", GAMM-Seminar: Unilaterale Probleme, Stuttgart, 1984.

18. Powell, J.J.D.: "Introduction to constrained optimization", in: Gill, P.E., Murray, W. (eds.): Numerical methods for constrained optimization, Acad. Pr., London (1974), 1-28.

19. Bischoff, D., Mahnken, R.: "Projizierte Newton- bzw. Quasi-Newton-Verfahren für Kontaktprobleme", to appear.

CONFRONTO TRA ESPERIMENTO E MODELLO TEORICO PER UN PROBLEMA DI STRUTTURE IN MURATURA

Briccoli Bati S., Di Pasquale S., Paradiso M., Tempesta G., Tonietti U.
Dipartimento di Costruzioni, Università di Firenze

SOMMARIO

In questa nota viene proposta l'analisi teorica e sperimentale relativa a semplici elementi strutturali costituiti da materiale scarsamente reagente a trazione. Lo scopo principale è quello di rendere ragione di alcuni aspetti caratteristici resi evidenti dalle indagini sperimentali [1] dei quali non è noto agli autori sia già stata data giustificazione. Infatti i diagrammi carico-spostamento, ottenuti con procedimenti a deformazione guidata, registrano, nelle fasi iniziali del processo di carico, la presenza di fenomeni tipo snap-through in corrispondenza dell'apertura delle fratture.

Nelle pagine che seguono si mostra come l'assunzione di un modello di comportamento del materiale di tipo elasto-plasto-fragile permetta di descrivere qualitativamente e quantitativamente il comportamento rilevato per via sperimentale: ciò garantisce della buona consistenza del modello teorico proposto.

1. INTRODUZIONE

Il problema della determinazione dello stato di tensione all'interno di un solido murario è stato affrontato negli ultimi anni da diversi autori mediante formulazioni basate sull'ipotesi che il materiale muratura presenti un comportamento di tipo unilatero, nel senso che sia incapace di reagire a trazione [2], [3], [4], [5], [6], [7], [8].

Occorre precisare che le teorie basate sull'ipotesi di materiale a comportamento unilatero non riescono a spiegare i fenomeni rilevati per via sperimentale; nè, viceversa, l'ipotesi suddetta sembra suscettibile di conferme sperimentali. Infatti sperimentando con le murature non è possibile eliminare la loro pur modesta resistenza a trazione. Il richiesto confronto teorico sperimentale può essere quindi effettuato solo postulando un materiale con rottura fragile e prefissata resistenza a trazione. E' intenzione degli autori dimostrare che, in questa ipotesi, la struttura reagente ultima cui si perviene al crescere del carico coincide, con buona approssimazione, con la parte di struttura effettivamente reagente nel caso di materiale rigorosamente non reagente a trazione.

Si prendono in esame strutture elementari, convenzionalmente schematizzabili come monodimensionali, costituite, appunto, da materiale poco reagente a trazione: ci si pone come obiettivo quello di controllare l'evolvere della struttura reagente al crescere di un fattore di carico λ .

A tale scopo sono stati esaminati due modelli strutturali per i quali si ritiene valida l'ipotesi di Navier delle sezioni piane. Inoltre si suppone ininfluente la deformazione dovuta al taglio. L'analisi viene con

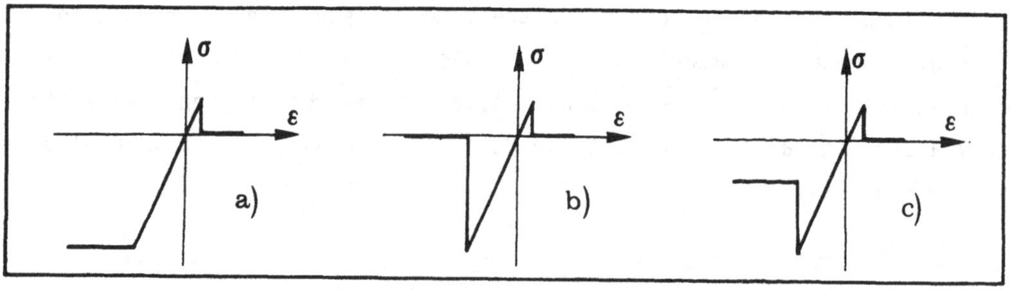

Fig. 1

dotta in regime di piccoli spostamenti supponendo in ogni caso che i fenomeni analizzati non siano soggetti ad instabilità di tipo euleriano.

Il primo modello esaminato è costituito da un pilastro caricato eccentricamente. Come è ben noto questa struttura ha costituito da sempre il banco di prova di tutte le teorie sui solidi poco reagenti a trazione (a tale proposito basti ricordare le celebri memorie di A. Signorini [9]).

L'analisi del secondo modello, costituito da una trave-arco incastrata agli estremi, ha come scopo quello di rendere ragione, mediante una simulazione analitica del fenomeno strutturale, di alcuni aspetti caratteristici dei diagrammi carico-spostamento ottenuti con indagini sperimentali.

In ambedue i casi il modello di comportamento assunto per il materiale tiene conto della, seppure scarsa, resistenza a trazione. La resistenza a compressione viene espressa in funzione di quella a trazione; per la scelta del rapporto 'n' tra i moduli delle due tensioni limite si è fatto riferimento a Résal [10] che indica in $10 \leqslant n \leqslant 20$ i valori comunemente accettati per il materiale muratura.

In ambedue le analisi il modello costitutivo assunto per il materiale, pur conservando l'ipotesi di rottura fragile a trazione, tiene conto della possibile differenza del comportamento a compressione in corrispondenza del valore limite: sono infatti ipotizzati sia un comportamento di tipo plastico che uno di tipo fragile (Fig. 1a, 1b).

La formulazione analitica descrive il problema secondo fasi contraddistinte dal raggiungimento del valore della tensione limite in corrispondenza dei bordi delle sezioni reagenti. Si tratta infatti di descrivere, in termini di carico e spostamento, il comportamento di strutture la cui configurazione è variabile in funzione di carichi crescenti monotonicamente. Si tiene conto della variazione della struttura effettivamente reagente per effetto della comparsa delle fratture. Il processo viene seguito fase per fase, simulando un procedimento a deformazione guidata con lo scopo di poter cogliere anche gli eventuali rami instabili del percorso di equilibrio.

2. IL PILASTRO CARICATO ECCENTRICAMENTE

Si analizza il problema di un pilastro soggetto a carico eccentrico crescente monotonicamente secondo un moltiplicatore λ , costituito da materiale con diversa tensione limite a trazione e compressione (Fig.2). Vengono considerati modelli di comportamento a compressione di tipo elasto-plastico (Fig. 1a) ed elasto-fragile (Fig. 1b)conservando l'ipotesi di rottura fragile a trazione in ambedue i casi.

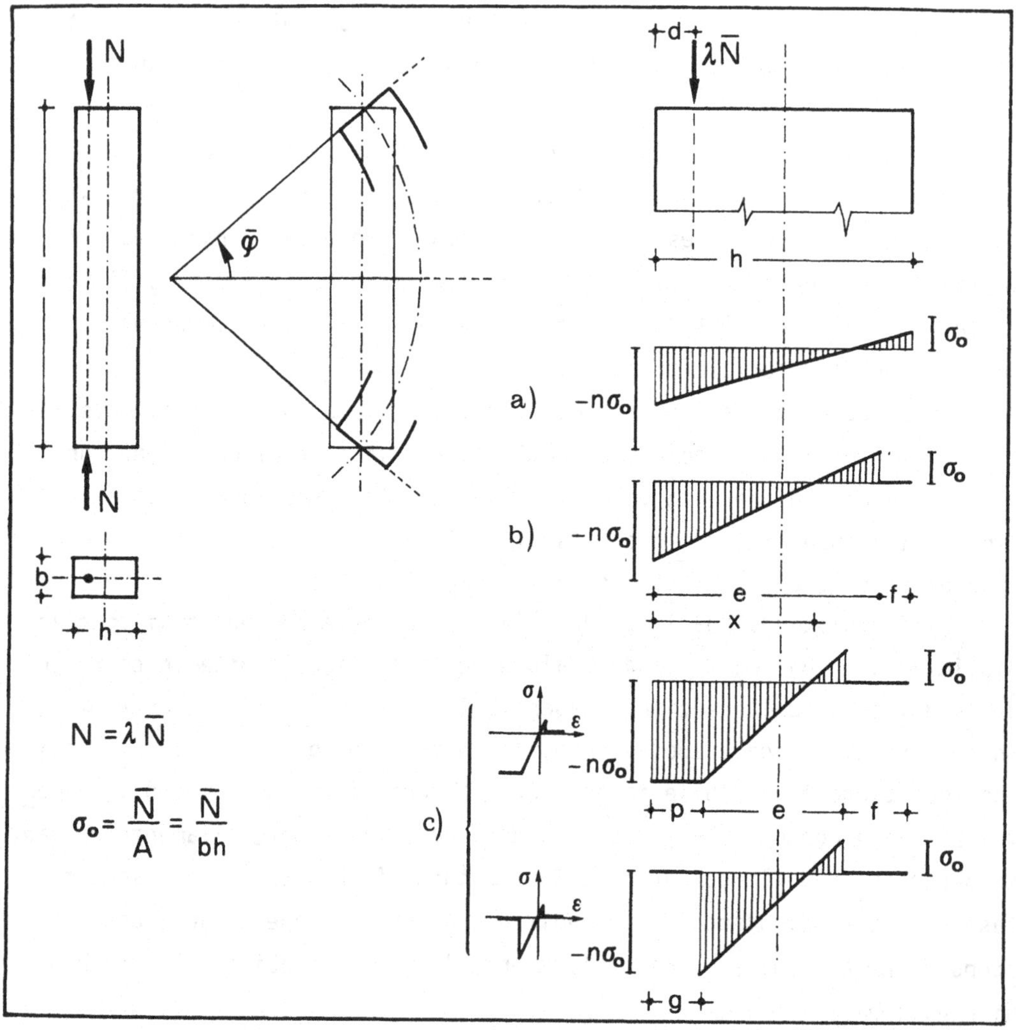

Fig. 2

Sia dunque σ_0 il valore della tensione limite a trazione(con $\sigma_0=\bar{N}/A=$ $=\bar{N}/bh$) e $-n\sigma_0$ il valore della tensione limite a compressione. Le fasi del comportamento della struttura sono contraddistinte da valori del moltiplicatore λ compresi in tre intervalli contigui:

$$0 \leqslant \lambda \leqslant \lambda_1$$
$$\lambda_1 \leqslant \lambda \leqslant \lambda_2$$
$$\lambda_2 \leqslant \lambda$$

dove λ_1 è il valore cui corrisponde il raggiungimento della tensione limite a trazione σ_0 al bordo teso della sezione del pilastro e λ_2 quello cui corrisponde il raggiungimento della tensione limite a compressione $-n\sigma_0$.

La condizione $\lambda_1 < \lambda_2$ impone che la distanza h/2-d (Fig.2) del centro di pressione dal baricentro sia sufficientemente grande; più precisamente deve essere:

$$d < \frac{n-2}{3(n-1)}\, h$$

La figura 3 mostra i percorsi di equilibrio $\lambda, \bar{\varphi}$ per due diversi valori di 'n'. Il parametro $\bar{\varphi}$, che rappresenta la rotazione della sezione a meno della costante $\frac{2\bar{N}l}{EAh}$, in questa fase è fornito da :

$$\bar{\varphi} = 3\lambda\,(1-2\bar{d}) \quad \text{con } \bar{d}=d/h$$

Appare evidente che i tratti OA, relativi alla fase elastica, non dipendono da 'n': il valore $\lambda = \lambda_1$ discende esclusivamente dalla condizione limite al bordo teso della sezione (Fig. 2a); in particolare si ha:

$$\lambda_1 = 1/(2-6\bar{d})$$

I tratti AB sono quelli relativi alla fase elastica fessurata. Tale fase comporta una riduzione della sezione resistente a causa della comparsa della frattura 'f'. Con riferimento alla figura 2b le equazioni che descrivono il problema possono essere poste nella forma:

$$\left\{\begin{array}{l} \dfrac{1}{2}\,\dfrac{x^2-(e-x)^2}{(e-x)h} = \lambda \\[2ex] \dfrac{x^3}{6h^2(e-x)} - \dfrac{(e-x)}{6h^2}\,(2e+x) = \lambda\,\dfrac{d}{h} \end{array}\right. \qquad (2.1)$$

Di esse, la prima esprime l'equilibrio verticale e la seconda l'equilibrio alla rotazione rispetto al baricentro della sezione effettivamente

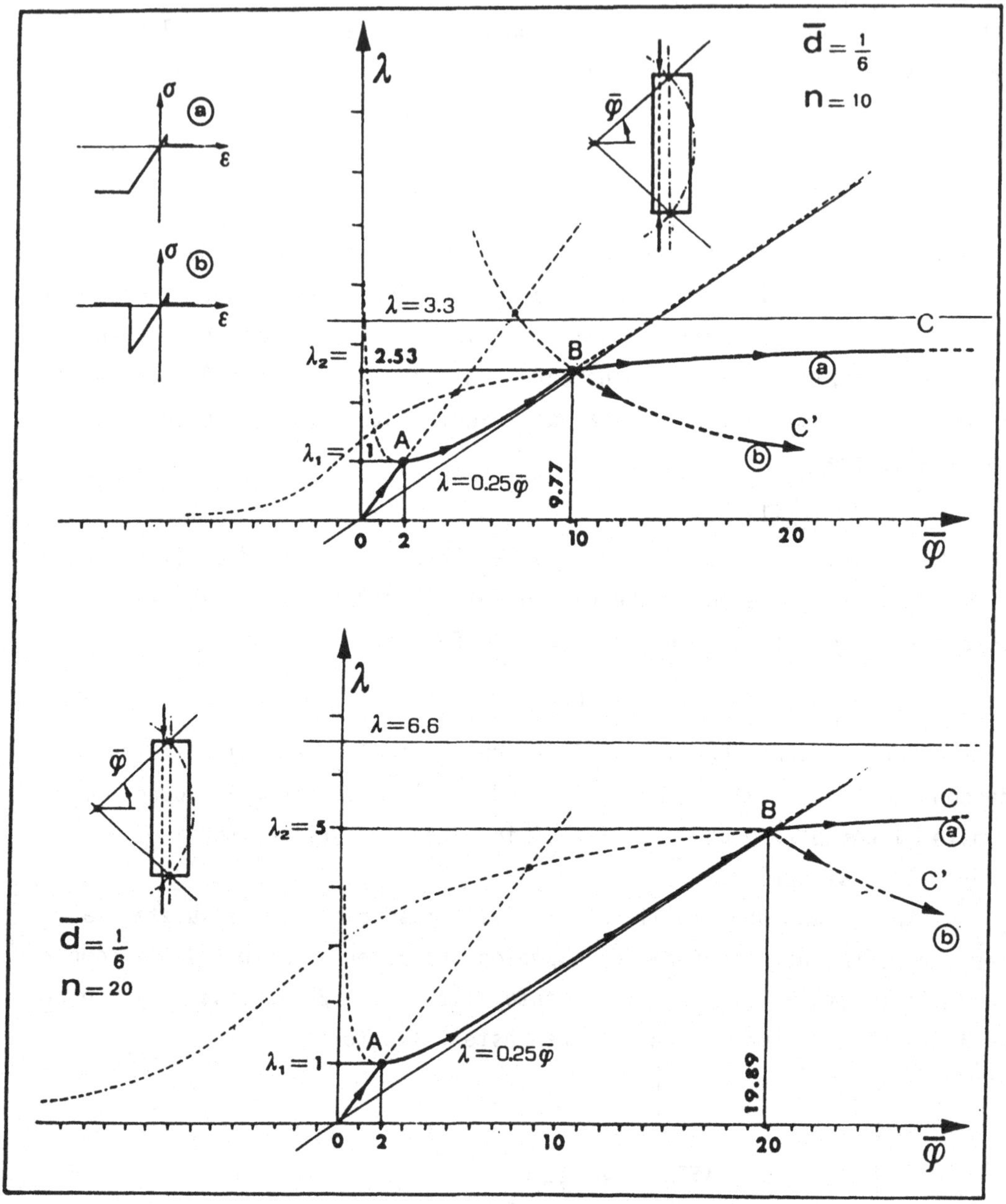
d̄ = 1/6
n = 10
λ
σ
ε
φ̄
λ=3.3
λ₂= 2.53
λ₁= 1
A
B
C
C'
λ=0.25φ̄
9.77
0
2
10
20
φ̄
λ=6.6
λ₂= 5
λ₁= 1
λ =0.25φ̄
19.89
d̄ = 1/6
n = 20

Fig. 3

reagente, di altezza 'e'. Dalle (2.1) è possibile ottenere in forma adimensionalizzata ($\bar{e}=e/h$) la relazione che lega 'e' al moltiplicatore λ :

$$\bar{e}=\lambda \pm \sqrt{\lambda^2-6\bar{d}\lambda} \qquad (2.2)$$

mentre $\bar{\varphi}$, tenendo conto del momento d'inerzia della parte reagente della sezione, vale:

$$\bar{\varphi}=3\lambda\,(\bar{e}-2\bar{d})\frac{1}{\bar{e}^3}$$

I punti appartenenti ai percorsi AB corrispondono alla soluzione della (2.2) che utilizza il segno negativo (la soluzione con segno positivo definisce i punti, alla sinistra di A, rappresentati in figura 3 con linea tratteggiata).

La condizione $\lambda^2-6\lambda\bar{d}=0$ permette di determinare il valore di λ per il quale la funzione assume il valore minimo: $\lambda=6\bar{d}$; se $\bar{d}=1/6$ il punto di minimo coincide con il punto A. E' facile verificare che tale minimo si trova rispettivamente alla sinistra od alla destra del punto A a seconda che sia $\bar{d}>1/6$ oppure $\bar{d}<1/6$.

I tratti AB sono asintotici rispetto alla retta di equazione $\lambda=9\bar{d}^2\bar{\varphi}$ Essa rappresenta il percorso di equilibrio nell'ipotesi di resistenza a trazione nulla.

I tratti BC del diagramma $\lambda,\bar{\varphi}$ descrivono il comportamento indefinitamente plastico della struttura dopo che il limite a compressione $-n\sigma_o$ è stato superato. Le equazioni, scritte direttamente in forma adimensionale sono:

$$\begin{cases} n\bar{p}+\frac{1}{2}(n-1)\bar{e}=\lambda \\ \frac{1}{2}n\bar{p}^2+\frac{1}{2}\bar{p}\bar{e}(n-1)+\frac{1}{6}\bar{e}^2(n-2)=\lambda\bar{d} \end{cases} \qquad (2.3)$$

avendo posto $\bar{p}=p/h$.

Il parametro $\bar{\varphi}$ rappresentativo della rotazione della sezione reagente è fornito da:

$$\bar{\varphi}=3\lambda\,(2\bar{p}-\bar{e}-2\bar{d})\frac{1}{\bar{e}^3}$$

La posizione del punto B, funzione di 'n', si può ottenere dalle (2.3). Infatti, ponendo $\bar{p}=0$, si ottiene:

$$\lambda_2=\frac{3}{2}\frac{(n-1)^2}{(n-2)}\bar{d}$$

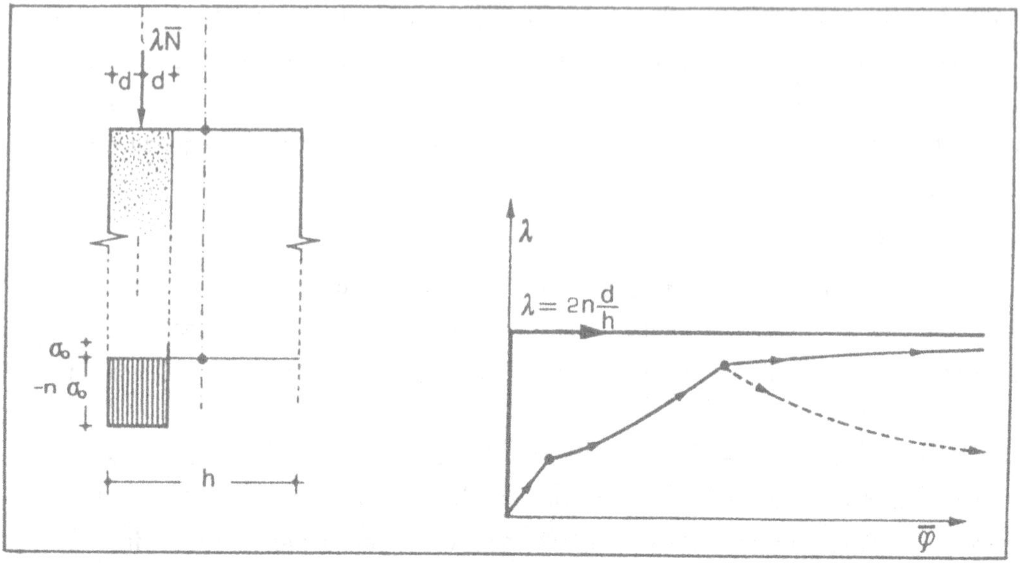

Fig. 4

Il parametro $\bar{\varphi}$ aumenta indefinitamente ed il percorso di equilibrio è descritto da una curva che tende asintoticamente ad una retta orizzontale di equazione λ =2n$\bar{d}$.

In questa situazione la sezione (Fig.4) di area '2db' è completamente plasticizzata. In particolare deve essere segnalato il fatto che, superata la fase delle fratture, la condizione di equilibrio limite della sezione richiede che l'asse di sollecitazione passi per il baricentro della parte reagente della sezione nella stessa situazione limite. Questa è la constatazione basilare che consente di affermare la validità delle teorie fondate sull'ipotesi di assenza di resistenza a trazione del materiale in virtù delle quali la porzione reagente di sezione e la distribuzione delle tensioni risulta costante durate tutte le fasi del processo di carico. Infatti, detta ipotesi fornisce la configurazione di equilibrio limite della sezione, alla completa plasticizzazione, indipendentemente dalla precedente evoluzione della sezione reagente in funzione del carico.

Se si ipotizza che lo stesso pilastro sia costituito da materiale a comportamento elasto-fragile sia a compressione che a trazione (pur conservando i limiti σ_o e $-n\sigma_o$) le equazioni che descrivono il problema so no le stesse nelle prime due fasi, mentre nella terza fase, in forma adimensionale, si ha:

$$\begin{cases} \frac{1}{2}(n-1)\bar{e}=\lambda \\ \frac{1}{2} \quad \bar{g}\bar{e}(n-1)+\frac{\bar{e}^3}{3}(n-2) = \lambda\bar{d} \\ \bar{q}=3\,\lambda\,(2\bar{g}-\bar{e}-2\bar{d})1/\bar{e}^3 \end{cases} \tag{2.4}$$

Dalla prima delle (2.4) sostituendo nella seconda e nella terza si ottiene:

$$\bar{g}=\bar{d}-\frac{2}{3}\,\frac{(n-2)}{(n-1)^2}$$

$$\bar{\varphi}=\frac{1}{4\lambda}(n^2-1)$$

In figura 3, il tratto BC' rappresenta il comportamento elasto-fragile a compressione. Esso è rappresentato da una curva concava tendente asintoticamente alla retta $\lambda=0$.

3. IL SISTEMA ARCO-TRAVE

Si consideri ora il sistema arco-trave incastrato agli estremi; per maggiore semplicità si fa riferimento ad uno schema a deformabilità concentrata in tre soli conci deformabili (Fig.5). Lo spessore dei tre conci deformabili, con comportamento dapprima elasto-fragile sia a trazione che a compressione e successivamente elasto-fragile a trazione ed elasto-plastico a compressione (Fig. 1b, 1a), è stato scelto in modo da dare, nelle varie sezioni, un diverso rapporto tra momento e rotazione relativa.

Siano dunque σ_0 la tensione limite a trazione, $-n\sigma_0$, con $n>1$, la tensione limite a compressione e λ il moltiplicatore del carico. Anche in questo caso le fasi di comportamento della struttura sono contraddistinte ciascuna da valori del moltiplicatore del carico λ compresi in intervalli contigui.

Per il caso di comportamento elasto-fragile sia a trazione che a com

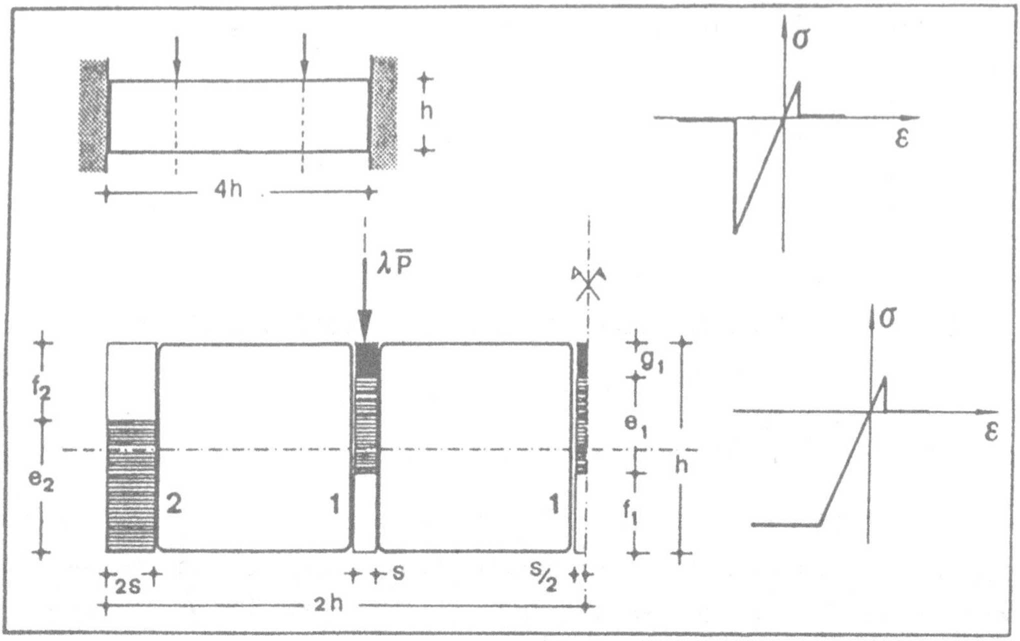

Fig. 5

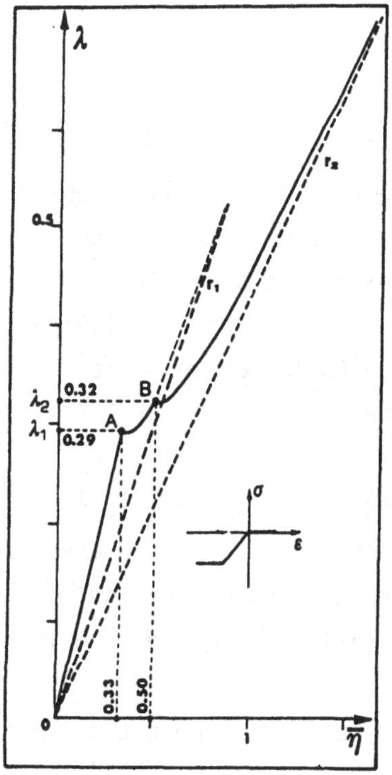

Fig. 6

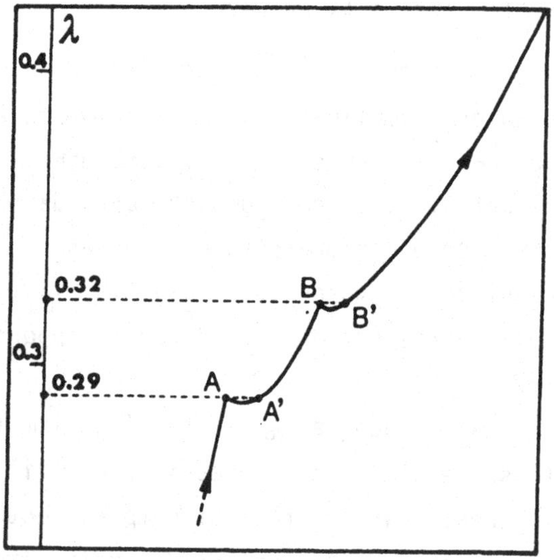

Fig. 7

pressione, gli intervalli sono:

$0 \leq \lambda \leq \lambda_1$;

$\lambda_1 \leq \lambda \leq \lambda_2$;

$\lambda_2 \leq \lambda \leq \lambda_3$

mentre per il caso elasto-fragile a trazione ed elasto-plastico a compressione essi sono:

$0 \leq \lambda \leq \lambda_1$;

$\lambda_1 \leq \lambda \leq \lambda_2$;

$\lambda_2 \leq \lambda \leq \lambda_3$;

$\lambda_3 \leq \lambda \leq \lambda_4$;

$\lambda_4 \leq \lambda$

dove λ_1, λ_2, λ_3, λ_4 sono i valori del moltiplicatore del carico cui corrispondono rispettivamente il raggiungimento della tensione:
limite a trazione al bordo teso dei conci 1;
limite a trazione al bordo teso del concio 2;
limite di rottura o di plasticizzazione al bordo compresso dei conci 1;
limite di rottura o di plasticizzazione al bordo compresso del concio 2.

La successione dei λ è determinata dalla scelta di 'n' e dalla geometria della struttura e dei carichi.

Le equazioni generali che reggono il problema in esame nella fase elastica non fessurata non presentano alcuna difficoltà di scrittura e pertanto non si ritiene necessario riportarle, mentre quelle valide in fase fessurata in ambedue i casi ma per $\lambda \leq \lambda_3$ ed opportunamente adimensionalizzate sono:

$$\begin{cases} a_{11}\bar{N}+a_{12}\bar{M}=b_1 \\ a_{21}\bar{N}+a_{22}\bar{M}=b_2 \\ -a_{31}\bar{N}+a_{32}\bar{M}+b_3=\bar{\eta} \end{cases} \tag{3.1}$$

I coefficienti b_i, a_{ij} funzioni delle fratture $\bar{f}_1$, $\bar{f}_2$ dipendono dalla fase di carico e valgono rispettivamente:

$$a_{11}=\frac{3}{2(1-\bar{f}_1)}+\frac{2}{(1-\bar{f}_2)}+\frac{6(\bar{f}_1+\bar{f}_2)^2}{(1-\bar{f}_2)^3}$$

$$a_{12} = a_{21} = \frac{12\bar{f}_1+12\bar{f}_2}{(1-\bar{f}_2)^3}$$

$$a_{22} = \frac{24}{(1-\bar{f}_2)^3} + \frac{18}{(1-\bar{f}_1)3}$$

$$a_{32} = \frac{4}{(1-\bar{f}_2)^3} + \frac{1}{(1-\bar{f}_1)^3} \; ; \; a_{31} = \frac{1}{6} b_1$$

$$b_1 = \frac{12\bar{f}_1+12\bar{f}_2}{(1-\bar{f}_2)^3} \quad ; \; b_2 = \frac{24}{(1-\bar{f}_2)^3} \quad ; \; b_3 = \frac{b_2}{12}$$

Le prime due impongono il rispetto delle condizioni di vincolo della sezione di simmetria nella quale agiscono le caratteristiche di sollecitazione $\bar{N}$ ed $\bar{M}$, mentre la terza fornisce il valore dell'abbassamento $\bar{\eta}$ del baricentro della sezione di simmetria.

Il valore limite $\lambda = \lambda_1$ è dedotto dalla condizione (3.2) scritta nella sezione 1 in presenza di fratture ponendo $\bar{f}_1=0$.

$$(1-\bar{f}_1)^2 = -\bar{N}(1-\bar{f}_1) + 6\bar{M} \tag{3.2}$$

mentre il valore limite $\lambda = \lambda_2$ è dedotto dalla condizione (3.3) scritta nella sezione 2 in presenza di fratture ponendo $\bar{f}_1$ 0, $\bar{f}_2=0$

$$(1-\bar{f}_2)^2 = -\bar{N}(1+2\bar{f}_2+3\bar{f}_1) - 6\bar{M} + 6 \tag{3.3}$$

Il tratto OA del percorso di equilibrio (Fig.6,7) rappresenta la relazione $\lambda, \bar{\eta}$ in fase elastica non fratturata. Il valore limite di $\bar{\eta}$ in fase elastica è :

$$\bar{\eta} = \frac{Ph^2 s}{3EJ}$$

Il tratto AB del percorso di equilibrio rappresenta la relazione λ, $\bar{\eta}$ in fase elastica e fratturata nelle sezioni 1; tale tratto è asintotico alla retta r_1 passante per l'origine che rappresenta il percorso di equilibrio della struttura con resistenza a trazione nulla nelle sezioni 1. Il minimo di $\lambda = \lambda(\bar{\eta})$ può trovarsi a destra o a sinistra del punto A;

nel secondo caso il tratto viene effettivamente percorso al crescere del parametro λ , mentre nel primo caso si assiste al fenomeno tipico dello snap-through per la nascita e propagazione istantanea delle fratture $\bar{f}_1$ (Fig.7). Al valore $\lambda = \lambda_1$ corrispondono dunque due valori distinti di per l'improvviso cambiamento della struttura reagente che da trave diviene arco.

Per $\lambda = \lambda_2$ si innescano le fratture agli incastri 2; il percorso di

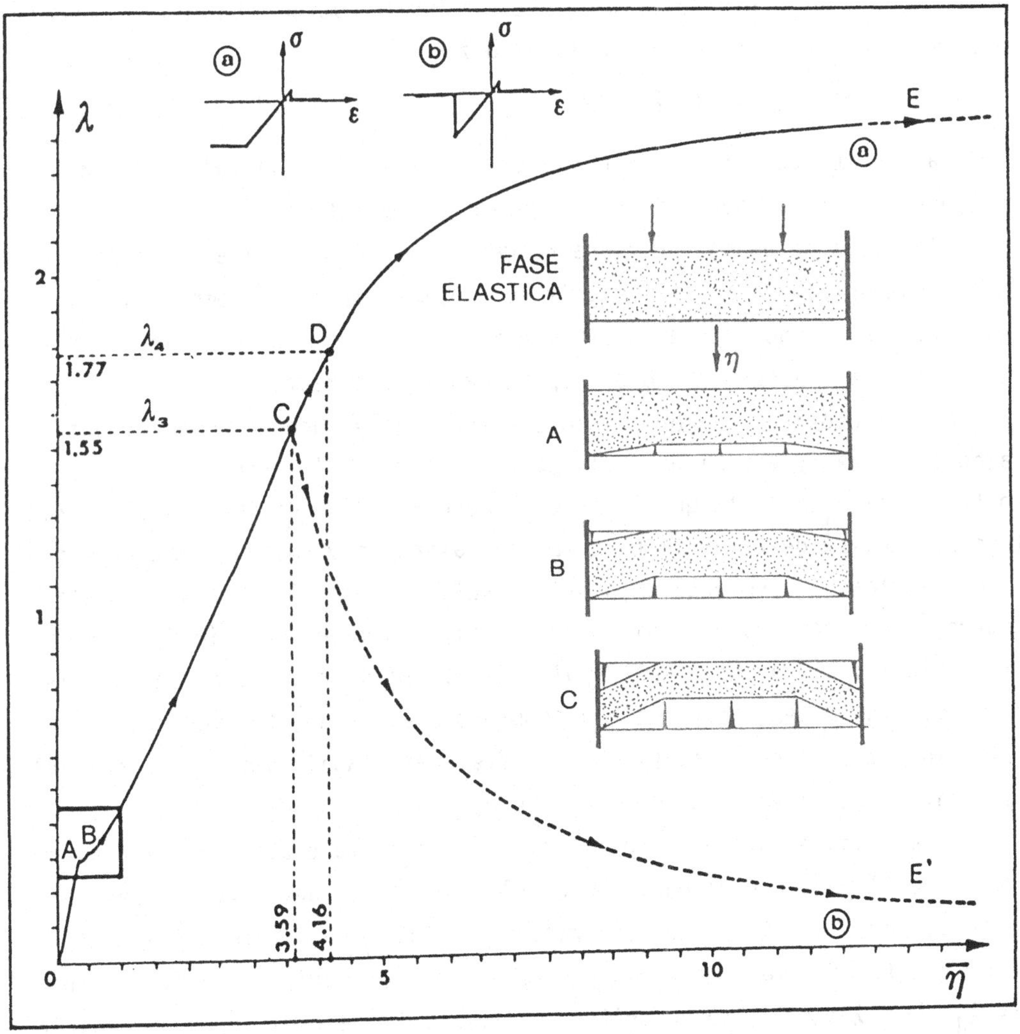

Fig. 8

equilibrio cambia bruscamente o con continuità passando al ramo AB al ramo BC, asintotico alla retta r_2 passante anch'essa per l'origine (Fig.7). Per il ramo BC valgono le stesse considerazioni fatte per il ramo AC. La retta r_2 rappresenta il percorso di equilibrio della struttura nell'ipotesi di resistenza a trazione nulla nelle sezioni 1 e 2 e risistenza a compressione illimitata. Le equazioni delle rette r_1 ed r_2 non sono di immediata determinazione e pertanto tali rette sono state tracciate con un procedimento puramente numerico.

Il raggiungimento della tensione limite di rottura o plasticizzazione a compressione è espresso dalla condizione:

$$n(1-\bar{f}_1)^2 = \bar{N}(1-\bar{f}_1)+6\bar{M} \tag{3.4}$$

scritta per le sezioni 1, che permette di individuare il valore $\lambda = \lambda_3$ e conseguentemente il punto C del percorso di equilibrio.

L'evoluzione della struttura reagente per $\lambda_3 < \lambda_4$ dipende dalla scelta tra comportamento elasto-fragile ed elasto-plastico a compressione. Nella figura 6 sono rappresentati i rami del percorso di equilibrio corrispondenti a ciascuno dei due comportamenti ipotizzati.

Nell'ipotesi di comportamento elasto-fragile anche a compressione valgono ancora le equazioni (3.1) salvo ovviamente le espressioni dei coefficienti b_i, a_{ij}. In questo caso il percorso di equilibrio della struttura per $\lambda_3 < \lambda$ descrive il ramo CE' con asintoto $\lambda = 0$. E' caratteristico in questa fase il comportamento delle fratture $\bar{f}_1$ ed $\bar{f}_2$ (Fig.9); infatti mentre $\bar{f}_2$ cresce fino a raggiungere il valore limite $\bar{f}_2 = 1$, la frattura $\bar{f}_1$ si richiude via via che viene espulso il materiale nelle zone in cui viene raggiunta la tensione limite a compressione. Questo fenomeno è stato colto in tutte le prove sperimentali effettuate dagli autori e di cui nel paragrafo successivo viene mostrato un episodio (Fig.10).

Se si assume invece l'ipotesi di comportamento elasto-plastico a compressione il percorso di equilibrio della struttura descrive successivamente il ramo CDE. Al punto D corrisponde il raggiungimento della tensione di plasticizzazione in corrispondenza delle sezioni 1: ciò avviene per $\lambda = \lambda_4$. Per $\lambda > \lambda_4$ viene descritto il tratto DE che evolve verso l'asinto

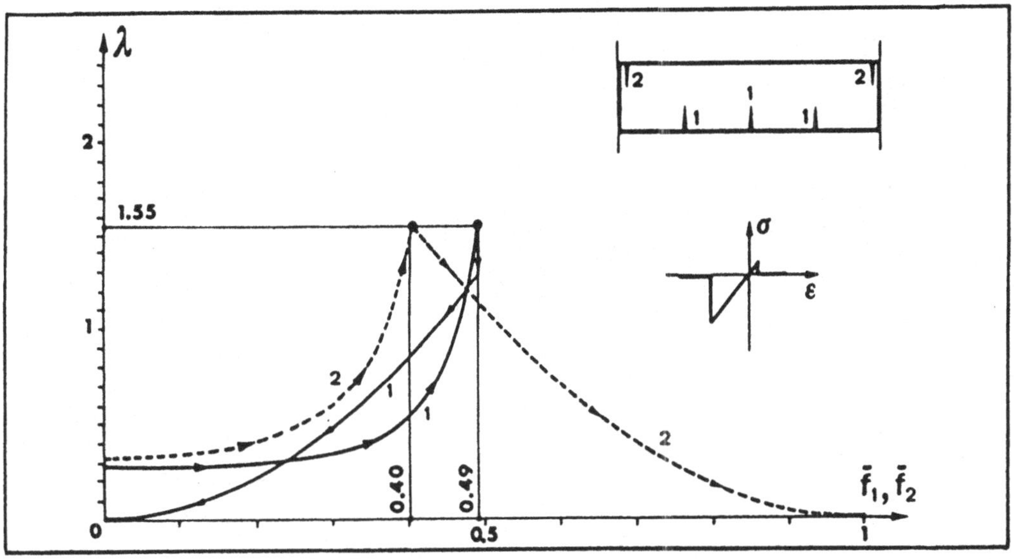

Fig. 9

to $\lambda = \lambda_{ultimo} = 0{,}5n$ che rappresenta il massimo moltiplicatore ammissibile del carico per la struttura esaminata.

E' significativa l'evoluzione della struttura che passa dallo schema di trave incastrata allo schema di arco; la situazione limite fornisce, in armonia con la teoria della plasticità $\bar{M}=0$ in tutte le sezioni così da individuare una sorta di struttura ad uniforme resistenza.

4. RISULTATI SPERIMENTALI ED OSSERVAZIONI CONCLUSIVE

La convalida sperimentale alle ipotesi sopra avanzate è stata ottenuta mediante prove eseguite nel Laboratorio annesso al Dipartimento di Costruzioni dell'Università di Firenze. Più precisamente le indagini sperimentali sono state condotte su di una serie di cinquanta travetti di dimensioni cm 49x6x2 in malta plastica normalizzata stagionata a 28 giorni. Del materiale impiegato per la confezione dei travetti erano state determinate le seguenti caratteristiche meccaniche: carico di rottura a flessione, carico di rottura a compressione ed a trazione, modulo di elasti-

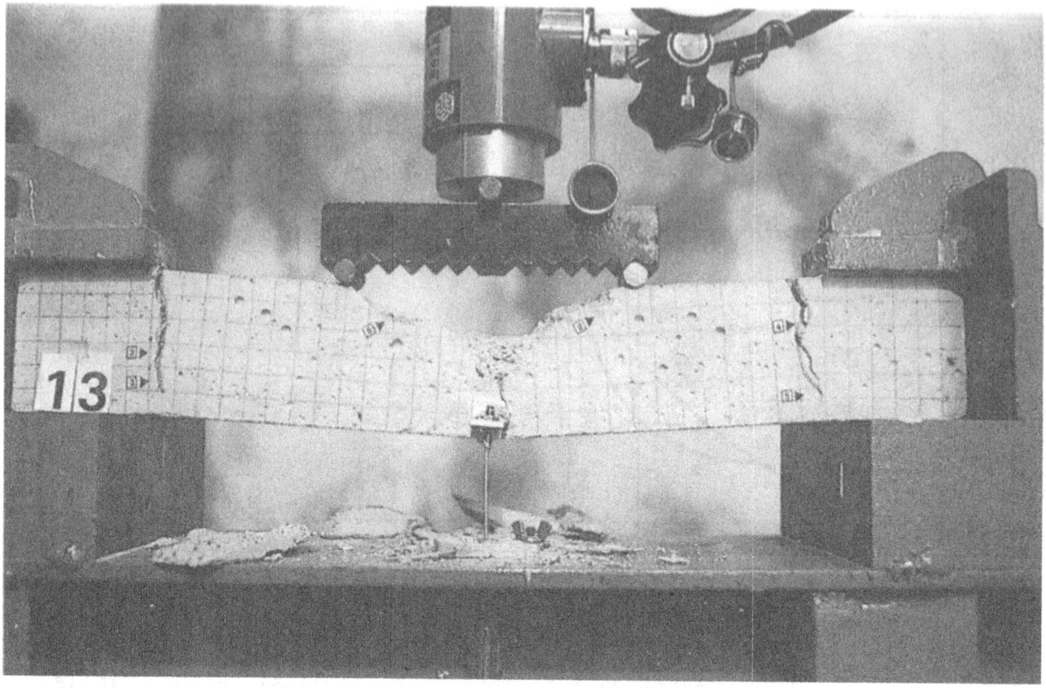

Fig. 10

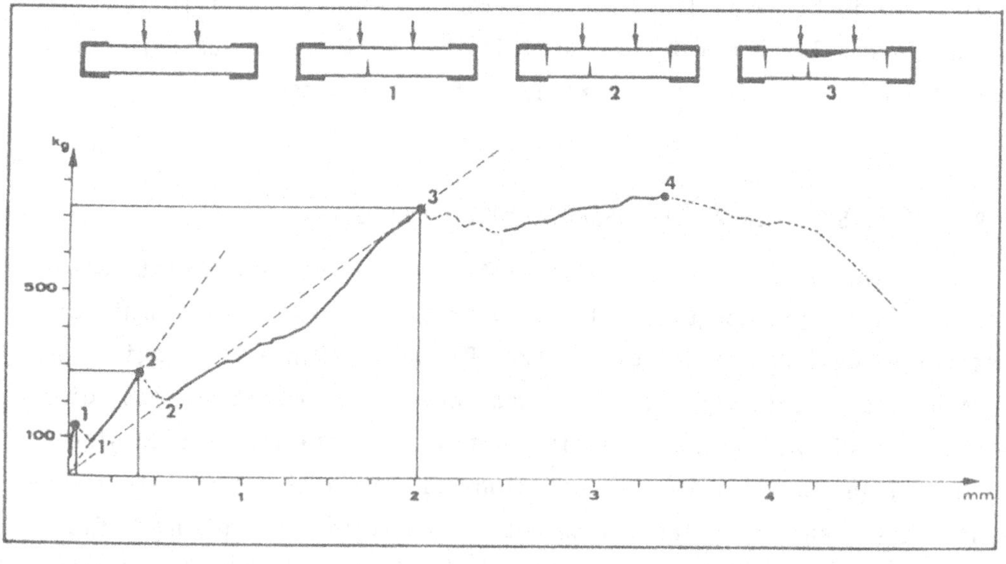

Fig. 11

cità longitudinale. Il processo di carico è stato a deformazione guidata. Tutte le prove eseguite hanno fornito gli stessi risultati.

Il percorso di equilibrio rappresentato in figura 11 ed ottenuto per via sperimentale su di un travetto campione incastrato agli estremi (Fig. 10), costituito da materiale scarsamente reagente a trazione, mostra con evidenza la presenza di fenomeni di snap-through perfettamente corrispondenti a quelli denunciati dal percorso di equilibrio ottenuto mediante simulazione analitica (Fig.6).

Infatti il raggiungimento del limite di rottura a trazione produce una frattura istantanea 1 cui corrisponde un improvviso aumento della deformazione con conseguente caduta del livello di carico (Fig.11). A partire dal punto 1' ed a crescere del moltiplicatore di carico λ la struttura manifesta il comportamento rappresentato dal tratto 1'-2 del diagramma, passante per l'origine: in altre parole il percorso di equilibrio corrisponde a quello di una trave fratturata fin dall'inizio. Il punto 2 del diagramma corrisponde alla brusca comparsa di fratture in corrispondenza delle sezioni di incastro, con il conseguente riproporsi del fenomeno prima descritto, caratterizzato da un aumento della deformazione e corrispondente caduta del livello di carico. Al successivo incremento del moltiplicatore di carico corrisponde un tratto del diagramma che può essere ancora assimilato ad una retta passante per l'origine ed il punto 3. In corrispondenza del punto 3 si ha la disgregazione del lembo superiore compresso. La prima frattura tende a richiudersi (Fig.9), dando luogo ad un arco con freccia decrescente; le tensioni di compressione crescono rapidamente sino al collasso totale in corrispondenza del punto 4.

Il confronto fra i percorsi di equilibrio relativi al caso teorico ed a quello sperimentale rivela come i fenomeni di snap-through si manifestino in entrambi i casi per valori del moltiplicatore del carico sufficientemente lontani dal valore ultimo. Inoltre anche nel percorso di equilibrio ottenuto per via sperimentale è possibile individuare le rette r_1 ed r_2 che rappresentano i percorsi di equilibrio della stessa struttura ma con resistenza a trazione nulla. Queste circostanze permettono di attribuire scarsa influenza alla resistenza a trazione sul comportamento globale della struttura e ciò potrebbe costituire una prima indiretta conferma all'ipotesi di considerare le murature incapaci di reagire a trazione.

Il modello teorico proposto permette dunque una edeguata descrizione del comportamento della struttura sia per quanto riguarda le fasi iniziali che quelle precedenti al collasso. Infatti è possibile rilevare che la curva sperimentale è compresa tra le due messe in evidenza dal modello teorico relative alle due ipotesi limite del materiale perfettamente plastico e perfettamente fragile a compressione. Ciò permette di congetturare che assumendo un modello di comportamento intermedio fra i due (Fig. 1c) i due percorsi di equilibrio sperimentale e teorico risulterebbero congruenti per tutte le fasi del percorso di equilibrio.

BIBLIOGRAFIA

1. Alessandri,C.; Briccoli Bati,S.; Tempesta,G.; Tonietti,U.: Indagine sperimentale sul comportamento di edifici in muratura: l'esperienza di Salvitelle, Atti del Dip. di Costruzioni, n.13/83, Firenze 1983.
2. Baratta,A.: Metodi di calcolo per le strutture non reagenti a trazione, in: Atti del VII Congresso Nazionale AIMETA, Trieste 1984.
3. Romano,G.; Sacco,E.: Sul calcolo di strutture non reagenti a trazione, in: Atti del VII Congresso Nazionale AIMETA, Trieste 1984.
4. Romano,G.; Romano,M.: Sulla soluzione dei problemi strutturali in presenza di legami costitutivi unilateri, in: Atti dell'Accademia Nazionale dei Lincei, serie VIII, LXVII, 1979.
5. Baratta,A.; Toscano,R.: Stati tensionali in pannelli di materiale non resistente a trazione, in: Atti del VI Congresso Nazionale AIMETA, Genova 1982.
6. Di Pasquale,S.: Questioni di meccanica dei solidi non reagenti a trazione, in: Atti del VI Congresso Nazionale AIMETA, Genova 1982.
7. Di Pasquale,S.: Statica dei solidi murari. Teoria ed Esperienze, in: Atti del Dipartimento di Costruzioni, n.27/84, Firenze 1984.
8. Villaggio,P.: Stress Diffusion in Masonry Walls: Journal of Structural

Mechanics, 9 (1981).

9. Signorini,A.: Sulla pressoflessione del cemento armato, Annali di Matematica, Serie Quarta, V (1926).

10. Résal,J.: Ponts an Maçonnerie, prem. partie, Stabilité des voutes, Parigi 1887.

APPLICATIONS OF THE MASONRY UNILATERAL MODEL TO THE LIMIT ANALYSIS OF BUILDINGS LOADED BY HORIZONTAL FORCES

M. Como, A. Grimaldi
Department of Civil Engineering, 2nd University of Rome

ABSTRACT

The Paper firstly recalls the unilateral elastic model of the masonry material and the formulation of the limit analysis theory of the masonry structures.

In this framework the collapse load of multistory masonry walls and of masonry buildings, loaded by gradually increasing horizontal forces, are given.

1. INTRODUCTION

The limit analysis theory of the masonry structures is of worthy interest for many important applications that it can receive, particularly for the evaluation of the seismic strength of masonry buildings.

A formulation of the limit analysis of masonry structures, by assuming as constitutive equation of the masonry the unilateral model of the elastic material with zero tensile strength, has been given in [1].

In this framework this Paper recalls the essential features of the proposed theory and gives some results obtained for the evaluation of the limit strength of masonry walls and buildings subjected to horizontal loads.

2. LIMIT ANALYSIS THEORY OF MASONRY BODIES ACCORDING TO THE ASSUMED UNILATERAL MODEL

The elastic material without tensile strength is the model commonly used in the static analysis of masonry structures [2, 3, 4].

The constitutive equations of this model are:

$$\sigma = C(\epsilon - \epsilon_c) \tag{1}$$

$$\sigma \leqslant 0 \tag{2}$$

$$\epsilon_c \geqslant 0 \tag{3}$$

$$\sigma \cdot \epsilon_c = 0 \tag{4}$$

where σ is the stress tensor whose principal values, according to condition (2), cannot be positive, ϵ the total strain tensor, ϵ_c the fracture strain tensor, whose principal values, on the contrary, cannot be negative. Eq. (1) represents the elastic relation between stress and the elastic strain, and the orthogonality condition (4) the physical assumption that the fracture dilatations occur only on planes where stress is zero.

The elastostatic analysis of the masonry body, that occupies the region Ω, has the boundary $\partial\Omega$, and is subjected to the loads p, is centered on the evaluation, if they exist, of the displacement u, the stress σ and the fracture strain ϵ_c.

These unknown quantities have to satisfy:

— the constitutive equations. (1-4)

— the compatibility equation

$$\epsilon = Du \tag{5}$$

where D is the operator that associates to the displacements u the strains ϵ

— the equilibrium equation

$$\langle \sigma, \delta\epsilon \rangle = \langle p, \delta u \rangle \qquad \forall\, \delta u \in V\,. \tag{6}$$

The unknown quantities

$$u, \sigma, \epsilon_c \tag{7}$$

must be defined in suitable functional spaces. In eq. (6) V is the space of the displacements fields.

The existence of the solution of the elastostatic problem of the masonry body, because of its unilateral aspect, is strictly connected to a restriction on the loads p. In this framework many results have been obtained in these last years, namely in [2] where many mathematical questions of the problem were analyzed, or in [3] where some variational formulations were given.

In the next some results, essential for the statement of the limit analysis theory of the masonry bodies in the context of the above quoted unilateral model of the material, will be reported.

First af all in [2] the existence of solutions of the elastostatic problem is proven under some suitable restrictions on the loads. It is also shown that the solution stress field is unique and that minimizes the complementary energy functional

$$E_c = 1/2 \langle C^{-1}\sigma, \sigma\rangle \tag{8}$$

in the set S of the admissible stress fields, i.e. the stress fields satisfying the equilibrium eq. (6) and the constraint inequality (2).

In [3] it is shown, under suitable assumptions, for instance in the case of systems with a finite degree of freedom, that the solution σ of the problem can be obtained minimizing the complementary energy functional in the set S. It is also shown that the condition that the set S is not empty is sufficient to the existence.

It is worthwhile to remark that in the case of rigid materials with zero tensile strength the previous difficulties to prove the sufficient condition of existence vanish. In this case the assumption $S \neq \phi$ is equivalent to admit the existence of the solution.

For applications it is more suitable, on the other hand, to formulate existence conditions that involve directly the loads p. Then it is useful to introduce the concept of «mechanism», i.e. of the displacement fields u that at any point of the masonry body never produce contractions.

The set M of the «mechanism» fields is therefore defined as

$$M = \{u \in V : \epsilon = Du \geqslant 0\} \tag{9}$$

and the useful result follows: to the existence of solution of the elastostatic problem is necessary that

$$\langle p, u\rangle \leqslant 0 \qquad \forall u \in M \tag{10}$$

We can immediately obtain this condition from the equilibrium eqs. (6) by assuming δu HM and taking into account the conditions (2).

The above result, of course, is equivalent to the statement that a sufficient condition for the nonexistence of a solution is that at least a mechanism exists such that:

$$\langle p, u\rangle > 0 \tag{11}$$

On the basis of these results the problem of the collapse of the masonry body can be easily established.

With reference to fig. 1 we will particularly examine the loading condition

$$p = g + \lambda q \tag{12}$$

with λ the loading parameter that defines the loading process, characterized by dead loads g and increasing loads λq.

It is evident the mechanical interest of such loading condition (fig. 1). In the examples sketched the solution of elastostatic problem exists at $\lambda = 0$ while, for increasing λ, existence can be lost during the loading, and the collapse occurs.

As proven in [1], if a solution exists at $\lambda = \lambda'$, a solution also does exist for any λ such that $0 \leqslant \lambda \leqslant \lambda'$.

On the contrary if for $\lambda = \lambda''$ solution does not exist, solution never exists for any $\lambda \geqslant \lambda''$.

Of special interest therefore is the value λ_c of the loading parameter, called the collapse multiplier, that during the loading process marks the passage from existence to non existence of solutions.

The evaluation of the so defined collapse multiplier can be performed, as proven in [1], by using the static and the kinematic approaches, similar to the well known procedures of the limit analysis of elasto-plastic bodies.

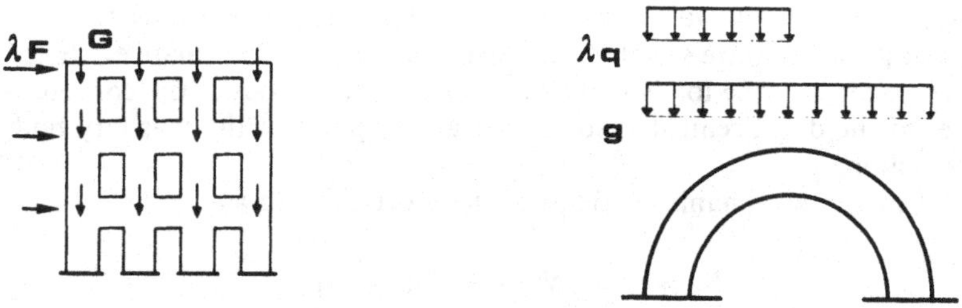

Fig. 1, Masonry structures subjected to dead load g and increasing load λq

Statical approach

Let us assume that under the loads

$$g + \lambda^{-} q$$

the corresponding set S of the statically admissible stress fields is not empty or, equivalently, that at least one stress field σ satisfying conditions (6) and (2) exists.

The multiplier λ^{-} is called statically admissible multiplier of the loads and we have, taking into account the previous statements and the definition of λ_c,

$$\lambda_c \geqslant \lambda^{-}(\sigma) \,. \tag{15}$$

Kinematical approach

With reference to an assigned «mechanism» displacement $u \in M$ the kinematical multiplier $\lambda^{+}(u)$ is defined as

$$\lambda^{+}(u) = -\frac{\langle g, u \rangle}{\langle q, u \rangle} \,. \tag{16}$$

The multiplier $\lambda^{+}(u)$ marks the change of sign of the work $\langle g + \lambda q, u \rangle$ of the loads $g + \lambda q$ along the mechanism u. For $\lambda > \lambda^{+}(u)$ the existence of the solution is not possible and therefore

$$\lambda_c \leqslant \lambda^{+}(u) \,. \tag{17}$$

A first direct application of the above formulation of the limit analysis of masonry bodies is the evaluation of the collapse thrust of the masonry panel of fig. 2.

As shown in [1] the collapse value of the horizontal force λF is

$$\lambda_c F = \frac{GB}{2H} \tag{18}$$

where G is the vertical dead load acting at the top section of the panel, B the width and H the height of the panel.

The collapse mechanism corresponds to the overturning of the panel.

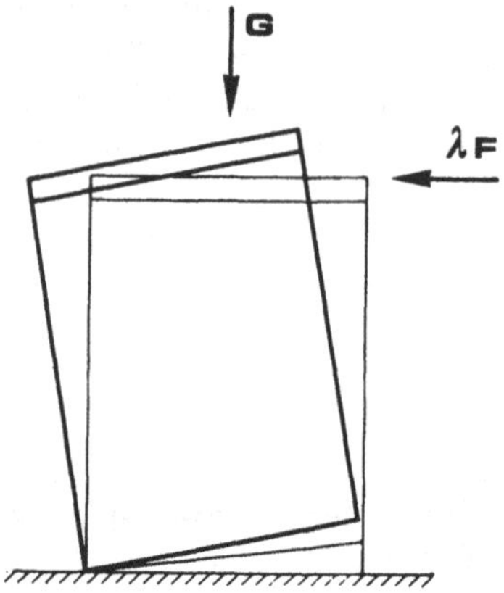

Fig. 2, Collapse mechanism of the masonry panel

This last result can be also obtained by using, for sake of simplicity, the resultant forces, i.e. the axial load N and the bending moment M, acting along the cross sections of the panel, together with the kinematical assumption of plane sections [1]. This last formulation meets the results obtained by Kooharian [5] and Heyman [6] in the case of the masonry arch made of rigid stones.

3. COLLAPSE OF MASONRY BUILDINGS UNDER HORIZONTAL FORCES

In the framework of the previous results, the limit analysis of masonry buildings subjected to horizontal loads has been developed. This analysis, of course, has many technical interesting applications, especially in the study of the seismic behaviour of masonry dwellings under earthquake shaking.

Essential elements of the masonry buildings are the multistory walls that, connected at any story with the floors, characterize the resistant structure of the building.

Let us consider the plane multistory masonry wall with n_p stories and n_c spans (fig. 3). The considered wall has a regular scheme with piers of width b and architraves of height a. The wall is subjected to vertical fixed dead loads G and to the gradually increasing horizontal forces λF, with λ the loading parameter, applied at the various stories along the architraves axes.

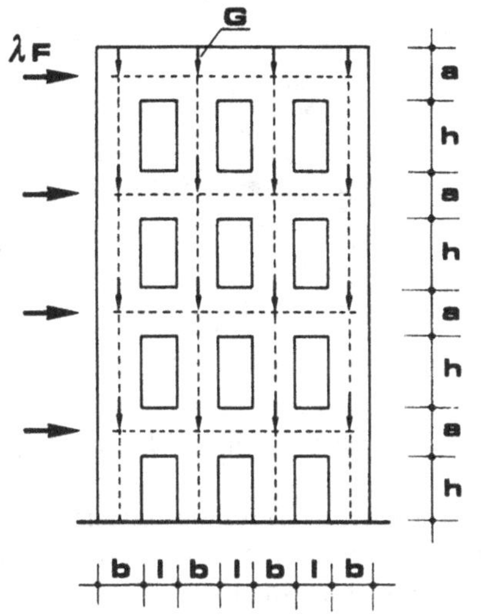

Fig. 3, The model of the multistory masonry wall with openings

We will assume that the mean compression stress at the base of the piers, produced by the vertical dead loads G, is small with respect to the compression strength. This assumption, as a rule, is satisfied in unreinforced masonry buildings.

The lateral strength of the wall depends on two different effects. The first is due to the lifting of the vertical dead loads during the pier rotation. The second effect is due to the plastic dissipation developing in the architraves, that we assume reinforced.

According to the assumption of elastic material with zero tensile strength and to the analysis previously developed, the collapse condition of the single pier can be represented in term of the resulting bending moment M and axial force N, by means of the failure domain of fig. 4. The strength of the reinforced architrave is represented, on the other hand, by the plastic bending moment M_0.

The corresponding shear force is therefore

$$T_0 = \frac{2M_0}{\ell} \tag{19}$$

where ℓ is the span length of the architrave. As shown in [8], according to the values of the characteristic ratio

$$Z = \frac{T_0 \ell}{G\, b} \tag{20}$$

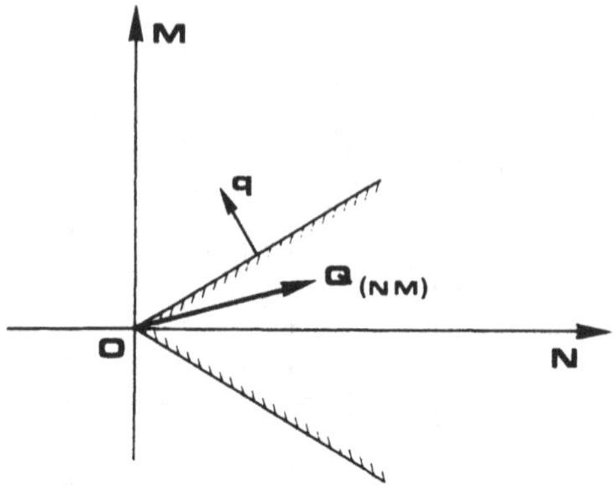

Fig. 4, The collapse domain M-N of the pier

collapse of multistory walls occurs with various mechanisms. Fig. 5a depicts the «architrave sidesway» mechanism, fig. 5b the «pier sidesway» mechanism and fig. 5c the «turn over» mechanism. For small values of Z, i.e. when

$$Z < Z_3 = \frac{1}{2\,(1 + K)} \tag{21}$$

6

with

$$K = \frac{b}{\ell} \tag{22}$$

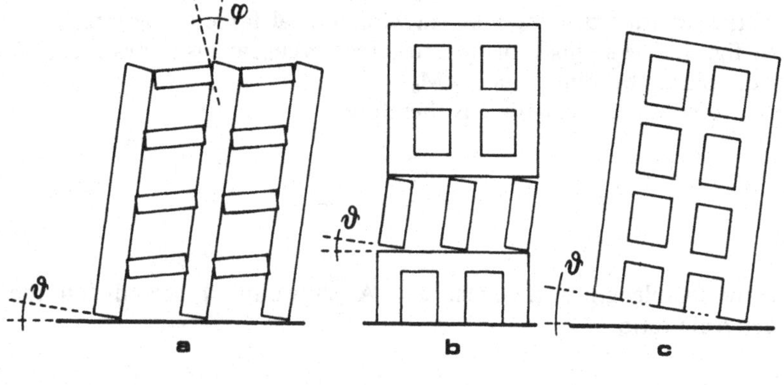

Fig. 5, Failure mechanisms of the wall (a = architrave sidesway, b = pier sidesway, c = turn over)

the collapse of the multistory wall develops with the «architrave sidesway» mechanism of fig. 5a. The corresponding collapse multiplier λ_c in adimensional form

$$\Lambda_c = \frac{\lambda_c F H}{G b} \tag{23}$$

is then given by

$$\Lambda_c = (n_c + 1) \frac{1 + 2Z (1 + K) \dfrac{n_c}{n_c + 1}}{1 + n_p (1 + 2t)} \tag{24}$$

with

$$t = \frac{a}{2H} \tag{25}$$

It is also rare to meet large values of the ratio Z, i.e. values of Z larger than Z_3, given in eq. (21). As a rule, in fact, the shear T_0 is very lower than G, that is the increment, at the story, of the vertical dead loads, while the span length ℓ and the width b of the pier are, more or less, of the some order of magnitude.

In the seismic analysis of walls it is usual to make reference to a conventional mean shear at the base section of the piers. According to (24), this ultimate shear is then given by [9]

$$\tau_c = \frac{G}{bHs} b \frac{n_p}{1 + n_p (1 + 2t)} [1 + 2Z (1 + K) \frac{n_c}{n_c + 1}] \tag{26}$$

where s is the thickness of the piers. It can be recognized that:

i — the number of the stories and of the openings has not relevant influence on τ_c.

ii — the ultimate shear is proportional to the piers width b and to the mean specific gravity G/bHs of the wall

iii — noteworthy influence on τ_c is due to the bending strength of the architraves by means the ratio Z.

A quantity more sharping to represent the seismic lateral strength of the multistory wall is the specific lateral strength r_c, i.e. the ratio between the total horizontal limit force and the weight of the whole wall. This quantity can represent the mean value of the horizontal seismic acceleration corresponding to the horizontal collapse load for the building. We have [9]

$$r_c = \frac{b}{H} \frac{1}{1 + n_p (1 + 2t)} [1 + 2Z (1 + K) \frac{n_c}{n_c + 1}] \tag{27}$$

Of course, the higher is the wall, i.e. larger the number of stories, the smaller is r_c.
Fig. 6 depicts the variation of τ_c and r_c with the architrave strengths, the story number etc., of some dwellings of simple plan.

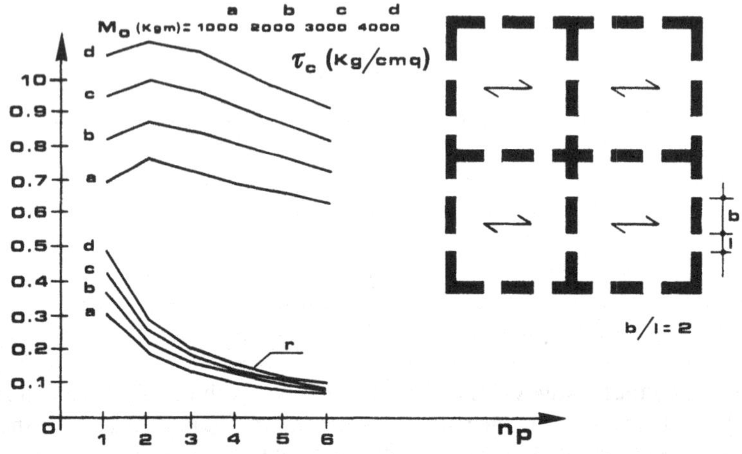

Fig. 6, Values of τ_c and r_c for symmetric masonry buildings.

In the presence of irregular plans, or asymmetric locations of the walls, the collapse mechanism of the whole building will be characterized by both translations and torsional rotations of the floors. The floors will be represented by rigid diafhragms that tie together around the walls. Furthermore, we will assume that suitable reinforcements, in the architraves, at the floors, at the corners etc., will be able to prevent local failures so that the whole resistant structure of the masonry building will be able to exhibit at the collapse its lateral ultimate strength. According to the analysis given in [10], we will assume that the building structure is composed of ν_x plane walls directed along y and by ν_y walls directed along the x axis. The strength of the single wall will be evaluated at a reference story and, once the vertical distribution of the seismic loads is assumed, the values of the horizontal forces at the remaining stories will be obtained. If a wall is not symmetric with respect to its vertical central axis, two different values of the lateral strength have to be considered.

The so defined masonry structure of the building will be subjected to:

i — vertical fixed dead loads G acting, at any story, along the axes of the piers;

ii — gradually increasing horizontal forces λF, acting at the centers C of the various floors (fig. 7).

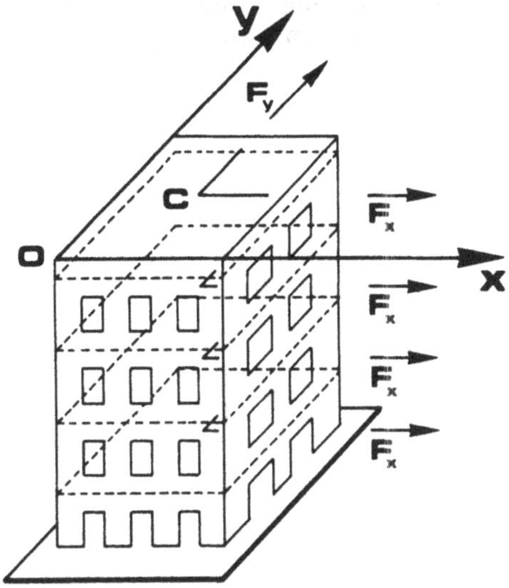

Fig. 7, The model of the multistory masonry building

The vector of the external horizontal forces applied at the floor is

$$Q = \begin{vmatrix} F_x \\ F_y \\ M \end{vmatrix} \tag{28}$$

The forces F_x and F_y are applied at the origin 0 of the axes and M is the moment of F_x and F_y with respect to the point 0.

By increasing the loading parameter, at a value λ_c of λ it will be attained the collapse of the building. The failure displacements of the floors will be defined as

$$D = \begin{vmatrix} u \\ v \\ \phi \end{vmatrix} \tag{29}$$

where u, v represent the horizontal displacements of the point 0 in the x and y directions and ϕ the torsional rotation of the floors. According to the assumed architrave sidesway mechanism, the vertical distribution of the horizontal displacements will be linear with the height.

The inplane displacements of the single walls, corresponding to the floor deformation D, can be casily obtained. They are represented by the vector

$$d = \begin{vmatrix} d_{x_1} \\ \cdot \\ \cdot \\ \cdot \\ d_{x_{v_y}} \\ \\ d_{y_1} \\ \cdot \\ \cdot \\ \cdot \\ d_{y_{v_x}} \end{vmatrix} \tag{30}$$

where d_{x_1}, d_{x_2}, ... d_{y_1}, d_{y_2}, ... are the inplane displacements of the walls, directed in the directions y and x respectively. Then the displacement vector d of the walls, that corresponds to the deformation vector D of the floors, can be obtained as

$$d = C\ D \tag{31}$$

where C is the compatibility matrix of the internal deformation of the building

$$C = \begin{vmatrix} 1 & 0 & y_1 \\ 1 & 0 & y_2 \\ \cdot & \cdot & \cdot \\ \cdot & \cdot & \cdot \\ 0 & 1 & -x_1 \\ \cdot & \cdot & \cdot \\ \cdot & \cdot & \cdot \\ 0 & 1 & -x_{v_y} \end{vmatrix} \tag{32}$$

The internal limit work D_r (D) done by the limit horizontal strengths of the walls for their inplane displacements, corresponding to the floor deformation D, is

$$D_r\ (D) = \phi_1\ (D)\ u + \phi_2\ (D)\ v + \phi_3\ (D)\ \phi \tag{33}$$

where ϕ_1 (D), ..., step functions of the floor deformation D, represent the limit response of the whole masonry structure, to the translations u, v and to the rotation ϕ.

In the space of the collapse mechanism vectors D, the function D_r (D) represents a positive plane faced cone with vertex at the origin. Points of each face of this cone represent the internal limit work evaluated for wall displacements that keep the same sign.

For a given collapse mechanism the kinematical multiplier

$$\lambda^{+}(D) = \frac{\phi_1(D)\,u + \phi_2(D)\,v + \phi_3(D)\,\phi}{F_x u + F_y v + M_y\,\phi} \tag{34}$$

can be casily evaluated. The collapse multiplier

$$\lambda_c \tag{35}$$

is the greatest lower bound of the kinematical multipliers $\lambda^{+}(D)$ for D varying in the space M of the collapse mechanism D. Hence we get

$$\lambda_c = \inf_{D \in M} \lambda^{+}(D) \tag{36}$$

The evaluation of λ_c can be simplified by considering the set S* of the mechanisms D* for which the horizontal forces Q make unit work, i.e.

$$S^{*} = \{D \in M : Q^T D = 1\} \tag{37}$$

The collapse multiplier is then given as

$$\lambda_c = \inf_{D^{*} \in S^{*}} (\phi_1(D)\,u + \phi_2(D)\,v + \phi_3(D)\,\phi) \tag{38}$$

It is also easy to prove that λ_c can be more simply obtained as

$$\lambda_c = \min_{D^{*} \in V^{*}} (\phi_1(D)\,u + \phi_2(D)\,v + \phi_3(D)\,\phi) \tag{39}$$

where V* is the set of the failure mechanisms D* ∈ S* corresponding only to translations of the floors in the x or y directions or to rotations of the floors around the crossing points of the walls. Fig. 8 sketches the set V* and the research of λ_c for the masonry house with simple plan [10].

The mean shearing stress τ_c at the base of piers and the corresponding specific lateral strength r_c of numerous examples of masonry houses are displayed in fig. 9. These numerical computations shown, that only in the presence of large excentricities between the center of gravity of masses and the center of the ultimate strengths of the walls a collapse mechanism accompanied by a torsional rotation of the floors appears.

The influence of the number of stories, of the dead loads G etc., is similar to the simpler case of symmetric buildings.

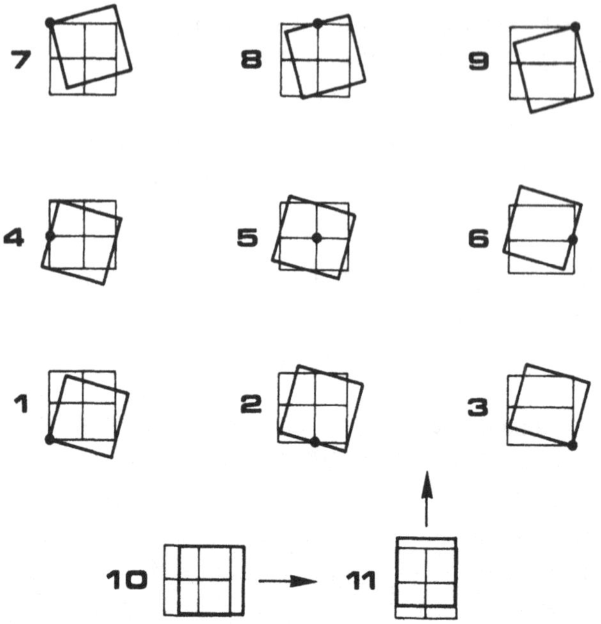

Fig. 8, Floor displacements corresponding to the building collapse mechanisms

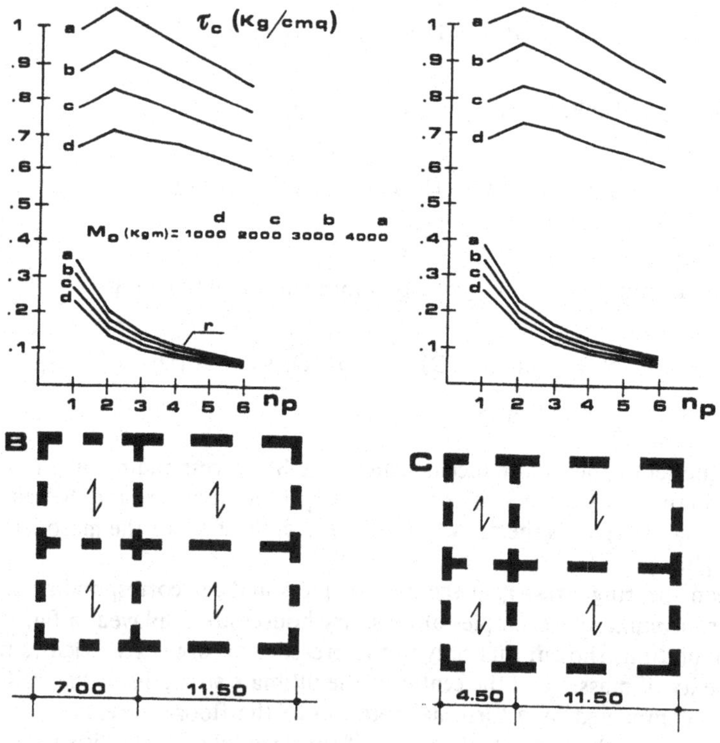

Fig. 9, Values of τ_c and r_c for asymmetric masonry buildings

REFERENCES

[1] M. Como, A. Grimaldi: «An unilateral model for the Limit Analysis of masonry walls». Meeting on «Unilateral problems in Structural Analysis», Ravello, 22-24 Sept. 1983, CISM courses and Lectures n. 288, Springer-Verlag, 1985

[2] M. Giaquinta, E. Giusti: «Researches on the statics of masonry structures». Arch. Rat. Mech. Anal., p. 359-392, 1985

[3] G. Romano, M. Romano: «Elastostatics of structures with Unilateral condition on Stress and displacement fields». Meeting on «Unilateral problems in Structural Analysis», Ravello, 22-24 Sept. 1983, CISM courses and Lectures n. 288, Springer-Verlag, 1985

[4] S. Di Pasquale: «Questioni di meccanica dei solidi non reagenti a trazione», AIMETA VI Congr. Naz.le, Genova 7-9 ott. 1982

[5] A. Kooharian: «Limit analysis of voussoir and concrete arches» Journ. Amer. Concrete Inst., 24, 1952

[6] J. Heyman: «The stone skeleton» Int. J. of Solids and Structures, 2, 1966

[7] J. Heyman: «The masonry arch», Cambridge Press, 1982

[8] M. Como, A. Grimaldi: «Analisi limite di pareti murarie sotto spinta», Università di Napoli, Atti Ist. Tecnica delle Costruzioni n. 546, Novembre 1983

[9] D. Abruzzese, M. Como, A. Grimaldi: «Sulla valutazione della resistenza sismica al collasso degli edifici in muratura». Atti del 2° Convegno Naz.le «L'Ingegneria Sismica in Italia», Rapallo, 6-9 Giugno 1984

[10] D. Abruzzese, M. Como, A. Grimaldi: «Analisi limite degli edifici murari sotto spinte orizzontali». Atti del Dipartimento di Ingegneria Civile Edile, 2ª Univ. di Roma, to be published.

TACT: A CONTACT ANALYSIS PROGRAM

A. Curnier
Laboratoire de Mécanique Appliquée
Ecole Polytechnique Fédérale de Lausanne

ABSTRACT

The computer analysis program **TACT** is applicable to wide classes of problems of **solid and fluid thermomechanics** and in particular to **dry and lubricated contact-impact problems.** The program can be utilized to solve one-, two- and **three**-dimensional problems which may be linear or **nonlinear,** steady state or **transient** . Both small and large deformation (total Lagrangean) formulations are available for the modelisation of solids as well as a spatial description (Eulerian) for liquids. Material laws include linear elastic, nonlinear elastic (rubber) and elastic-plastic (Von Mises) models for solids and linear viscous (Newtonian) liquids.

Both small and large gap and slip simulations are possible, along contact surfaces. Interface laws include adherence (stick), sliding (slip) and nonassociated (Coulomb's) friction. Lubricants are separately treated as a third body.

The program is based on the **finite element** method for the spatial discretization, successive **linearizations** to overcome the non-linearities and on a choice of **finite difference** schemes for the time integration. The **penalty** and the **duality** methods are used for dry contacts with special impact-release conditions in dynamics. The friction algorithm is directly inspired from the **radial return** algorithm of plasticity, with a consistently derived nonsymmetric tangent matrix.

1. INTRODUCTION

With the advent of electronic computers, **numerical** techniques have emerged as valuable alternatives to the traditional analytical and experimental methods for the analysis of mechanical systems. Indeed, several general purpose computer programs are already available to study rather intricate problems of continuum mechanics [e.g. 1,2].

However it is more difficult to find programs offering thorough **contact** analysis capabilities. This scarcity is unfortunate because the global behaviour of a mechanical system is often determined by the contact conditions prevailing at the interfaces between its different components rather than by the intrinsic deformations or strengths of each of these components.

The purpose of this paper is to present a **contact analysis program,** called TACT, which fulfills part of this shortage.

The program is applicable to wide classes of problems of solid and fluid thermomechanics but more specifically to dry and lubricated contact-impact problems. In principle, these problems may be linear or nonlinear, steady state or transient and in one, two or three dimensions.

The program is based on the combination of three classical numerical methods :

- the **finite element method** (FEM) for the spatial discretization of the bodies,
- the **linear iteration method** (LIM) to overcome the geometric and material nonlinearities,
- the **finite difference method** (FDM) to integrate the equations of motion in time.

In addition, a selection of specific techniques are implemented in the program to deal with contact problems :

- the **penalty** and the **duality** methods for dry quasi-static contacts, together with impact-release conditions for dynamic impacts;
- a special **finite element** for lubricated contacts.

The principle and the implementation of these various methods are reviewed in this paper. The general numerical methods are grouped in section 2 to give an overview of the architecture of the program. The specific contact techniques are discussed in section 3 and provide additional insight into the numerical analysis of contact problems. Finally a typical application is displayed in the conclusion to illustrate the capabilities of the program.

The material summarized in this paper is neither complete nor new. The interested reader is referred to [3] for a more exhaustive description of the program and to [4] and [5] for the origin of the contact techniques utilized.

2. GENERAL NUMERICAL ANALYSIS METHODS [3]

The Finite Element Method (FEM), the Linearization Method (LIM) and the Finite Difference Method (FDM) constitute a powerful combination of numerical techniques to obtain approximate solutions to Non Linear Initial Boundary Value Problems (NLIBVP).

The FEM performs the spatial discretization, the LIM the successive linearizations and the FDM the time step integration. The order of application of these three methods to go from the exact continuous problem to its approximate discrete counterpart is quite interchangeable. Tradition and common sense suggest however to apply the FEM first or "inner", the LIM next or "inter" and the FDM last or "outer". The following diagram clarifies these three steps :

$$\textbf{NLIBVP} \xrightarrow{\textbf{FEM}} \textbf{NL-ODE} \xrightarrow{\textbf{LIM}} \textbf{ODE} \xrightarrow{\textbf{FDM}} \textbf{ALGEQ}$$

where NL-ODE stands for Non Linear - Ordinary Differential Equation (in time) and the resulting system of linear algebraic equations (ALGEQ) is solved by standard methods of linear algebra. The use of numerical quadrature formulas during the FE phase must also be mentioned in order to complete this overview of the numerical techniques utilized in the program.

In this section, the formalism of each method is briefly reviewed together with a compact way to combine them. To this end, let the "exact" continuous problem, to be discretized, linearized and integrated in time, be characterized by a general **(weak) statement** of the form :

$$\int_V f(z,t) \cdot \check{z} \, dV = 0 \tag{1}$$

where f(z,t) denotes a differential operator governing the problem and acting on the unknown field variable, z(x), defined over the domain V and where $\check{z}(x)$ is a test function often interpreted as the virtual counterpart of z.

Typical examples of equation (1) are the principle of virtual work in mechanics :

$$\int_V \nabla\tau \cdot \check{u} \, dV = - \int_V \tau \cdot \nabla\check{u} \, dV + \int_A \tau_n \cdot \check{u} \, dA = \int_V \rho\ddot{u} \cdot \check{u} \, dV$$

where z=u is the displacement and τ is the stress; and the weak form of the Laplace equation of heat conduction :

$$\int_V \nabla^2 T \cdot \check{T} \, dV = - \int_V \nabla T \cdot \nabla\check{T} \, dV + \int_A \nabla T_n \cdot \check{T} \, dA = \int_V \rho c \dot{T} \cdot \check{T} \, dV$$

where z=T is the temperature and ∇T the heat flow.

It must be emphasized that whereas equation (1) is inherently **linear in $\check{z}$** by construction, it is in general **nonlinear in z** by nature of the operator f(z) which embeds the material and geometric nonlinearities of the problem. It must also be recognized that V does not need to be homogeneous or even connected in which case (1) could be expanded into :

$$\sum_s \int_{{}^sV} {}^s f(z,t) \cdot \check{z} \, dV = 0$$

where the left superscript s would denote connected homogeneous subregions. Finally, V may vary with time if the current configuration is utilized to describe the motion and even its connectivity may change with time as in **contact problems,** creating sources of non linearity in their own right.

2.1 Formalism of the Finite Element Method (FEM)

The FEM replaces the exact continuous problem (1) by an approximate discrete problem more amenable to computations.

Basically the domain V is subdivided into elements e , connected at N nodes. This subdivision provides a systematic procedure to evaluate the weak form (1), element per element, along the rule : $\int_V = \sum_e \int_{{}^eV}$, where ${}^eV$ is the element volume.
Within each element the real and virtual field variables, z and $\check{z}$, are discretized at the n nodes delimiting the element and approximated within the element subregion by means of interpolation functions, often called shape functions, S and $\check{S}$, e.g. $z(x) = S^a(x)\, Z^a$, a = 1,n, where Z^a are the nodal values of z.

Enforcement of the fact that the weak statement (1) must be satisfied for **all** virtual admissible fields $\check{z}$, and thus $\check{Z}$, yields the system of N nonlinear dynamic ordinary differential equations (or simply N nonlinear static algebraic equations)

$$F(Z,t) = \sum_e \int_{{}^eV} f(S^b Z^b, t) \cdot S^a \, dV = 0 \tag{2}$$

where the N-dimensional vector, Z, is called the discrete solution.

A typical example of equation (2) is provided by the discrete equation of non linear (thermo-)dynamics :

$$F(Z,t) = H(a) + G(v) + F(u) = B \tag{3}$$

where $H(a) \simeq C.a$ represent the forces of inertia; $a = \ddot{u}$ being the acceleration vector and C the mass matrix.

$G(v) \simeq D.v$ represent the viscous (or convection) forces; $v=\dot{u}$ being the velocity (or temperature rate) vector and D the damping (or capacity) matrix.

$F(u) \simeq E.u$ represent the internal mechanical forces (or thermal flows), u being the displacement (or temperature) vector and E the stiffness (or conductivity) matrix.

B represents the vector of external forces (or flows)

The FEM provides great flexibility in modeling complicated geometries with the possibility of increasing the accuracy locally. It handles elegantly inhomogeneities, like fluid-solid interfaces, at marginal effort.

All phases of the FEM are so systematic that they can be fully automated. They are implemented in the program presently discussed. The programmer is left with the task of designing one master element for each class of problem of interest to him. Several important classes of problems and associated elements are already developed and implemented. They are discussed in the theoretical manual [3].

2.2 Formalism of the Linearization Method (LIM)

The LIM replaces the resulting (discrete) nonlinear problem (2) by a sequence of linear problems directly solvable by standard methods of linear algebra.

The algorithm may be summarized as follows :

$$Z^o = 0, \quad 0 = t^o < \ldots t^k \ldots < t^* = 1 \qquad \text{(a)}$$

$$DF(Z^k) \cdot \Delta Z = - F(Z^k, t^{k+1}) \qquad \text{(b)}$$

$$Z^{k+1} = Z^k + \Delta Z \qquad \text{(c)} \qquad (5)$$

where t is a (normalized) time or incrementation parameter, DF is the derivative of the (dual) function F with respect to the (primal) variable Z and ΔZ is the variable increment. One pass "k" through (5b-c) is called a **step** or increment if $t^{k+1} > t^k$ and an **iteration** if $t^{k+1} = t^k$.

A typical example is provided by the equation of equilibrium of nonlinear statics, $F(u) = B$, obtained from (3) after deletion of the inertia and viscous forces H and G :

$$DF(u^k) \cdot \Delta u = B - F(u^k, t^{k+1}) \qquad (6)$$

The standard LIM resorting to a single large increment (obtained by setting $t^k=1$ for all k in (5a)) is only locally convergent and tends to be rather restricted in the choice of the initial guess Z^o. Continuation is a standard technique to ensure global convergence. The name of the method comes from the existence of a continuous curve, Z(t), connecting a trivial or known initial solution, Z^o, of F(Z,0) = 0 to the desired final solution, Z*, of F(Z,1) = 0.

Its success entirely depends upon the partition, t^k, of the interval [0,1]. Adaptation constitutes a rational procedure to construct an optimal partition by adjusting the time step in accordance with the predicted local radius of convergence. The process starts at Z^o and marches along the curve Z(t) as fast as possible to converge towards Z* in a final descent.

A mixed termination criterion involving both the norms of the variable increment, ΔZ, and of the equation, F, is utilized.

In some applications it may turn out more economical to keep the same derivative DF over several increments or iterations. All these variations of the LIM are readily available in this program.

2.3 Formalism of the Finite Difference Method (FDM)

The FDM replaces the discrete linearized ordinary differential equation (5), whenever the parameter t plays the role of real time, by a temporal sequence of algebraic equations.

This is achieved by replacing the time derivatives by finite difference quotients e.g.: $Z = (Z_{n+1}-Z_n)/\tau$ where $Z_n=Z(t_n)$ and τ is the time step. The solution is pursued step by step starting at the initial time, t=0 .

There exist two basic types of finite difference methods to integrate equations (3) in time :

Explicit or predictor methods use some forward difference formula to predict the new value of the unknown variable in terms of its past own and rate values alone, quite independently of the differential equation. This differential equation is then used to update the rates :

$$u_{n+1} = u_n + \tau v_n$$
$$v_{n+1} = D^{-1} \cdot (B_{n+1} - F(u_{n+1})) \qquad (7)$$

Since a lumped mass or capacity matrix D is tacitly retained in this case, the computational effort per time step is reduced to a minimum. Unfortunately, explicit methods are conditionally stable i.e. they are restricted to small time steps : some fraction of the smallest period of the system.

Implicit or predictor-corrector methods use some backward difference formula to correct either the past value or more often an otherwise predicted value of the unknown by means of its current rate. This procedure requires the simultaneous solution of the difference and the differential equations :

$$\begin{aligned} u_{n+1} &= u_n + \tau\, v_{n+1} \,, \\ D \cdot v_{n+1} &+ F(u_{n+1}) = B_{n+1} \,. \end{aligned} \tag{8}$$

The nonlinear equation, obtained by substitution of the difference equation into the differential one, requires considerable computational effort for its solution by iterations at each time step. Fortunately, "sufficiently correcting" implicit methods are unconditionally stable i.e. exempt of any restriction on the time step other than to accurately integrate the lower modes of course : some fraction of the largest period of the system often proves adequate.

Explicit methods are well suited for wave propagation problems where the higher modes of the system govern its response (accuracy as stringent as stability) or for bulk problems characterized by a narrow modal spectrum (soft ODE). Conversely implicit methods are well suited to analyse inertial problems where only the lower modes are excited (stability much more stringent than accuracy) or shell like structures characterized by a wide spread spectrum (stiff ODE).

In this program, only one explicit scheme and one range of implicit schemes belonging to the same family are available. These are :

Explicit : forward difference for thermics, centered difference for dynamics

Implicit : generalized trapezoidal rule } for thermics and dynamics

They have been selected for their well established individual performances as well as their compatibility for thermo-dynamics. They are defined by :

For nonlinear thermics

$$D \cdot v_{n+1} + F(u_{n+1}) = B_{n+1} \qquad \text{(a)}$$
$$u_{n+1} = u_n + (1-\alpha)\tau\, v_n + \alpha\tau\, v_{n+1} \qquad \text{(b)} \tag{9}$$

For nonlinear dynamics

$$C \cdot a_{n+1} + D \cdot v_{n+1} + F(u_{n+1}) = B_{n+1} \qquad \text{(a)}$$
$$u_{n+1} = u_n + (1-\alpha)\tau\, v_n + \alpha\tau\, v_{n+1} + \left(\frac{1}{2} - \alpha\right)\tau^2\, a_n \qquad \text{(b)} \tag{10}$$
$$v_{n+1} = v_n + (1-\beta)\tau\, a_n + \beta\tau\, a_{n+1} \qquad \text{(c)}$$

The implicitness parameter α is restricted to the range $0 \leq \alpha \leq 1$ whereas the dissipation parameter is confined to $1/2 \leq \beta \leq 3/4$ in practice.

2.4 Compact combination of the FE-LI-FD methods

It is traditional in numerical analysis to solve a nonlinear problem by a sequence of linear problems and a dynamic problem by another sequence of static problems. Therefore a nonlinear thermodynamics analysis program consists essentially of a **basic linear analysis package** embedded into an **inner iteration loop,** to overcome the nonlinearity, itself embedded **into an outer load incrementation or time integration loop,** to advance the solution, both loops being accompanied by appropriate update steps. It can be repeated for clarity that the linear analysis package performs the finite element spatial discretization of the continuous problem, whereas the inner and outer loops with their associated updates represent the linearization phase and finite difference temporal discretization, respectively.

A nonlinear thermodynamics solution algorithm can be applied to any problem subclass like linear thermics or nonlinear statics, for instance, with minor adjustment. Moreover, all the members of a family of time integration schemes applicable to first order ODEs of thermics and second order ODEs of dynamics and ranging from explicit-diagonal predictors to implicit-unsymmetric predictor-correctors can be programmed simultaneously in a compact scheme.

The result is a code offering
- linear and nonlinear
- thermal and mechanical
- explicit and implicit

options that are performant.

The basic idea is to split the FDM into a **predictor** followed by a **corrector** and to insert the FEM and the LIM between the two. The result is a scheme of the form [3],

- **a predictor** **(FDM)**
- **a former** **(FEM)**
- **a solver** **(ALG)**
- **a corrector** **(LIM + FDM)**

A more explicit version of this algorithm is summarized in table 1. Three formulations are offered in option in order to accomodate non linear thermics and nonlinear dynamics as well.

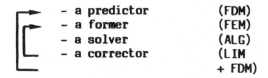

TABLE 1. NONLINEAR THERMODYNAMIC ALGORITHM

1. predict $u^{o}_{n+1} = u_{n} + \tau\, v_{n} + \frac{1}{2}\,\tau^{2} a_{n}$, $v^{o}_{n+1} = v_{n} + \tau\, a_{n}$

$a^{o}_{n+1} = a_{n}$. (i=0)

2. form $F^{i}_{n+1} = F(u^{i}_{n+1})$; $G^{i}_{n+1} = D \cdot v^{i}_{n+1}$; $H^{i}_{n+1} = C \cdot a^{i}_{n+1}$

$E^{i}_{n+1} = \frac{dF}{du}(u^{i}_{n+1})$; $D = \frac{dG}{dv}$; $C = \frac{dH}{da}$.

3. solve $[cC + dD + eE^{i}_{n+1}] \cdot h = B_{n+1} - F^{i}_{n+1} - G^{i}_{n+1} - H^{i}_{n+1}$

4. correct $u^{i+1}_{n+1} = u^{i}_{n+1} + e\,h$

$v^{i+1}_{n+1} = v^{i}_{n+1} + d\,h$

$a^{i+1}_{n+1} = a^{i}_{n+1} + c\,h$

5. adapt <NOT CONV.> i = i + 1 , GO TO 2.
<CONVERGED> n = n + 1 , GO TO 1.

Legend :	Formulation	Increment	Constants		
	DISP/TEMP	$h = \Delta u$	$c=1/\alpha\beta\tau^{2}$	$d=1/\alpha\tau$	$e=1$
	VELO/RATE	$h = \Delta v$	$c=1/\beta\tau$	$d=1$	$e=\alpha\tau$
	ACCE	$h = \Delta a$	$c = 1$	$d=\beta\tau$	$e=\alpha\beta\tau^{2}$

3. SPECIFIC CONTACT ANALYSIS TECHNIQUES [4-6]

Contact mechanics is such a wide subject that a selection had to be exercised at the origin of the program development. Currently, the program is limited to two classes of contact problems, namely

- **dry friction contacts,** which involve two bodies, (usually two solids), in quasi-static (contact) or dynamic (impact) interaction.
- **lubricated contacts,** characterized by the presence of a third body (usually a liquid), in between the two solids.

Both classes are discussed in this section. In the discussion, the modelisation of the contacting bodies is assumed to be known, [e.g. 2]. For the sake of completeness, it is mentioned, however, that the program possesses a library of elements which includes in particular standard isoparametric elements for the discretisation of continuum bodies in 2D and in 3D, [3].

For **solids,** a large deformation total Lagrangean formulation and a choice of elastic plastic material models are available. Selective numerical integration schemes extend the range of application of the linear options of the elements to incompressible materials and shell like configurations.

For **liquids,** the spatial description is applied to a linear viscous fluid (Navier Stokes equations). Discontinuous test functions extend the range of application of the elements to all Reynolds numbers and selective integration accomodates nearly incompressible liquids.

3.1 Dry friction contacts [4,5]

A dry contact occurs when two bodies coalesce over a certain portion of their boundaries for a certain period of time. By hypothesis, the two bodies can come in contact, deform each other and then separate. Along the contact surface, they may stick, slip or even rub against each other, but they cannot penetrate each other across the interface.

Quasi-static **contacts** are governed by two complementary unilateral constraints : the kinematic condition of impenetrability and the static principle of action and reaction in compression only. A variational formulation of these statements constitutes a good and concise basis for the development of numerical solution schemes.

To this end consider two bodies, one called the **striker** and the other the **target,** bound to contact one another within a surface A characterized by the unit outward normal N to the target.

Basically, to the standard weak statements of equilibrium, e.g. the principle of virtual work, of the two deformable bodies, needs to be added a contact term of the form [4] :

$$\int_A F \cdot \dot{D}\, dA = \int_A (F_N \cdot \dot{D}_N + F_T \cdot \dot{D}_T)\, dA \quad (\geq 0) \tag{11}$$

In the above the vector D represents the **distance** of contact (or gap) separating each point S on the striker from a point T on the target, i.e. $D = S - T$, which are (were or will be) in contact with one another (depending on the instant). A superposed dot denotes a rate or an increment (or even a variation) depending on the temporal (or functional) context. It proves convenient to resolve the relative velocity $\dot{D}$ into a normal and a tangential components $\dot{D}_N = (\dot{D} \cdot N)\, N$ and $\dot{D}_T = \dot{D} - \dot{D}_N$. In general this decomposition is not applicable to the contact distance D itself due to the variations in the definition of the unit normal N with the curvature and deformation of the contact surface. There exist two exceptions when it remains valid, namely for small amounts of slip and nearly flat and rigid contact surfaces.

The vector F is the **force** of contact per unit area (also called stress or traction vector) acting on the target at the point $S = T$ whenever contact occurs, i.e. when $D_N = \dot{D}_N = 0$. $F = (F \cdot N)\, N$ and $F_T = F - F_N$ are the normal and tangential components of the contact force.

It is found convenient to attach the notion of point of contact with a material point S on the striker and to regard the target surface as a sliding rink oriented by its outward normal N . The actual surface of contact C defined by the set of striker points in contact with the target at a given time, must be distinguished from the potential (or candidate) surface of contact A which includes the actual one at all times as an envelope.

The contact integral (11) may be indifferently defined over the reference contact surface A with unit outward normal N (in material Lagrangean description), F representing the first Piola-Kirchhoff stress vector, or over the deformed contact surface a with normal n (in spatial Eulerian description), F becoming the Cauchy stress vector. Because the amounts of slip and deformation are assumed to remain small in this presentation, the two descriptions are undistinguishable.

The contribution of the contact constraint term (11) is non-negative. Indeed it is null whenever a positive normal gap exist ($D_N > 0$, $\dot{D}_N \neq 0$, no contact) since then $F = F_N = F_T = 0$. When the normal gap closes ($D_N = 0$, $\dot{D}_N = 0$, contact), the normal contribution $F_N \cdot \dot{D}_N$ remains zero, which accounts well for the condition of impenetrability, but the tangential term $F_T \cdot \dot{D}_T$, which represents the energy dissipated by friction, must only be non-negative according to the entropy inequality.

Of course penetration ($D_N < 0$, $\dot{D}_N \neq 0$) is implicitly excluded whereas tension ($F_N > 0$, $D_N = 0$) due to adhesion may be retained.
The dissipation due to friction vanishes in the two limiting cases of perfect stick ($\dot{D}_T = 0$) and perfect slip ($F_T = 0$) . In between those extremes, a **law of friction** relating the force of friction to the amount of slip (and other internal variables) of the form $F_T = F_T\,[D_T, \dot{D}_T , \ldots]$ is necessary.

Enforcement of the principle of action and reaction across the surface of contact is achieved by taking a variation of the contact distance or velocity D , i.e. separate variations of S and T , like in a classical displacement formulation. To the contrary kinematic compatibility is enforced by taking an independent variation of the contact force F according to the force method. More explicitly the contact term (11) can be expanded into the sum of two terms

$$\int_A (F \cdot \check{D} + \check{F} \cdot \dot{D})\, dA \tag{12}$$

where a "vee" denotes a variation and overrides the "dot" whenever superposed. The first term in (12) is adapted to the implementation of penalty methods whereas the second term is suitable for a treatment by dual Lagrange multipliers. Recent formulations resorting to "penalty-duality augmented Lagrangians" provide a unified treatment of these alternative approaches.

The spatial discretisation of the contact term (11) is facilitated by the absence of any derivative in the integrand. Thereby, concentrated contact forces are legitimate provided the contact distance is continuous, once contact is established, which is usually the case.
Accordingly, the integral (11) is replaced by a finite element sum of the form

$$\sum_e {}^eF \cdot {}^eD \tag{13}$$

where ${}^eF$ are the contact element concentrated forces and ${}^eD$ the corresponding contact distances. The geometry of the contact element is determined by the meshes and the kinematics of the striker and target. If a one to one correspondance between the boundary nodes of the striker and the ones of the target in regard can be established and maintained throughout the contact duration (adherent or small slip contact), then a node on node element is quite adequate. Otherwise a node on facet element must be used to account for initial mismatching as well as subsequent sliding.

For the **node on node element** the contact distance represents the length of the element

$$D = S - T = x^1 - x^2 \tag{14}$$

where $S = x^1$ and $T = x^2$ denote the deformed coordinates of node 1 attached to the striker and node 2 attached to the target, respectively. The element is similar to a bar element as far as its geometry is concerned.

For the **node on facet element,** the contact distance keeps its signification, of course, but now the target point T is expressed in terms of the facet corner coodinates $T^a = x^a$ by means of standard shape functions $N^a(\xi)$

$$T = N^a(\xi)T^a \ , \quad a = 1,N \tag{15}$$

where ξ are the natural coordinates of the facet and N is the number of nodes per facet. The element can be visualized as a bipode in 2D and a quadripode in 3D in connection with standard linear isoparametric continuum elements.

Once a penetration is detected at the end of a time or load step, the location and the instant of impact (or incipient contact) are determined by solving the complete contact equation

$$D(\xi,\tau) = S(\tau) - N^a(\xi)T^a\,(\tau) = 0 \tag{16}$$

for both the natural coordinates ξ and the time parameter τ .(For the node on node element, it is sufficient to solve the normal gap equation $D_n(\tau) = 0$, since the contact location is known a priori).

In the **penalty** formulation, the normal component of the contact force is related to the normal contact distance by the condition of near impenetrability

$$\begin{aligned} F_n &= 0 && \text{if} \quad D_n > 0 \quad \text{(gap)} && \text{(a)} \\ F_n &= P\,D_n && \text{if} \quad D_n \leqslant 0 \quad \text{(contact)} && \text{(b)} \end{aligned} \tag{17}$$

where P is a large coefficient (a penalty) in comparison to the stiffness of the adjacent solids.

In the **duality** formulation, the contact force is taken as an independent variable with the contact distance for conjugate function. Exact impenetrability may be achieved with this approach, namely

$$\begin{aligned} F_n &= 0 && \text{if} \quad D_n > 0 \quad \text{(gap)} && \text{(a)} \\ F_n &< 0 && \text{if} \quad D_n = 0 \quad \text{(contact)} && \text{(b)} \end{aligned} \tag{18}$$

The additional force variable is fetched into an extra node for convenience.

Mixed Penalty-duality formulations based on a so-called augmented Lagrangian combine these two techniques.

A friction model, inspired from the theory of plasticity, governs the tangential component of the normal force [5]. This model is based on four hypotheses :

- an additive decomposition of the contact distance into adherence (reversible) and slip (irreversible) components

$$D_T = D_T^A + D_T^S \tag{19}$$

- an exclusive dependence of the contact force on the adherence

$$F_T = \mathbf{P}\, D_T^A \tag{20}$$

- Coulomb's criterion to delimit adherence from slip

$$|F_T| + \mu F_N \leq 0 \tag{21}$$

- a non associated slip rule governing the slip increment

$$dD_T^S = \lambda F_T \;. \tag{22}$$

A friction algorithm, inspired from the radial return algorithm of plasticity, generates the path dependent friction-force/slip-distance relationship,

$$F_T = F_T \left[D_T, D_T^S\right] .$$

Dynamic impacts require a specific treatment. When two bodies collide, the contact force, which prevents penetration, and the relative velocity, which characterizes their approach, suffer abrupt changes, assimilable to discontinuities. Standard finite difference schemes like the one used in this program are unable to capture these jumps. Special **impact-release conditions** must be applied to avoid spurious oscillations in the post impact and post release calculations.

The jump conditions may be derived either by applying the elementary theory of shocks between rigid bodies to the discrete masses located at the periphery of the contacting bodies, or by discretizing the formulas provided by the theory of wave propagation in elastic media.

According to the first approach, for instance, the post impact values for the node on node element are given by

$$V = \frac{M_S\,\dot{S}_- + M_T\,\dot{T}_-}{M_S + M_T} \qquad \text{(a)}$$

$$A = \frac{M_S\,\ddot{S}_- + M_T\,\ddot{T}_-}{M_S + M_T} \qquad \text{(b)} \qquad (23)$$

$$F = \frac{M_S\,M_T}{M_S + M_T}\,(\ddot{S}_- - \ddot{T}_-). \qquad \text{(c)}$$

In the above formulas M denotes a concentrated mass at a node, V and A the compatible values of the velocity and acceleration of the nodes in contact. A subscript "-" indicates a pre-impact value.

The corresponding post-release conditions are

$$\dot{S}_+ = V + \frac{\tau F}{2M_S} \qquad \text{(a)}$$

$$\dot{T}_+ = V - \frac{\tau F}{2M_T} \qquad \text{(b)} \qquad (24)$$

$$\ddot{S}_+ = A + \frac{F}{M_S} \qquad \text{(c)}$$

$$\ddot{T}_+ = A - \frac{F}{M_T} \qquad \text{(d)}$$

where τ denotes the time step and a subscript "+" post-release values.

For the node on facet element, the different exchanges of impulse momentum and forces of inertia must be consistently distributed over the facet corners by means of the shape functions. For instance in 2D, the counterpart of equation (23) is obtained by solving the system

$$\begin{bmatrix} M_T^1 + N^1N^1M_S & N^1N^2M_S \\ N^2N^1M_S & N^2N^2M_S + M_T^2 \end{bmatrix} \begin{Bmatrix} \dot{T}^1 \\ \dot{T}^2 \end{Bmatrix} = \begin{Bmatrix} M_T^1\dot{T}^1_- + N^1M_S\dot{S}_- \\ M^2\dot{T}^2_- + N^2M_S\dot{S}_- \end{Bmatrix}$$

for the target node velocities $\dot{T}^1$ and $\dot{T}^2$.

3.2 Lubricated contacts [6]

The main function of lubrication is to decrease the friction between two solids in contact. The lubricant is a viscous liquid which is inserted between the two rubbing surfaces to form a film and carry the load. Frequently, the deformations of the two adjacent solids play a significant role in the lubrication process. The two deformable solids separated by the fluid film constitute a **lubricated contact.**

The coupled analysis of a lubricated contact problem involves the simultaneous solution of the lubrication equation for the fluid and the equations of equilibrium of the two solids. In the following, attention is focussed on the lubricant contribution since the formulations for the adjacent solids is classical [e.g. 2].

To fix the ideas, let the film of lubricant be spread over an area of contact A, in the plane $X_3 = 0$ spanned by the two coordinates X_α, $\alpha = 1,2$. Thus, the thickness of the film $h=h(X_\alpha,t)$ is measured along the direction X_3. Basically, to the standard weak formulations of the equilibrium of the two deformable solids, needs to be added a lubrication term of the form

$$\int_A w,_\alpha Q_\alpha \, dA + \int_{\partial A} w \, Q_n \, dL \, . \tag{25}$$

In the above expression, $w=w(X_\alpha)$ is a weighting function, best interpreted as a virtual pressure, with partial derivatives $w,_\alpha$. Q is the **mean flow** across the thickness of the film, a vector with cartesian components defined by

$$Q_\alpha = hv_\alpha - \frac{h^3}{12\mu} P,_\alpha \tag{26}$$

where $v_\alpha = (v_\alpha^1 + v_\alpha^2)/2$ is the mean (rolling) velocity of the flow in the direction α, μ the viscosity of the liquid and P the internal pressure.

The mean flow (26) is recognized as the primitive of the Reynolds equation $Q_{\alpha,\alpha} = 0$ and therefore (25) as a weak form of this equation.

It is recalled for completeness, that the Reynolds equation results from the combination of the very simplified Navier Stokes equations (by means of a dimensional analysis based on the fundamental hypothesis that the thickness of the film remains small in comparison to the size of the contact) and the averaged continuity equation (by integration over the thickness). A crucial result of the dimensional analysis is that the pressure remains constant across the thickness i.e. $P=P(X_\alpha,t)$ thus reducing the dimension of the problem.

A generalized version of the theory of lubrication including thermal effects and pressure variations across the thickness is available in [6].

The mean flow Q depends **nonlinearly** on the film thickness which varies with the deformations of the adjacent solids in a coupled analysis. Moreover, if the viscosity of the fluid is assumed to depend on the pressure according to some empirical model $\mu = \mu(P)$, then the mean flow becomes a nonlinear function of the pressure as well :

$$Q = Q\,[h,P]\ .$$

The mean flow (26) is linearized in order to construct a convergent algorithm to overcome these nonlinearities.

Formally

$$\tilde{Q}_\alpha(\Delta h, \Delta P) = Q_\alpha(h,p) + \frac{\partial Q_\alpha}{\partial h}\,\Delta h + \frac{\partial Q_\alpha}{\partial P}\,\Delta P$$

and more explicitly

$$\frac{\partial Q_\alpha}{\partial h}\,\Delta h = \left(v_\alpha - \frac{h^2}{4\mu}\,P,_\alpha\right)\Delta h$$

$$\frac{\partial Q_\alpha}{\partial P}\,\Delta P = \frac{h^3}{12\mu^2}\,\frac{\partial \mu}{\partial P}\,P,_\alpha\,\Delta P - \frac{h^3}{12\mu}\,\Delta P,_\alpha\quad .$$

The spatial discretization of the lubrication term (25) follows the general finite element pattern outlined in subsection 2.1. The lubricant element developed is an isoparametric element with linear and quadratic shape functions in two dimensional analysis and bilinear ones in three dimensions. Consequently it is entirely compatible with the solid element adopted besides. The unconventional geometry of the element consists of a surface to represent the midsurface of the film equipped with nodal bipodes to account for its thickness. Note that this geometry supposes an alignement of the boundary nodes of the meshes of the adjacent solids.

The pressure interface conditions are simply treated as any distributed load acting over a solid boundary.

4. CONCLUSION

The simulation of the impact of a liquid jet against an elastic plastic wall displayed on Fig. 1. gives an idea of the program applicability.

Several developments are anticipated to improve and extend the current potential of the program. They include : improved continuum and structural elements, quasi-Newton linearisation methods, explicit-implicit schemes; non-Newtonian lubrication and third body modelisation.

References

[1] Taylor R.L. "Computer Procedures for Finite Element Analysis", Ch.24 of Zienkiewicz O.C., **"The Finite Element Method"**, 3rd Ed., Mc Graw Hill, London, 1978.

[2] Bathe K.J. **"Finite Element Procedures in Engineering Analysis"**, Prentice Hall, 1982.

[3] Curnier A., "TACT, A Contact Analysis Program", Program manual, Lab. de Méca. Appliquée, Ecole Polytechnique Féd. de Lausanne, 1983.

[4] Hughes & al., "A Finite Element Method for a Class of Contact - Impact Problems", Comp. Methods in Applied Mech. and Engng, Vol.8, 1976.

[5] Curnier A., "A Theory of Friction", Int. Journal of Solids and Struct., Vol. 20, pp.637-647, (1984).

[6] Curnier A. and Taylor R.L., "A Thermomechanical Formulation and Solution of Lubricated Contacts between Deformable Solids", J. of Lubrication Technology, Trans ASME Series E, Vol.104, 1982.

liquid jet (water)

compressible (Tait)

$$p = B\left[\left(\frac{\rho}{\rho_0}\right)^n - 1\right]$$

B = 3000 bar

n = 7.5

velocity
600 m/s

solid wall (aluminum)

elastic – plastic (V.Mises)

E = 700 000 bar

ν = 0.25

ρ = 2.5 g/cm^3

σ_e = 200 bar

P = 5000 bar

Q = 2500 bar

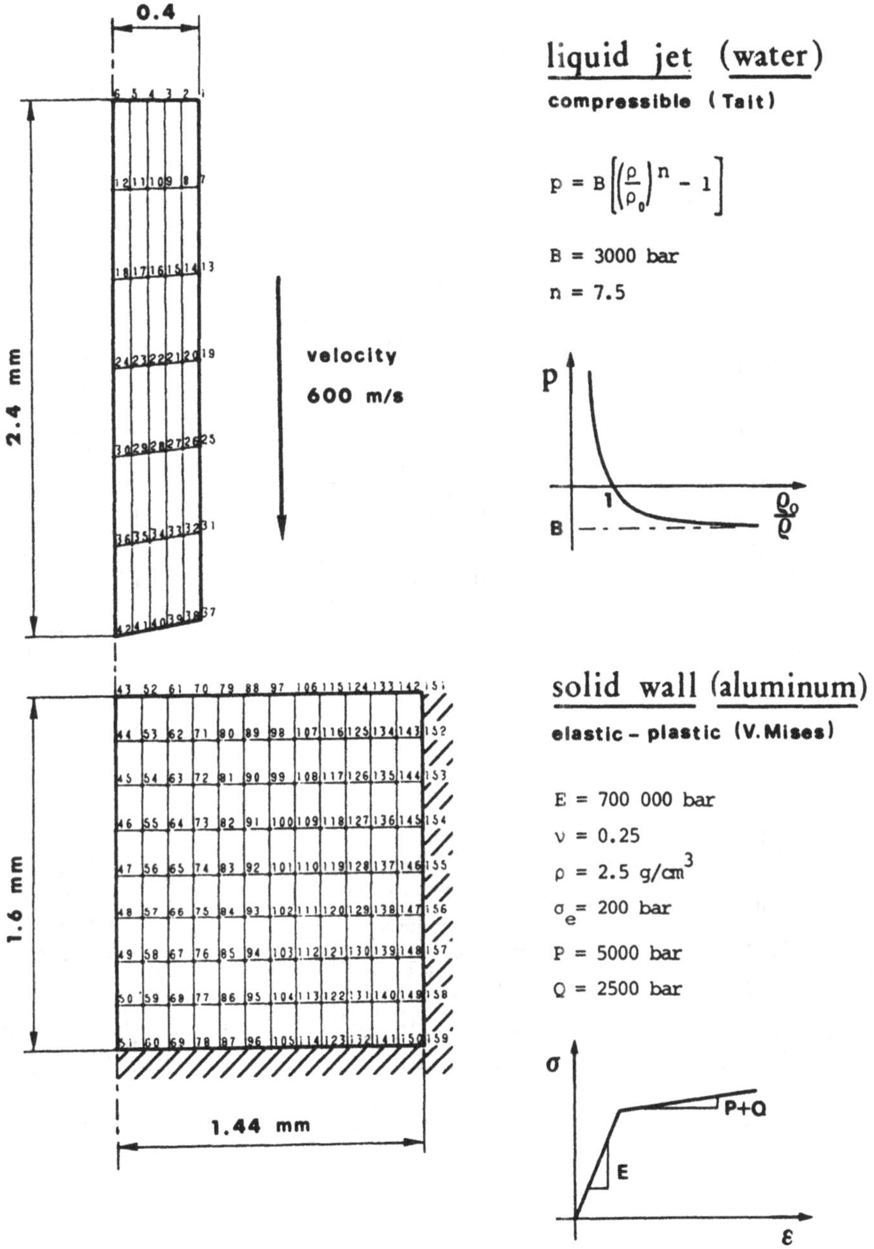

Fig. 1a   Impact of a liquid jet against an elastic-plastic wall, geometric and material data

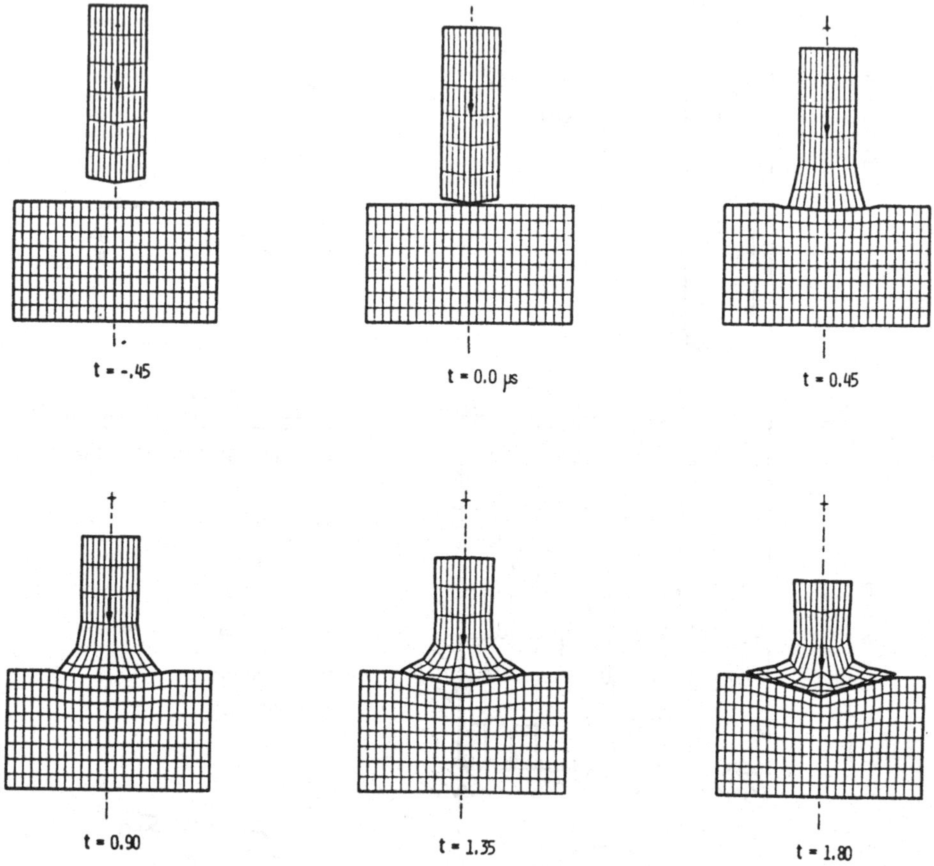

Fig. 1b Impact of a liquid jet against an elastic-plactic wall, time sequence of deformed configurations

CONTACT UNILATERAL AVEC ADHERENCE
UNE THEORIE DU PREMIER GRADIENT

M. Fremond
Laboratoire Central des Ponts et Chaussees, Paris

ABSTRACT

Using the methods of continuum mechanics, an adhesion theory is described. A scalar variable, the adhesion intensity or proportion of active links β describes the degreee of contact. A first gradient theory (i.e. the free energy depends on grad β) and a pseudo-potential of dissipation give the constitutive laws. The equations show that, during the the separation process, far from the separation line the contact is bilateral, in the vicinity of the separation line only a small reaction is allowed, and on the separation line the contact becomes unilateral.

1. INTRODUCTION

On a déjà présenté une théorie de l'adhésion en introduisant une nouvelle variable β, l'intensité d'adhésion ou proportion de liaisons actives entre une structure et le socle sur lequel elle repose avec adhérence [2], [3]. Dans cette théorie l'énergie libre Ψ dépend de β et conduit à une théorie dite du zéro gradient puisque Ψ ne dépend d'aucune dérivée de β. De ce fait toute influence mutuelle réversible entre des points voisins de la surface de contact est ignorée. Des influences ne sont cependant pas exclues mais sont reportées sur les effets dissipatifs. La faiblesse de ce modèle n'est qu'apparente puisqu'elle permet de retrouver la théorie habituelle de l'adhérence et divers résultats expérimentaux [3]. Nous présentons ici une théorie plus fine, celle du premier gradient, qui permet cette fois une interaction réversible entre les points voisins de la surface de contact. L'énergie libre devient une fonction de β et grad β . La présence de grad β traduit les influences mutuelles.

Parmi les conséquences de cette influence mutuelle, notons par exemple que lors du décollement la ligne de décollement est entourée d'une zone où la détérioration des liaisons est progressive : loin de la ligne de décollement la réaction du socle est quelconque, plus près la réaction ne peut être trop grande, en particulier elle ne peut être une traction trop importante, et lorsque l'on atteint la ligne de décollement le contact devient unilatéral.

Comme dans la théorie du zéro gradient la principale idée que nous mettons en oeuvre est que les champs de déplacements habituels sont insuffisants pour décrire le contact. En effet lorsque deux solides sont en contact ils peuvent être plus ou moins liés (il y a adhérence) ou bien simplement juxtaposés sans aucune liaison (il n'y a pas adhérence). On introduit donc la nouvelle variable cinématique β, intensité d'adhésion (paragraphe 2). La théorie est construite en utilisant le principe des puissances virtuelles (paragraphes 3,4,5). Le second principe de la thermodynamique donne les lois de comportement (paragraphes 6,7,8,9). On examine ensuite le comportement non dissipatif et l'on donne une formulation variationnelle de la recherche des positions d'équilibre (paragraphe 10). On termine par les problèmes dissipatifs (paragraphes 11,12,13). La dissipation a dans cette théorie une structure habituelle alors que dans la théorie du zéro gradient elle a une forme inusuelle faisant intervenir des interactions entre points voisins [3].

2. VARIABLES CINEMATIQUES

Soit une structure occupant à l'instant t un domaine $\overline{\Omega}\subset\mathbb{R}^3$, fixée sur une partie Γ_0 de sa frontière et reposant par une autre partie Γ_1 sur un socle plan S (figure 1).

Le système considéré est $\overline{\Omega}$. Sur la partie Γ_1 le système inclut les points matériels de la structure et les points matériels du socle. La description cinématique est assurée dans $\overline{\Omega}$ par des champs de déplacement fonction du point x et du temps t. Sur Γ_1 sont définis deux champs de

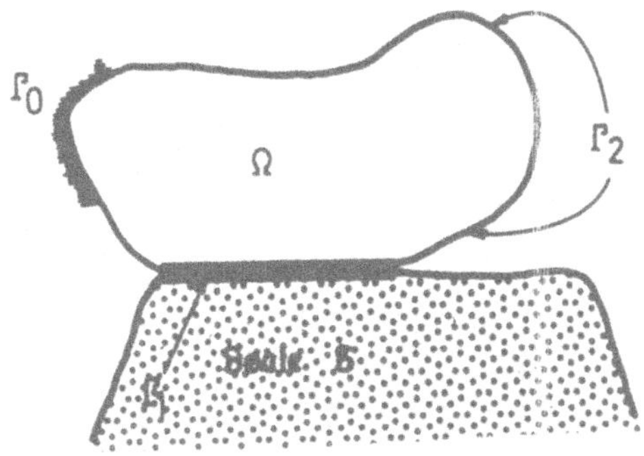

- Figure 1 -

La structure Ω adhère au socle S.

déplacement, d'une part le déplacement des points de la structure qui est la trace sur Γ_1 du déplacement $\vec{u}$ des points de Ω, d'autre part le déplacement des points du support $\vec{u}_{ext}$. Le déplacement $\vec{u}$ est une fonction de $x \in \overline{\Omega}$ et de t. Le déplacement $\vec{u}_{ext}$ est une fonction de $x \in \Gamma_1$ et de t.

Nous avons déjà exposé que ces déplacements étaient insuffisants pour décrire l'état du contact sur Γ_1 [2], [3]. En effet en un point de contact, on doit préciser s'il y a adhérence ou pas avec le socle S, ou encore si les liaisons socle-structure sont actives ou rompues. On a défini pour cela une nouvelle variable cinématique β, appelée intensité d'adhérence. Elle représente la proportion de liaisons actives entre la structure et le socle,

si $\beta = 0$ toutes les liaisons sont rompues, il n'y a pas adhésion ;
si $\beta = 1$ toutes les liaisons sont actives, il y a adhésion totale ;
si $0 < \beta < 1$ une partie des liaisons est active, l'autre partie $1-\beta$ est rompue, l'adhésion est partielle.

3. PUISSANCE VIRTUELLE DES EFFORTS INTERIEURS

Considérons un mouvement tel que $\dot{\vec{u}} = 0$, $\dot{\vec{u}}_{ext} = 0$ et $\dot{\beta} \neq 0$. Dans ce mouvement des liaisons microscopiques entre la structure et le socle se font et se défont ($\dot{\beta} \neq 0$). Des déplacements microscopiques provoquent ces modifications (figure 2). La puissance de ces mouvements doit apparaître dans la puissance des efforts intérieurs. L'expression usuelle de cette puissance

$$-\int_{\Omega} \sigma D(\dot{\vec{u}})\,d\Omega - \int_{\partial\Omega\cap\Gamma_1} \vec{R}(\dot{\vec{u}}_{ext} - \dot{\vec{u}})\,d\Gamma$$

nulle pour $\dot{\vec{u}} = 0$, $\dot{\vec{u}}_{ext} = 0$, ne prend pas en compte la puissance des mouve-

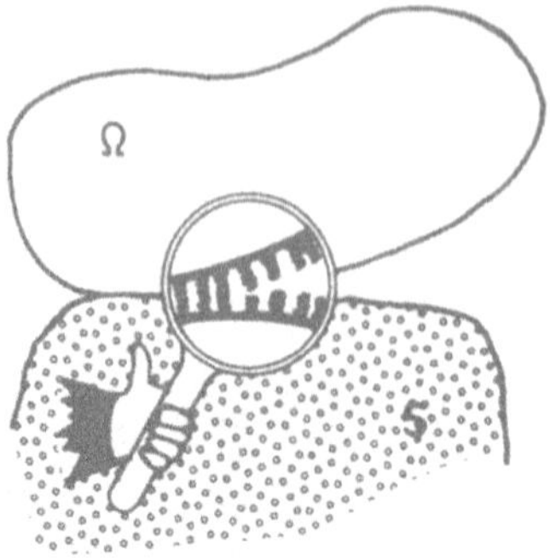

- Figure 2 -

Les liaisons se font et se défont entre S et Ω.

ments microscopiques (σ représente les contraintes, $D(\dot{\vec{u}}) = \left(\frac{1}{2}(\dot{u}_{i,j}+\dot{u}_{j,i})\right)$ le taux de déformation, $\vec{R}$ l'effort intérieur de liaison). Cette puissance doit donc apparaître dans un autre terme ayant la structure habituelle d'une puissance : c'est-à-dire une forme linéaire sur l'espace des vitesses. Comme nous souhaitons faire une théorie du premier gradient nous choisissons

$$-\int_{\Gamma_1 \cap \partial\Omega} F\dot{\beta}\, d\Gamma - \int_{\Gamma_1 \cap \partial\Omega} \vec{J}\, \mathrm{grad}\, \dot{\beta}\, d\Gamma.$$

Le premier terme a déjà été introduit et étudié [2], [3]. Le second prend en compte les influences réversibles entre les points voisins. La quantité F a la dimension d'un travail par unité de surface. Le vecteur $\vec{J}$ a la dimension d'un travail par unité de longueur. Les équations d'équilibre donneront une interprétation physique du vecteur $\vec{J}$.

Pour définir la puissance virtuelle $\mathcal{P}_i$ des efforts intérieurs, forme linéaire sur l'espace $\mathcal{B}\times\mathcal{V}\times\mathcal{V}_{ext}$ des vitesses vérifiant l'axiome des puissances virtuelles [4] : $\mathcal{P}_i = 0$ dans tout mouvement rigidifiant le système à l'instant t, il convient de préciser ce que l'on entend par mouvement rigidifiant. On dit que $(\gamma, \vec{v}, \vec{v}_{ext})$ est rigidifiant si

$$D(\vec{v}) = 0 \text{ dans } \Omega,\ \vec{v}_{ext} - \vec{v} = 0,\ \gamma = 0 \text{ sur } \Gamma_1.$$

Les deux premières relations sont classiques. La dernière est soutenue par l'expérience qui montre la nécessité d'avoir des déformations pour modifier l'adhérence (c'est-à-dire pour avoir $\gamma \neq 0$).

Ce choix étant fait, on retient comme puissance des efforts intérieurs à un domaine $\mathcal{D}$:

$$\forall \mathcal{D} \subset \Omega, \quad \forall (\vec{\gamma}, \vec{v}, \vec{v}_{ext}) \in \mathcal{B} \times \mathcal{V} \times \mathcal{V}_{ext},$$

$$\mathcal{T}_i = - \int_{\mathcal{D}} \sigma D(\vec{v}) d\Omega - \int_{\partial \mathcal{D} \cap \Gamma_1} \{ \vec{R}(\vec{v}_{ext} - \vec{v}) + F\gamma + \vec{J} \text{ grad } \gamma \} d\Gamma,$$

expression qui vérifie évidemment l'axiome des puissances virtuelles.

4. PUISSANCE DES EFFORTS EXTERIEURS

Soit un domaine $\mathcal{D} \subset \Omega$. Les efforts extérieurs se classent en deux catégories : les actions locales (la force surfacique $\vec{T}$ sur $\partial\mathcal{D}$, le travail a sur la frontière $\partial\delta$ dans Γ_1 de l'intersection δ de Γ_1 et $\partial\mathcal{D}$ (figure 3)) et les actions à distance (la force volumique $\vec{f}$ et éventuellement A et $\vec{h}$ sur Γ_1).

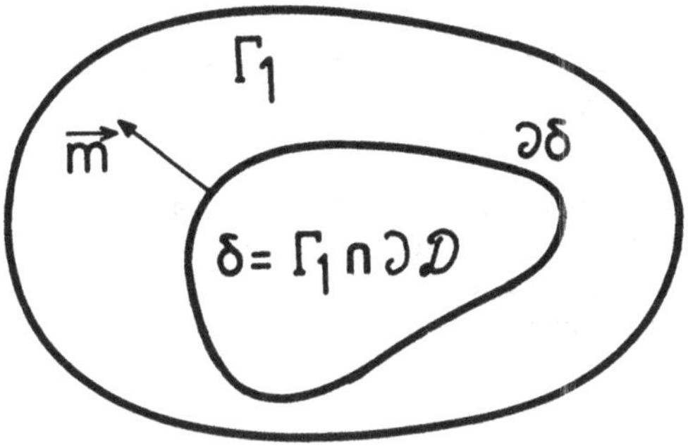

- Figure 3 -

L'intersection de $\partial\mathcal{D}$ et de Γ_1

On a déjà donné des exemples d'actions A et $\vec{h}$ [3]. Le travail par unité de longueur a est celui qu'il faut fournir à δ pour que son état ne change pas lorsque l'on enlève la partie $\Omega - \mathcal{D}$.

La puissance virtuelle des efforts extérieurs est alors

$$\forall \mathcal{D} \subset \Omega, \quad \forall (\gamma, \vec{v}, \vec{v}_{ext}) \in \mathcal{B} \times \mathcal{V} \times \mathcal{V}_{ext},$$

$$\mathcal{T}_e = \int_{\mathcal{D}} \vec{f}\,\vec{v}\, d\Omega + \int_{\partial\mathcal{D} - \Gamma_1} \vec{T}\,\vec{v}\, d\Gamma + \int_{\delta} \{ \vec{T}\,\vec{v}_{ext} + \vec{h}(\vec{v}_{ext} - \vec{v}) + A\gamma \} d\Gamma + \int_{\partial\delta} a\,\gamma\, d\ell$$

où $d\ell$ est l'élément de longueur sur $\partial\delta$.

5. PRINCIPE DES PUISSANCES VIRTUELLES. EQUATIONS D'EQUILIBRE

Nous nous limitons au cas quasi-statique pour lequel le principe est

$$\forall \mathcal{D} \subset \Omega, \quad \forall (\gamma, \vec{v}, \vec{v}_{ext}) \in \mathcal{B} \times \mathcal{V} \times \mathcal{V}_{ext}, \quad \mathcal{T}_e + \mathcal{T}_i = 0.$$

En faisant des choix convenables pour $(\gamma, \vec{v}, \vec{v}_{ext})$ on obtient les équations classiques :

$$\sigma_{ij,j} + f_i = 0 \text{ dans } \mathcal{D}, \quad \sigma_{ij} n_j = T_i \text{ sur } \partial\mathcal{D} - \Gamma_1,$$

où les n_i sont les cosinus directeurs de la normale extérieure $\vec{n}$ à $\mathcal{D}$,

$$\sigma_{ij} n_j = R_i - h_i, \quad T_i + h_i = R_i, \text{ sur } \Gamma_1 \cap \partial\mathcal{D}.$$

On obtient ensuite

$$F - \operatorname{div} \vec{J} = A \text{ sur } \delta = \Gamma_1 \cap \partial\mathcal{D}, \tag{1}$$

$$J_\alpha m_\alpha = a \text{ sur } \partial\delta, \tag{2}$$

où m_α sont les cosinus directeurs de la normale extérieure $\vec{m}$ à δ dans Γ_1 (figure 3) (la sommation des indices α va de 1 à 2).

Les deux dernières équations sont nouvelles. La deuxième donne une interprétation du vecteur $\vec{J}$, $\vec{J}\vec{m}$ est la quantité de travail par unité de longueur qui est imposée par l'extérieur de $\mathcal{D} \cap \Gamma_1$ à $\mathcal{D} \cap \Gamma_1$. Le vecteur $\vec{J}$ peut être appelé vecteur courant de travail.

L'équation (1) donne

$$- \operatorname{div} \vec{J} = A - F, \tag{3}$$

A-F peut être interprétée comme une production surfacique de travail (source si $A-F \geqslant 0$ et puits si $A-F \leqslant 0$). Les équations (1) et (2) peuvent se comprendre comme une équation de conservation de travail.

L'équation (1) remplace l'équation $F = A$ de la théorie du zéro gradient [3]. Elle a la structure habituelle des équations aux dérivées partielles résultant de lois de conservation. La connaissance des efforts extérieurs n'est plus suffisante pour connaître les efforts intérieurs, comme c'est le cas dans la théorie du zéro gradient. L'intervention d'une loi de comportement est ici nécessaire pour leur détermination.

6. SECOND PRINCIPE DE LA THERMODYNAMIQUE

Bien que l'on puisse partir de la forme classique conduisant à la relation de Clausius-Duhem [4], nous posons a priori cette relation et supposons pour simplifier les évolutions isothermes

$$\frac{d\Psi}{dt} \leq \int_\Omega \sigma D(\vec{\dot{u}})\,d\Omega + \int_{\Gamma_1} \{\vec{R}(\vec{\dot{u}}_{ext} - \vec{\dot{u}}) + F\dot{\beta} + \vec{J}\,\mathrm{grad}\,\dot{\beta}\}\,d\Gamma$$

où Ψ est l'énergie libre. L'inégalité de Clausius-Duhem doit être vérifiée par toute évolution réelle du système. La structure du second membre montre que Ψ devra contenir, outre un terme volumique, un terme surfacique sur Γ_1. Cette idée est développée ci-dessous dans le cadre de l'hypothèse des petites perturbations et de l'hypothèse de l'état local [4].

7. ENERGIE LIBRE DANS L'HYPOTHESE DES PETITES PERTURBATIONS

Nous supposons que la température (supposée constante), les petites déformations $\varepsilon_{ij}(\vec{u}) = D_{ij}(\vec{u})$, l'écart $\bar{u} = \vec{u}_{ext} - \vec{u}$ entre les déplacements réels $\vec{u}_{ext}$ du socle et $\vec{u}$ de la structure ainsi que l'intensité réelle d'adhésion β et son gradient, grad β, forment un système de variables normales dont l'énergie libre est fonction.

On doit remarquer que l'écart $\bar{u}$ et l'intensité d'adhésion β ne sont pas indépendants : ils vérifient les liaisons internes

$$0 \leq \beta \leq 1, \tag{4}$$

$$\beta\bar{u} = 0, \quad (5) \qquad \bar{u}\cdot\vec{n} \geq 0, \quad (\bar{u} = \vec{u}_{ext} - \vec{u}). \tag{6}$$

La deuxième exprime que lorsqu'il y a décollement ($\bar{u} \neq 0$), il n'y a pas d'interaction entre la structure et le support ($\beta = 0$). La troisième relation exprime qu'il n'y a pas interpénétration de la structure et du socle. D'autres hypothèses sont possibles, on peut, par exemple, autoriser des interactions à distance pour modéliser des filets de colle se rompant progressivement.

Les évolutions réelles qui vérifient l'inégalité de Clausius-Duhem sont liées par les liaisons internes (4) à (6). On peut donc définir à loisir l'énergie libre si les liaisons internes ne sont pas vérifiées sans affecter l'inégalité de Clausius-Duhem. Nous profitons de cette liberté pour inclure les liaisons internes dans le potentiel Ψ en lui donnant la valeur $+\infty$ si (4), (5) ou (6) n'est pas vérifiée. La dérivée $\frac{d\Psi}{dt}$ doit toujours s'entendre comme dérivée particulaire, c'est-à-dire en suivant les points matériels qui vérifient à chaque instant les liaisons internes. Sa valeur n'est donc pas modifiée. La fonction Ψ a là où les liaisons internes sont vérifiées toutes les propriétés des potentiels thermodynamiques : Ψ est convexe sur toute partie convexe où elle est finie. Plus précisément, les fonctions $\varepsilon(\vec{u})$, β, $\mathrm{grad}\,\beta \rightarrow \Psi(\varepsilon(\vec{u}),\bar{u},\beta,\mathrm{grad}\beta)$ et $\varepsilon(\vec{u}),\bar{u} \rightarrow \Psi(\varepsilon(\vec{u}),\bar{u},\beta,\mathrm{grad}\beta)$ sont convexes. Par contre Ψ considérée comme fonction de tous ses arguments n'est pas convexe. En effet, l'ensemble K des $\bar{u}$, β qui vérifient (4),(5) et (6) n'est pas convexe.

On choisit comme énergie libre

$$\Psi\left(\varepsilon(\vec{u}),\bar{u},\beta,\ \text{grad}\,\beta\right) = \int_\Omega \frac{1}{2}\{\lambda\left(\varepsilon_{kk}(\vec{u})\right)^2 + 2\mu\varepsilon_{ij}(\vec{u})\varepsilon_{ij}(\vec{u})\}d\Omega$$
$$+ \int_{\Gamma_1} \{\frac{k}{2}\,\bar{u}^2 - w\beta + \frac{\bar{k}}{2}(\text{grad}\,\beta)^2\} + I_K(\beta,\bar{u})$$

où

$$K = \{(\gamma,\vec{v}) \mid \gamma\in\mathcal{B}\ ;\ \vec{v}\in\mathcal{V}\ ;\ 0\leq\gamma\leq 1\ ;\ \gamma\bar{v}=0\ ;\ \bar{v}\vec{n}\geq 0 \text{ sur } \Gamma_1\},$$

I_K est la fonction indicatrice de l'ensemble K $\left(I_K(x) = 0 \text{ si } x\in K,\ I_K(x) = +\infty \text{ si } x\notin K\right)$, λ et μ sont les paramètres de Lamé de la théorie de l'élasticité, $w>0$ l'énergie d'adhésion de Dupré [1], k et $\bar{k}$ des paramètres positifs.

La présence de la fonction indicatrice de l'ensemble non convexe K rend le calcul de $\frac{d\Psi}{dt}$ délicat. On peut le faire en supposant ψ localement sous-différentiable [2]. Voici la définition de la sous-différentiabilité locale :

Soit Φ une application d'un espace vectoriel topologique X en dualité avec X^* dans $\mathbb{R}\cup\{+\infty\}$. On dit que Φ est localement sous-différentiable au point $x\in X$ si elle est finie au point x et s'il existe un voisinage $\mathcal{W}(x)$ et $x^*\in X^*$ tels que

$$\forall z\in\mathcal{W}(x),\quad \Phi(z)\geq\Phi(x) + \langle x^*, z-x\rangle .$$

On note $\partial\Phi(x)$ l'ensemble des sous-gradients x^* qui vérifient la relation précédente. Remarquons que cette notion ne suppose pas que Φ soit convexe. On a donné dans [2] et [3] des exemples de fonctions localement sous-différentiables.

8. PROPRIETES DE L'ENERGIE LIBRE

L'essentiel de ce paragraphe porte sur la dérivation de Ψ par rapport à ses arguments. C'est l'étude du sous-différentiel local. Les propriétés les plus intéressantes portent sur les dérivées généralisées de Ψ par rapport à $\bar{u}$ et β puisque ce sont les limitations sur ces deux variables qui créent des difficultés.

8.1. SOUS-DIFFERENTIEL LOCAL DE Ψ.

Nous nous plaçons dans une situation qui n'altère pas la généralité des résultats en fixant le socle. Il est à partir de maintenant rigide et fixe, $\vec{u}_{ext} = 0$. Ce qui fait que $\bar{u} = -\vec{u}$ sur Γ_1. On pose

$$\mathcal{V}_0 = \{\vec{v} \mid \vec{v}\in\mathcal{V}\ ;\ \vec{v}|\Gamma_0 = 0\}.$$

L'espace $\mathcal{B}\times\mathcal{V}_o\times\varepsilon(\mathcal{V}_o)$ est en dualité avec l'espace des efforts intérieurs σ, $\vec{R}$, F, $\vec{J}$ par la forme

$$\int_\Omega \sigma\varepsilon(\vec{v})\,d\Omega + \int_{\Gamma_1}\{\vec{R}\vec{v} + F\gamma + \vec{J}\,\mathrm{grad}\,\gamma\}d\Gamma .$$

Comme le suggère un exemple déjà donné [2], [3], on suppose que le potentiel Ψ est localement sous-différentiable en tout point où il est fini. Soit $(\tau,\vec{G}_1,G_2,\vec{G}_3)$ un sous-gradient de Ψ au point $(\varepsilon(\vec{u}),-\vec{u},\beta,\mathrm{grad}\beta)$ avec

$$\tau = \frac{\partial\Psi}{\partial\varepsilon},\quad \left(\tau_{ij} = \lambda\varepsilon_{kk}(\vec{u})\delta_{ij} + 2\mu\varepsilon_{ij}(\vec{u})\right), \tag{7}$$

où τ est la contrainte thermodynamique ou réversible classique,

$$\vec{G}_3 = \frac{\partial\Psi}{\partial(\mathrm{grad}\,\beta)} = \bar{k}\,\mathrm{grad}\,\beta . \tag{8}$$

Examinons maintenant les propriétés de $\vec{G}_1$ et G_2 liées à la présence de l'ensemble non convexe K.

Nous allons montrer que

– <u>là où il y a contact ($\vec{u}=0$)</u>, on a

$$G_2 = -w \text{ là où } 0<\beta<1 ; \tag{9}$$

$$G_2 \geqslant -w \text{ là où } \beta = 1 ; \tag{10}$$

$$G_2 \leqslant -w \text{ là où } \beta = 0 . \tag{11}$$

Ces relations signifient que le travail intérieur est assez grand là où il y a adhésion totale, assez faible là où il n'y a pas adhésion.

On montre d'autre part que la partie positive de la force normale $(G^1_n)^+$, $(G^1_n = \vec{G}^1\vec{n})$ et la force tangentielle, $\vec{G}^1_T = \vec{G}_1 - G^1_n\vec{n}$, ne sont pas trop grands là où β est assez petit, en particulier près de la ligne de décollement où β s'annule.

Si l'on suppose les comportements non dissipatifs ($F = G_2$, $\vec{R} = \vec{G}_1$), la relation (3) montre que la source de travail $A - F = A - G_2$ n'est pas trop grande là où il y a adhésion totale et n'est pas trop faible là où il n'y a pas adhésion. Là où β est assez petit, en particulier près de la ligne de décollement où β s'annule, la réaction normale ne peut être une traction trop grande et la réaction tangentielle n'est elle aussi pas trop grande.

– <u>là où il y a décollement ($\beta = 0$)</u> on a

$$G^1_n + ku_n \leqslant 0,\quad (G^1_n + ku_n)u_n = 0,\quad \vec{G}^1_T + k\vec{u}_T = 0, \tag{12}$$

qui sont les relations du contact unilatéral classique avec force de rappel proportionnelle à l'écart. Là où $\vec{u}$ est assez petit, en particulier près de la ligne de décollement où $\vec{u}$ s'annule la partie positive de $w+G_2$, $(w+G_2)^+$, n'est pas trop grande. Si l'on suppose le comportement élastique $(F = G_2)$ la relation (3) montre que la source de travail $A-F = A-G_2$ n'est pas trop faible là où il y a décollement.

Les relations précédentes vont être démontrées en utilisant la définition de la sous-différentiabilité locale. Nous donnons d'abord deux propriétés utiles pour la suite.

8.2. NOTATIONS ET PREMIERES PROPRIETES DE $\partial\Psi$.

Soit $(\tau,\vec{G}_1,G_2,\vec{G}_3) \in \partial\Psi\left(\varepsilon(\vec{u}),-\vec{u},\beta, \text{grad}\,\beta\right)$. Il existe alors un voisinage $\mathcal{W}$ de $(\beta,\vec{u})$ tel que

$$\forall(\gamma,\vec{v}) \in \mathcal{W},\ \tilde{\Psi}(\gamma,\vec{v}) - \tilde{\Psi}(\beta,\vec{u}) \geqslant \int_\Omega \tau\varepsilon(\vec{v}-\vec{u})d\Omega + \int_{\Gamma_1} \left\{-\vec{G}_1(\vec{v}-\vec{u}) + G_2(\gamma-\beta) + \vec{G}_3\,\text{grad}\,(\gamma-\beta)\right\}d\Gamma, \tag{13}$$

en notant $\tilde{\Psi}(\gamma,\vec{v}) = \Psi\left(\varepsilon(\vec{v}),-\vec{v},\gamma,\text{grad}\,\gamma\right)$.

Donnons quelques définitions,

$$a(\vec{v},\vec{v}) = \int_\Omega \left\{\lambda\left(\varepsilon_{kk}(\vec{v})\right)^2 + 2\mu\varepsilon_{ij}(\vec{v})\varepsilon_{ij}(\vec{v})\right\}d\Omega + \int_{\Gamma_1} k\vec{v}^2\, d\Gamma,$$

$$b(\gamma,\gamma) = \int_{\Gamma_1} \bar{k}(\text{grad}\,\gamma)^2\, d\Gamma.$$

En utilisant les relations (7) et (8), la relation (13) donne

$$\forall(\gamma,\vec{v}) \in \mathcal{W}\cap K,\ \frac{1}{2}a(\vec{v}-\vec{u},\vec{v}-\vec{u}) + \frac{1}{2}b(\gamma-\beta,\gamma-\beta) + \int_{\Gamma_1}(\vec{G}_1+k\vec{u})(\vec{v}-\vec{u})d\Gamma - \int_{\Gamma_1}(w+G_2)(\gamma-\beta)d\Gamma \geqslant 0. \tag{14}$$

Soit enfin $\mathcal{E}(\beta)$ l'ensemble des fonctions $\gamma \in \mathcal{B}$ sont le support est contenu dans celui de β et $\mathcal{E}(\vec{u})$ l'ensemble des fonctions $\vec{v} \in \mathcal{V}_0$ dont le support est contenu dans celui de $\vec{u}$. En remarquant que sur les espaces $\mathcal{E}_\beta = \mathcal{E}(\beta)\times\{\vec{u}\}$ et $\mathcal{E}_{\vec{u}} = \{\beta\}\times\mathcal{E}(\vec{u})$, la fonction Ψ est convexe, on constate que la relation (14) est vraie non seulement sur $\mathcal{E}_\beta \cap \mathcal{W} \cap K$ et $\mathcal{E}_{\vec{u}} \cap \mathcal{W} \cap K$ mais encore sur $\mathcal{E}_\beta \cap K$ et $\mathcal{E}_{\vec{u}} \cap K$.

8.2. TOPOLOGIE SUR $\mathcal{B}\times\mathcal{V}_0$

Pour utiliser la notion de sous-différentiel, il faut préciser ce que l'on entend par voisinage pour les espaces $\mathcal{B}\times\mathcal{V}_0$.

Soit un élément $(\gamma,\vec{v}) \in \mathcal{B}\times\mathcal{V}_0$. Dire qu'il est voisin de $(\beta,\vec{u})$, c'est d'abord dire que les déplacements $\vec{v}$ et $\vec{u}$, les intensités γ et β sont voisins aux sens des normes naturelles des espaces $\mathcal{B}$ et $\mathcal{V}_0$ (norme de l'espace de Sobolev H^1). Ces normes ne sont cependant pas suffisantes : prenons ($\vec{v}_1 = 0$, $\gamma_1 = \varepsilon$) (contact sur tout Γ_1), {$\vec{v}_2 = \vec{\varepsilon}$ (de l'ordre de ε), $\gamma_2 = 0$}, (décollement sur tout Γ_1). La norme $\|\vec{v}_1 - \vec{v}_2\| + \|\gamma_1 - \gamma_2\|$, qui est de l'ordre de ε, est donc petite alors que les états mécaniques sont fort différents puisque dans un cas il y a collage, dans l'autre décollement total. Il est donc nécessaire de dire que si $(\gamma,\vec{v})$ est voisin de $(\beta,\vec{u})$, les zones collées et décollées sont en plus voisines (on appelle zone collée le support de β : supp β).

L'espace $\mathcal{B}\times\mathcal{V}_0$ est muni d'une structure d'espace métrique par

$$d\left((\beta,\vec{u}),(\gamma,\vec{v})\right) = \|\vec{v}-\vec{u}\| + \|\gamma-\beta\| + \tilde{d}\left(\text{support}\,\gamma,\ \text{support}\,\beta\right),$$

où $\tilde{d}$ est une distance sur les ensembles fermés dont on peut donner une définition précise. La distance $\tilde{d}$ assure que si $(\gamma,\vec{v})$ et $(\beta,\vec{u})$ sont voisins, l'ensemble des points qui changent d'état, c'est-à-dire qui passent de l'état collé à l'état décollé ou l'inverse, est petit quand on passe de $(\beta,\vec{u})$ à $(\gamma,\vec{v})$.

Pour démontrer les relations de comportement, on se place en un point x où l'une des deux situations examinées est réalisée. On fait des variations sur des $(\gamma,\vec{v})$ qui ne changent pas la nature du contact au voisinage du point x. On utilise alors les tests classiques des formulations variationnelles qui donnent des renseignements sur $\vec{G}_1$ et G_2. On fait aussi des variations sur des $(\gamma,\vec{v})$ qui changent la nature du contact dans un voisinage $\mathcal{O}$ du point considéré (situation collée remplacée par situation décollée et vice-versa). On obtient alors des limitations sur $\vec{G}_1$ et G_2.

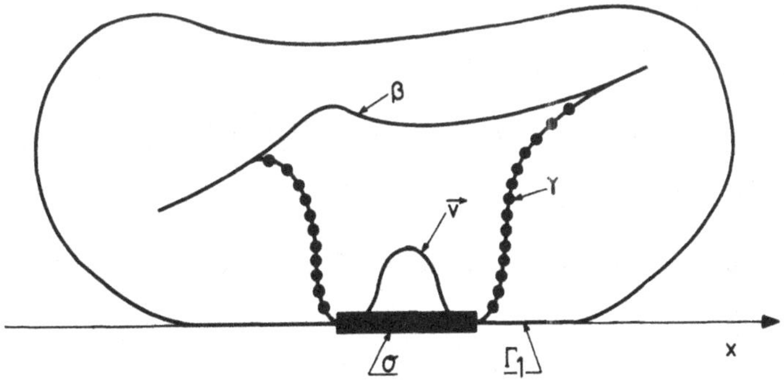

- Figure 4 -

La nature du contact est changée sur $\mathcal{O}$.

8.4. CONTACT

Soit $\mathcal{O}$ un voisinage dans Γ_1 d'un point x dans lequel la structure est en contact avec le socle {$\forall y \in \mathcal{O}$; $\vec{u}(y) = 0$}. Prenons d'abord $(\gamma,\vec{u}) \in \mathcal{E}_\beta$

avec $\gamma = \beta$ hors de $\mathcal{O}$. On a d'après (14)

$$\frac{1}{2}b(\gamma-\beta,\gamma-\beta) - \int_{\mathcal{O}}(w+G_2)(\gamma-\beta)d\Gamma \geq 0.$$

On en déduit de façon classique que

$$G_2 = -w \quad \text{là où} \quad 0<\beta<1, \tag{9}$$

$$G_2 \geq -w \quad \text{là où} \quad \beta = 1, \tag{10}$$

$$G_2 \leq -w \quad \text{là où} \quad \beta = 0. \tag{11}$$

Ces relations signifient que le travail intérieur est assez grand là où il y a adhésion totale et qu'il est assez faible là où il n'y à pas adhésion. Si l'on fait l'hypothèse qu'il n'y a pas de dissipation sur β $(F = G_2)$, l'équation d'équilibre (3) montre que la source de travail $A - F = A - G_2$ n'est pas trop grande là où il y a adhésion totale et n'est pas trop faible là où il n'y a pas adhésion.

Choisissons maintenant $(\gamma,\vec{v})$ tels que sur Γ_1 :

$\gamma = 0$ sur $\mathcal{O}$ et $\vec{v} = \vec{u}$ hors de $\mathcal{O}$ (figure 4).

On suppose que $\mathcal{W}$ est assez grand pour que $(\gamma,\vec{v}) \in \mathcal{W} \cap K$. D'après la relation (14) on a

$$\frac{1}{2}a(\vec{v}-\vec{u},\vec{v}-\vec{u}) + \int_{\mathcal{O}}\vec{G}_1\vec{v}\, d\Gamma + \frac{1}{2}b(\gamma-\beta,\gamma-\beta) - \int_{\Gamma}(G_2+w)(\gamma-\beta)d\Gamma \geq 0.$$

En posant $c^2(\mathcal{O})$ la borne inférieure sur tous les γ possibles de $\frac{1}{2}b(\gamma-\beta,\gamma-\beta) - \int_{\Gamma}(w+G_2)(\gamma-\beta)d\Gamma$ (expression qui est positive d'après les relations (9) (10 et (11)), on a

$$c^2(\mathcal{O}) + \frac{1}{2}\, a(\vec{v}-\vec{u},\vec{v}-\vec{u}) + \int_{\mathcal{O}}\vec{G}_1\vec{v}\, d\Gamma \geq 0.$$

La relation précédente montre que $\vec{G}_1$ n'est pas trop grand. En effet

$$\mathrm{Inf}\{a(\vec{w}-\vec{u},\vec{w}-\vec{u}) \mid \vec{w} = \vec{v} \text{ sur } \Gamma_1\} = \|\vec{v}\|^2,$$

est le carré d'une norme de la trace de $\vec{v}$ sur $\mathcal{O}$ (c'est la norme $H^{1/2}(\mathcal{O})$) et l'on a

$$c^2(\mathcal{O}) + \frac{1}{2}\|\vec{v}\|^2 + \int_{\mathcal{O}}\vec{G}_1\vec{v}\, d\Gamma \geq 0.$$

Supposons que G_n^1 soit positif sur $\mathcal{O}$ (G_n^1 est une traction). Il existe alors $\hat{\varphi}$ avec $\hat{\varphi}_n \leq 0$ tel que

$$\|\hat{\varphi}\|\,\|\vec{G}_1\| = -\int_{\mathcal{O}}\vec{G}_1\hat{\varphi}\, d\Gamma,$$

où la norme de $\vec{G}_1$ est la norme duale de celle de $\vec{v}$.

Ce qui donne avec $\vec{v} = \alpha\hat{\varphi}$ $(\alpha > 0)$,

$$\|\vec{G}_1\| \leq \frac{c^2(\mathcal{O})}{\alpha\|\hat{\varphi}\|} + \frac{\alpha}{2}\|\hat{\varphi}\| .$$

En prenant la valeur minimale sur α (α ne peut être trop grand car $(\gamma,\vec{v})$ doit appartenir à $\mathcal{W}$), on voit bien que $\vec{G}_1$ n'est pas trop grand.

Cette propriété, en supposant le comportement élastique $(\vec{R} = \vec{G}_1)$, montre que la liaison socle structure avec adhérence peut supporter une traction sans décollement. Cette dernière ne doit cependant pas être trop grande. Cette importante propriété distingue fortement le contact avec adhérence du contact unilatéral classique qui ne peut supporter de traction.

Supposons maintenant que $G_n^1 \leq 0$ (G_n^1 est une compression). Il existe alors $\bar{\varphi}_T$ tel que

$$\|\bar{\varphi}_T\|\,\|\vec{G}_T^1\| = -\int_{\mathcal{O}} \vec{G}_T^1 \bar{\varphi}_T \, d\Gamma .$$

On a alors pour $\vec{v} = \alpha\bar{\varphi}_T + \psi\vec{n}$ $\quad(\psi \leq 0)$

$$\frac{1}{\alpha\|\bar{\varphi}_T\|}\left(c^2(\mathcal{O}) + \frac{1}{2}\|\alpha\bar{\varphi}_T + \psi\vec{n}\|^2 + \int_{\mathcal{O}} G_n^1\psi \, d\Gamma\right) \geq \|\vec{G}_T^1\| .$$

En prenant la borne inférieure sur α et ψ du membre de gauche (qui est positif) on en déduit que $\|\vec{G}_T^1\|$ ne doit pas être trop grand.

S'il n'est pas possible de trouver $\mathcal{O}$ voisinage de x et γ tels que $(\gamma,\vec{v}) \in \mathcal{W} \cap K$, la force $\vec{G}_1$ est quelconque. En supposant le comportement élastique $(\vec{R} = \vec{G}_1)$ le contact est bilatéral. Cette situation peut se produire puisque la norme $\|\gamma-\beta\|^2 = \int_\Gamma \{(\gamma-\beta)^2 + \text{grad}^2(\gamma-\beta)\} d\Gamma$ peut être grande lorsque $\mathcal{O} \to x$ [grad$(\gamma-\beta)$ peut en effet être grand]. On a d'autant plus de chances d'avoir $(\gamma,\vec{v}) \in \mathcal{W}$ que la norme $\|\gamma-\beta\|_{\mathcal{O}}$ est petite donc que β est petit sur $\mathcal{O}$. Comme la fonction β est continue (au moins en un certain sens) elle est petite au voisinage du bord du support de β ou encore près de la ligne de décollement. On en déduit que $\vec{G}_1$ (ou la réaction $\vec{R}$ si le comportement est élastique) n'est pas trop grand au voisinage de la ligne de décollement. La force $\vec{G}_1$ (ou la réaction $\vec{R}$ si le comportement est élastique) peut par contre être quelconque (contact bilatéral) loin de cette ligne. Ce comportement est celui que l'on imagine : détérioration progressive de la liaison au fur et à mesure que l'on se rapproche de la ligne de décollement où la rupture se produit.

8.5. DECOLLEMENT

Soit $\mathcal{O}$ un voisinage dans Γ_1 d'un point x dans lequel il y a décollement $(\forall y \in \mathcal{O}\ ;\ \beta(y) = 0)$. Prenons $(\beta,\vec{v}) \in \mathcal{E}_{\vec{u}}$ avec $\vec{v} = \vec{u}$ hors de $\mathcal{O}$. On a d'après (14)

$$\frac{1}{2}\, a(\vec{v}-\vec{u},\vec{v}-\vec{u}) + \int_{\mathcal{O}} (\vec{G}_1 + k\vec{u})(\vec{v}-\vec{u})\, d\Gamma \geq 0.$$

On en déduit classiquement que

$$G_n^1 + ku_n \leq 0,\quad (G_n^1 + ku_n)u_n = 0,\quad \vec{G}_T^1 + k\vec{u}_T = 0, \tag{12}$$

qui sont les équations du contact unilatéral classique avec force de rappel proportionnelle au déplacement (si l'on fait l'hypothèse du comportement élastique $\vec{R} = \vec{G}_1$).

Choisissons maintenant $(\gamma,\vec{v})$ tels que sur Γ_1 :

$\vec{v} = 0$ sur $\mathcal{O}$ et $\gamma = \beta$ hors de $\mathcal{O}$ (figure 5).

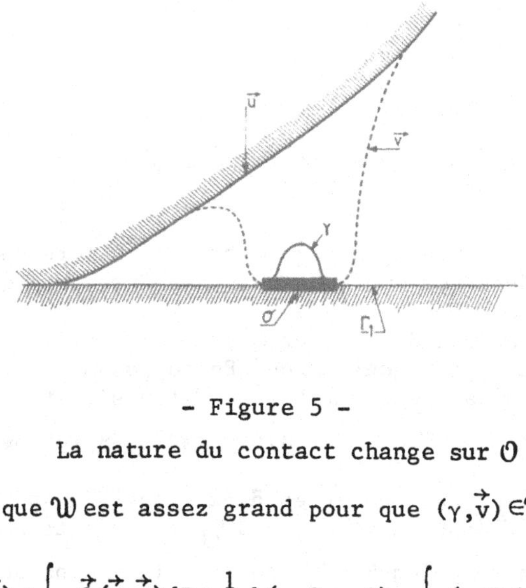

- Figure 5 -

La nature du contact change sur $\mathcal{O}$

On suppose que $\mathcal{W}$ est assez grand pour que $(\gamma,\vec{v}) \in \mathcal{W} \cap K$. D'après (14) on a

$$\frac{1}{2}\, a(\vec{v}-\vec{u},\vec{v}-\vec{u}) + \int_{\Gamma_1} \vec{G}_1(\vec{v}-\vec{u})\, d\Gamma + \frac{1}{2}\, b(\gamma-\beta,\gamma-\beta) - \int_{\mathcal{O}} (w+G_2)\gamma\ d\Gamma \geq 0.$$

Cette relation est toujours vérifiée si $w+G_2 \leq 0$. Supposons donc $w+G_2 \geq 0$ sur $\mathcal{O}$. En posant $c^2(\mathcal{O})$ la borne inférieure sur tous les $\vec{v}$ possibles de $\frac{1}{2}\, a(\vec{v}-\vec{u},\vec{v}-\vec{u}) + \int_{\Gamma_1} \vec{G}_1(\vec{v}-\vec{u})\, d\Gamma$ (expression qui est positive d'après les relations (12)), on obtient

$$c^2(\mathcal{O}) + \frac{1}{2}\, b(\gamma-\beta,\gamma-\beta) - \int_{\mathcal{O}} (w+G_2)\gamma\ d\Gamma \geq 0.$$

La quantité

$$b(\gamma-\beta,\gamma-\beta) = \overline{k}\int_{\mathcal{O}} \mathrm{grad}^2\gamma \; d\Gamma = \|\gamma\|^2,$$

est le carré d'une norme de la trace de γ sur $\mathcal{O}$ (c'est la norme de $H^1_o(\mathcal{O})$). Il existe alors $\hat{\gamma} \geqslant 0$ tel que

$$\int_{\mathcal{O}} (w+G_2)\hat{\gamma} \; d\Gamma = \|w+G_2\| \; \|\hat{\gamma}\|$$

où la norme de $w+G_2$ est la norme duale (c'est celle de $H^{-1}(\mathcal{O})$). On obtient alors pour $\gamma = \alpha\hat{\gamma}$ $(\alpha > 0)$

$$\frac{c^2(\mathcal{O})}{\alpha\|\hat{\gamma}\|} + \frac{\alpha}{2}\|\hat{\gamma}\| \geqslant \|w+G_2\| .$$

En prenant la borne inférieure sur α, on voit donc que $w+G_2$ ou encore la partie positive de $w+G_2$, $(w+G_2)^+$, n'est pas trop grande. Si l'on fait l'hypothèse qu'il n'y a pas de dissipation sur β, on trouve d'après la relation (3) que le travail fourni n'est pas trop faible là où il y a décollement.

Remarque

1) Les relations (11) et (12) sont simultanément vraies là où il y a à la fois contact $(\vec{u} = 0)$ et décollement $(\beta = 0)$.

2) Les relations (9), (10), (11) et la limitation sur $w+G_2$ dans la zone décollée sont à rapprocher des relations obtenues dans la théorie du zéro gradient [3]. Par exemple, la dernière limitation peut être comparée à la relation $w+G_2 \leqslant H$ de la théorie du zero gradient (H est le taux de restitution de l'énergie).

9. LES LOIS DE COMPORTEMENT. LES LOIS D'ETAT ET LES LOIS COMPLEMENTAIRES.

Calculons maintenant $\frac{d\Psi}{dt}$. Pour un accroissement de temps Δt assez petit, on a, en supposant $\vec{u}$ et β continus en temps, $\{\vec{u}(t+\Delta t),\beta(t+\Delta t)\} \in \mathcal{W}$ et du fait de la sous-différentiabilité de Ψ

$$\Psi(t+\Delta t) - \Psi(t) \geqslant \int_\Omega \tau\big(\varepsilon(\vec{u}(t+\Delta t)) - \varepsilon(\vec{u}(t))\big)d\Omega + \int_{\Gamma_1} \big\{\vec{G}_1\big(-\vec{u}(t+\Delta t)+\vec{u}(t)\big) + G_2\big(\beta(t+\Delta t) - \beta(t)\big) + \vec{G}_3 \; \mathrm{grad}\big(\beta(t+\Delta t) - \beta(t)\big)\big\}d\Gamma.$$

En divisant par Δt positif puis négatif et en passant à la limite, on a

$$\frac{d\Psi}{dt} = \int_\Omega \tau\varepsilon(\dot{\vec{u}})d\Omega + \int_{\Gamma_1} \{-\vec{G}_1\dot{\vec{u}} + G_2\dot{\beta} + \vec{G}_3 \; \mathrm{grad}\,\dot{\beta}\}d\Gamma.$$

La relation de Clausius-Duhem donne alors

$$\tilde{D} = \int_\Omega (\sigma-\tau)\varepsilon(\dot{\vec{u}})d\Omega + \int_{\Gamma_1} \{-(\vec{R}-\vec{G}_1)\dot{\vec{u}} + (F-G_2)\dot{\beta} + (\vec{J}-\vec{G}_3) \; \mathrm{grad}\,\dot{\beta} \; d\Gamma \geqslant 0,$$

qui doit être vérifiée pour toute évolution réelle du système, c'est-à-dire pour tous les champs de vitesse $\dot{\vec{u}}$ et $\dot{\beta}$.

La relation

$$(\tau,\vec{G}_1,G_2,\vec{G}_3) \in \partial\Psi\left(\varepsilon(\vec{u}),-\vec{u},\beta,\text{grad}\,\beta\right) \tag{15}$$

forme les lois d'état. On remarque que les liaisons internes (2) à (4) font partie des lois de comportement en assurant que Ψ est finie donc sous-différentiable.

Pour avoir toutes les lois de comportement, on doit donner les lois complémentaires.

Nous donnons deux exemples. L'un, le comportement élastique, a une dissipation nulle $\tilde{D} = 0$, l'autre, le comportement standard généralisé [6], a une dissipation non nulle.

9.1. COMPORTEMENT ELASTIQUE (dissipation nulle).
On dit que le comportement est élastique si

$$\sigma = \tau, \quad \vec{R} = \vec{G}_1, \quad F = G_2, \quad \vec{J} = \vec{G}_3.$$

Pour les contraintes, on retrouve la loi de comportement élastique classique.

9.2. COMPORTEMENT STANDARD GENERALISE (dissipation non nulle).
Pour définir le comportement standard généralisé on introduit un pseudo-potentiel de dissipation défini par des densités volumique $\varphi(\dot{\varepsilon})$ et surfacique $\Phi(\dot{\vec{u}},\dot{\beta},\text{grad}\dot{\beta})$. Les fonctions φ et Φ sont convexes, positives et nulles à l'origine.

Les lois de comportement complémentaires sont

$$(\sigma-\tau) \in \partial\varphi(\dot{\varepsilon}), \quad (\vec{R}-\vec{G}_1,F-G_2,\vec{J}-\vec{G}_3) \in \partial\Phi(-\dot{\vec{u}},\dot{\beta},\text{grad}\dot{\beta}),$$

où $\partial\varphi$ et $\partial\Phi$ sont les sous-différentiels classiques de φ et Φ.

Examinons maintenant l'équilibre de structures élastiques. Nous étudierons plus loin les problèmes d'évolution.

10. EQUILIBRE D'UNE STRUCTURE ELASTIQUE - FORMULATION VARIATIONNELLE

Considérons la structure chargée par les forces $\vec{f}$ de volume, $\vec{g}$ de surface sur Γ_2, en contact unilatéral avec adhérence sur Γ_1 et fixée sur Γ_0 (figure 6). Pour garder toute la généralité on suppose que les actions extérieures A,a et $\vec{h}$ sont non nulles.

Les équations que doivent vérifier une éventuelle position d'équilibre $(\beta,\vec{u}) \in \mathcal{B}\times\mathcal{V}_0$ sont

$$\sigma_{ij,j} + f_i = 0 \text{ dans } \Omega, \quad \sigma\vec{n} = \vec{g} \text{ sur } \Gamma_2,$$

$$F - \operatorname{div} \vec{J} = A, \quad \vec{R} = \sigma\vec{n} + \vec{h} \text{ sur } \Gamma_1, \quad \vec{J}\,\vec{m} = a \text{ sur } \partial\Gamma_1.$$

$$(\sigma, \vec{R}, F, \vec{J}) \in \partial\Psi\big(\varepsilon(\vec{u}), -\vec{u}, \beta, \operatorname{grad}\beta\big). \tag{16}$$

Les données de ces équations sont $\Psi, \vec{f}, \vec{g}, \vec{h}, A$ et a, les inconnues sont β et $\vec{u}$.

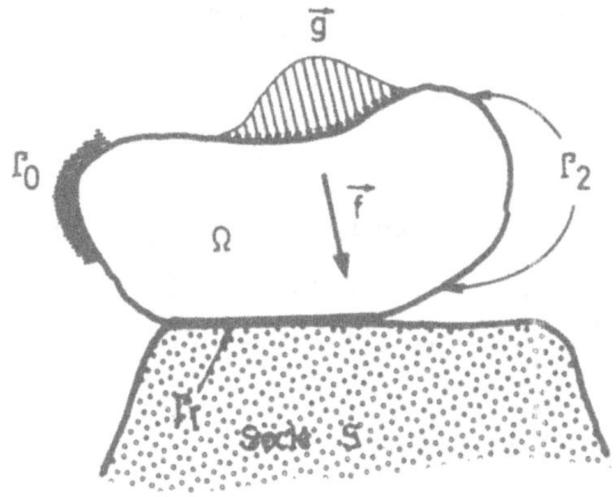

\- Figure 6 -

La structure chargée par les forces volumiques $\vec{f}$ et surfaciques $\vec{g}$.

On peut donner une formulation variationnelle des équations précédentes. On a d'abord

$$\forall(\gamma,\vec{v}) \in \mathcal{W}(\vec{u},\beta), \quad \tilde{\Psi}(\gamma,\vec{v}) - \tilde{\Psi}(\beta,\vec{u}) \geq \int_\Omega \sigma\big(\varepsilon(\vec{v}) - \varepsilon(\vec{u})\big)\,d\Omega \; +$$

$$+ \int_{\Gamma_1} \{F(\gamma-\beta) + \vec{J}\operatorname{grad}(\gamma-\beta) - \vec{R}(\vec{v}-\vec{u})\}\,d\Gamma.$$

En utilisant les équations d'équilibre

$$\tilde{\Psi}(\gamma,\vec{v}) - \tilde{\Psi}(\beta,\vec{u}) \geq \int_\Omega \vec{f}(\vec{v}-\vec{u})\,d\Gamma + \int_{\Gamma_2} \vec{g}(\vec{v}-\vec{u})\,d\Gamma - \int_{\Gamma_1} \vec{h}(\vec{v}-\vec{u})\,d\Gamma \; +$$

$$+ \int_{\Gamma_1} A(\gamma-\beta)\,d\Gamma + \int_{\partial\Gamma_1} a(\gamma-\beta)\,d\ell.$$

On a donc

<u>THEOREME 1</u>. Les équations (16) sont équivalentes à trouver $(\beta,\vec{u}) \in K \cap \mathcal{V}_0$ et un voisinage $\mathcal{W}$ de $(\beta,\vec{u})$ qui vérifient

$$\forall(\gamma,\vec{v}) \in \mathcal{W} \cap K, \quad \mathcal{F}[\gamma,\vec{v}] \geq \mathcal{F}[\beta,\vec{u}], \tag{17}$$

avec

$$\mathcal{F}[\gamma,\vec{v}] = \frac{1}{2}\int_\Omega \{\lambda[\varepsilon_{kk}(\vec{v})]^2 + 2\mu\varepsilon_{ij}(\vec{v})\varepsilon_{ij}(\vec{v})\}d\Omega +$$
$$+ \int_{\Gamma_1} \{\frac{k}{2}\vec{v}^2 - w\gamma + \frac{\bar{k}}{2}\text{grad}\,\gamma^2\}d\Gamma -$$
$$- \int_\Omega \vec{f}\vec{v}\,d\Omega - \int_{\Gamma_2} \vec{g}\vec{v}\,d\Gamma + \int_{\Gamma_1} \vec{h}\vec{v}\,d\Gamma - \int_{\Gamma_1} A\gamma - \int_{\partial\Gamma_1} a\gamma.$$

Les positions d'équilibre sont les minimums locaux de l'énergie potentielle.

En précisant les espaces on peut démontrer le théorème d'existence suivant.

THEOREME 2. Si Ω et Γ_1 sont réguliers, $\text{mes}(\Gamma_0) > 0$, $\text{mes}(\Gamma_1) > 0$, $\mathcal{V}_0 = \{\vec{v} \mid v_i \in H^1(\Omega);\ \vec{v}|\Gamma_0 = 0\}$, $\mathcal{B} = H^1(\Gamma_1)$, $\vec{g} \in L^2(\Gamma_2)$, A, $h_i \in L^2(\Gamma_1)$; $a \in L^2(\partial\Gamma_1)$; $\vec{f} \in L^2(\Omega)$, le problème (17) possède au moins une solution.

11. LOIS DE COMPORTEMENT DISSIPATIVES

L'essentiel des effets de dissipation est dû à la dissipation sur la variable β. Nous nous limitons donc à cette seule dissipation.

11.1. COMPORTEMENT VISQUEUX

On choisit

$$\varphi \equiv 0, \qquad \Phi_1(\dot{\beta}) = \frac{c}{2}\int_{\Gamma_1} \dot{\beta}^2\, d\Gamma, \ c > 0,$$

qui donne un effet dissipatif,

$$F - G_2 = c\dot{\beta},$$

les autres effets dissipatifs sont nuls

$$\sigma - \tau = 0, \quad \vec{R} - \vec{G}_2 = 0, \quad \vec{J} - \vec{G}_3 = 0.$$

On a ici un comportement visqueux linéaire classique.

11.2. COMPORTEMENT VISQUEUX AVEC RECOLLEMENT IMPOSSIBLE

Dans cette situation seul le décollement est possible. Une fois celui-ci effectué, il n'est plus possible de recoller quels que soient les efforts appliqués. C'est le cas du papier à tapisser collé sur un mur. On choisit

$$\varphi \equiv 0, \quad \Phi_2(\dot{\beta}) = \frac{c}{2}\int_{\Gamma_1} \dot{\beta}^2\, d\Gamma + I_-(\dot{\beta})$$

où I_- est l'indicatrice de la demi droite négative $\left(I_-(x) = 0 \text{ si } x \leq 0,\ I_-(x) = +\infty \text{ si } x > 0\right)$.

On a alors

$$F - G_2 = c\dot{\beta} \quad \text{si } \dot{\beta} < 0,$$

$$F - G_2 \geqslant 0 \quad \text{si } \dot{\beta} = 0.$$

12. EQUATIONS D'EVOLUTION DU COMPORTEMENT VISQUEUX

Les équations que doit vérifier une éventuelle solution dépendant du temps t, $\left(\beta(t),\vec{u}(t)\right) \in \mathcal{B}\times\mathcal{V}_0$ sont

$$\sigma_{ij,j} + f_i = 0 \text{ dans } \Omega, \quad \sigma\vec{n} = \vec{g} \text{ sur } \Gamma_2,$$

$$F - \operatorname{div}\vec{J} = A, \quad \vec{R} = \sigma\vec{n} + \vec{h} \text{ sur } \Gamma_1, \quad \vec{J}\vec{m} = a \text{ sur } \partial\Gamma_1 \tag{18}$$

$$(\sigma,\vec{R},G_2,\vec{J}) \in \partial\Psi\left(\varepsilon(\vec{u}),-\vec{u},\beta,\operatorname{grad}\beta\right) \quad ; \quad F - G_2 \in \partial\Phi(\dot{\beta}).$$

On complète ces équations par la condition initiale

$$\beta(x,0) = \beta_0(x), \quad 0 \leqslant \beta_0(x) \leqslant 1. \tag{19}$$

Les données des équations (18) et (19) sont $\vec{f},\vec{g},A,\vec{h},a,\beta_0$ les fonctions Ψ et Φ, les inconnues sont les fonctions $\vec{u}(x,t)$ et $\beta(x,t)$ fonction du point x et du temps t.

On peut donner de ces équations des formulations variationnelles.

Pour bien voir la structure des équations remplaçons le potentiel Ψ par un potentiel Ψ_η différentiable. La quantité $\eta > 0$ est destinée à tendre vers 0.

13. POTENTIEL APPROCHE. METHODE NUMERIQUE

On remplace Ψ par Ψ_η en pénalisant les conditions définissant l'ensemble K comme cela a été fait dans la théorie du zéro gradient [5].

$$\Psi_\eta\left(\varepsilon(\vec{v}),-\vec{v},\gamma,\operatorname{grad}\gamma\right) = \frac{1}{2}\, a(\vec{v},\vec{v}) + \int_{\Gamma_1} \left\{\frac{k}{2}\,\vec{v}^2 - w\gamma + \frac{\bar{k}}{2}(\operatorname{grad}\gamma)^2\right\} d\Gamma$$

$$+ \frac{1}{2\eta}\int_{\Gamma_1} \left\{\gamma^2 v^2 + \left((\gamma-1)^+\right)^4 + (\gamma^-)^4 + (v_n^+)^2\right\} d\Gamma.$$

D'un point de vue physique le nouveau potentiel modélise d'une part une force de réaction d'intensité $-\frac{1}{\eta}\gamma^2\vec{v} - \frac{1}{\eta}v_n^+\vec{n}$ qui s'oppose au décollement si $\gamma \neq 0$ et à l'interpénétration, d'autre part un travail de réaction $-\frac{2}{\eta}\left((\gamma-1)^+\right)^3 + \frac{2}{\eta}(\gamma^-)^3$ qui s'oppose à la sortie de l'intervalle (0,1).

Le potentiel Ψ_η étant dérivable, les équations sont alors

$$- (\lambda+\mu) \operatorname{grad} \operatorname{div} \vec{u} - \mu\Delta\vec{u} + \vec{f} = 0 \text{ dans } \Omega.$$

$$\sigma \vec{n} = \vec{g} \text{ sur } \Gamma_2,$$

$$c\dot{\beta} - \bar{k}\Delta\beta = w + A - \frac{1}{\eta}\beta\vec{u}^2 - \frac{2}{\eta}\left((\beta-1)^+\right)^3 + \frac{2}{\eta}(\beta^-)^3, \text{ sur } \Gamma_1$$

$$\sigma\vec{n} + \left(k + \frac{\beta^2}{\eta}\right)\vec{u} + \frac{1}{\eta} u_n^+ \vec{n} = 0, \text{ sur } \Gamma_1$$

$$\bar{k}\frac{\partial\beta}{\partial n} = a \text{ sur } \partial\Gamma_1 \qquad (20)$$

que l'on complète par la condition initiale

$$\beta(x,0) = \beta_0(x), \; 0 \leqslant \beta_0(x) \leqslant 1.$$

Ce système d'équations peut être résolu numériquement. On peut même retrouver certains faits expérimentaux. Montrons par exemple que le collage en bloc est possible alors que le décollage en bloc est impossible (on dit que le collage est en bloc si β est indépendant de x [3]).

Soit Ω simplement posé sur le socle et soumis à aucune force extérieure ($A = 0$, $a = 0$, $\vec{f} = \vec{g} = 0$). L'état initial est $\beta(x,0) = 0$. On a alors $\vec{u}(x,t) = 0$, $\beta(x,t) = \frac{w}{c}t$ pour $0 \leqslant t \leqslant \frac{c}{w}$ et ensuite $\vec{u} = 0$ et $\beta = 1$. On a un collage en bloc et l'on retrouve que du papier adhésif posé sur une surface lisse y adhère spontanément.

Considérons maintenant Ω collé sur le support, $\beta(x,0) = 1$. Supposons qu'aucun travail extérieur ne soit appliqué, $A = 0$, $a = 0$. Quelles que soient les forces $\vec{f}$ et $\vec{g}$ appliquées, le décollement ne peut se faire en bloc, c'est-à-dire de façon que $\beta(x,t) = \beta(t)$. Dans ce cas, en effet, les équations (20) donnent $c\dot{\beta} = w$ ce qui est contradictoire avec $\dot{\beta} < 0$.

REFERENCES

1. M. BARQUINS, D. MAUGIS. Fracture Mechanics and Adherence of Viscoelastic Solids dans Adhesion and Adsorption of Polymers, Plenum Publ. Corp. New York, 1980, part A, p. 203.

2. M. FREMOND. Contact unilatéral avec adhérence dans Unilateral Problems in Structural Analysis (G. Del Piero, F. Maceri éditeurs), Springer-Verlag, New York, 1985.

3. M. FREMOND. Adhérence des Solides, à paraître au Journal de Mécanique Théorique et Appliquée.

4. P. GERMAIN. Mécanique des milieux continus, Masson, Paris, 1973.

5. H. GHIDOUCHE, N. POINT.Unilateral contact with adherence in Free boundary problems : applications and theory, Volume IV, A. Bossavit, A. Damlamian, M. Frémond editors, Pitman, 1985.

6. J.J. MOREAU. Fonctions de résistance et fonctions de dissipation. Séminaire d'Analyse Convexe, Montpellier, 1971.

EXISTENCE RESULTS FOR MINIMA OF NON-COERCIVE FUNTIONALS AND APPLICATIONS TO UNILATERAL PROBLEMS IN ELASTICITY

F. Gastaldi
Istituto di Analisi Numerica del CNR, Pavia

ABSTRACT

The problem of minimizing a functional on a subset of a Hilbert space is considered in a non-coercive and non-convex framework. Sufficient conditions for the existence of minima are given, involving a suitable recession functional. Some particular functionals are detailed, notably the quadratic ones, for which the existence theorem is specialized.
Applications to the bending of a partially supported plate and to the classical Signorini problem are given. A modified interpretation of the unilateral condition for the Signorini problem is also introduced, suitable in finite elasticity. The abstract existence theorem is then applied to this concrete problem, providing sufficient conditions for the existence of an equilibrium configuration for an elastic body constrained to lie inside a box with rigid contour.

The aim of this lecture is to present some existence results for quite general minimum problems and their application to several unilateral problems in elasticity.

The outline of the lecture is the following.

1. Statement of the abstract minimum problem (P).
2. An example: the classical Signorini problem in linear elasticity.
3. Statement of an abstract existence result for problem (P) under convexity, non-coerciveness assumptions.
4. Application to the equilibrium problem of a partially supported plate.
5. Discussion of the unilateral condition in Signorini's problem and of a modified approach to it, resulting in possible non-convexity of the set of admissible displacements.
6. Statement of an abstract existence result for problem (P) under non-convexity, non-coerciveness assumptions.
7. Discussion of the abstract sufficient condition and application to the equilibrium problem of an elastic body subject to the unilateral condition introduced above.

Let us consider the following abstract minimum problem:

$$\begin{cases} \textit{to find } u \in \mathbf{K} \textit{ such that} \\ G(u) \leq G(v), \quad \forall v \in \mathbf{K}, \end{cases} \tag{P}$$

where

- $\mathbf{K}$ is a subset of a Hilbert space V;
- G is a functional defined on V, with values in $]-\infty,+\infty]$.

Many mechanical situations admit an abstract formulation as problem (P): a typical category is that of equilibrium problems in elasticity. In this case, the equilibrium configuration often [1] is a minimizer of the total energy of the system; possible constraints are taken into account by minimizing over a subset, instead of the whole space.

In many cases, the functional G has the following characteristic shape

$$G(v) = \frac{1}{2} a(v,v) - \langle L,v \rangle, \quad v \in V. \tag{1}$$

Here,

[1] A remarkable exception occurs, for instance, in presence of friction.

- $a(\bullet,\bullet)$ is a bilinear, symmetric form defined on $V\times V$;
- L is a linear functional on V, say an element of the topological dual V' of V;
- $\langle\bullet,\bullet\rangle$ denotes the pairing between V' and V.

As a concrete example, we consider the classical Signorini problem, which we briefly describe as follows.

EXAMPLE 1 Let Ω be an open subset of $\mathbf{R}^3$ with smooth boundary $\partial\Omega$. Let Γ be a smooth, connected piece of $\partial\Omega$. Consider the Sobolev space

$$V \equiv [\mathbf{H}^1(\Omega)]^3 \equiv \{v \in [\mathbf{L}^2(\Omega)]^3 \colon \frac{\partial v_i}{\partial x_j} \in \mathbf{L}^2(\Omega),\ i,j = 1,2,3\}, \tag{2}$$

endowed with the natural norm

$$\| v \| \equiv \left[\sum_{i=1}^{3} \| v_i \|^2_{\mathbf{L}^2(\Omega)} + \sum_{i,j=1}^{3} \| \frac{\partial v_i}{\partial x_j} \|^2_{\mathbf{L}^2(\Omega)} \right]^{\frac{1}{2}} \tag{3}$$

Introduce the bilinear form

$$a(u,v) = \sum_{i,j,h,k=1}^{3} \int_\Omega a_{ijhk}(x) \frac{\partial u_i}{\partial x_j}(x) \frac{\partial v_h}{\partial x_k}(x)\, dx \qquad \forall u,v \in V, \tag{4}$$

where

$$a_{ijhk} \in \mathbf{C}^1(\bar{\Omega}),\quad i,j,h,k = 1,2,3,$$

are the elements of a fourth order tensor satisfying

$$a_{ijhk}(x) = a_{jihk}(x) = a_{hkji}(x)$$

for all $i,j,h,k = 1,2,3$ and all x in $\bar{\Omega}$ and

$$\sum_{i,j,h,k=1}^{3} a_{ijhk}(x)\, \xi_i\, \xi_h\, \eta_j\, \eta_k > \alpha \mid \xi \mid^2 \mid \eta \mid^2, \quad \forall x \in \Omega \tag{5}$$

for some α strictly positive and all vectors ξ, η of $\mathbf{R}^3$.
Finally, consider the closed, convex, non-empty subset of V

$$\mathbf{K} \equiv \{v \in V : v \bullet \nu \leq 0 \ \ on\ \Gamma\}, \tag{6}$$

where ν is the outward unit vector orthogonal to $\partial\Omega$.

With the previous notations, the problem (P) for the functional G given in (1) is called *Signorini problem in linearized elasticity* (see [1]). It provides the mathematical model for the physical problem of equilibrium of an elastic body, whose reference configuration is Ω, not allowed to penetrate a rigid surface initially in contact with Ω along the so called *potential contact area* Γ. The total energy due to a virtual displacement v is expressed by the functional $G(v)$ and is given by the sum of the elastic energy $\frac{1}{2}a(v,v)$ and of the potential energy $\langle L,v\rangle$ due to the applied load L (2). The equilibrium configuration corresponds to a displacement which minimizes the total energy and is compatible with the constraint. ■

Let us discuss the solvability of (P), assuming (1) and

$$\mathbf{K} \text{ is closed, convex and non-empty.} \tag{7}$$

Under these assumptions, a well known theorem states that there is exactly one solution to (P) if the bilinear form is coercive, say

$$\exists \alpha > 0: \quad a(v,v) \geq \alpha \| v \|^2, \quad \forall v \in V. \tag{8}$$

Yet, example 1 shows that some problems in elasticity require a non-coercive setting, the bilinear form being just non-negative:

$$a(v,v) \geq 0, \qquad \forall v \in V. \tag{9}$$

In this framework, existence of a solution to (P) is proved in [2], [3], [4], [5] under additional hypotheses (of compatibility) on a, $\mathbf{K}$ and L. A recent result, which unifies and extends the previous ones, is given in [6], [7]. To state it, let us introduce the following notations.

$$Y \equiv \{ v \in V: \ a(v,v) = 0 \} \tag{10}$$

is the kernel of the form a;

$$rc\,\mathbf{K} \equiv \{ v \in V: \ v_0 + \lambda v \in \mathbf{K}, \ \forall v_0 \in \mathbf{K}, \ \forall \lambda \geq 0 \} \tag{11}$$

is the *recession cone* of $\mathbf{K}$ (well defined, because we assume (7)).

THEOREM 1 *Assume (1), (9), (7).* (P) *has a solution if all the following conditions are satisfied:*

$$Y \cap rc\,\mathbf{K} \text{ has finite dimension;} \tag{12}$$

$$\langle L,w \rangle \leq 0, \quad \forall w \in Y \cap rc\,\mathbf{K}; \tag{13}$$

(2) Allowing more regularity on L, we may assume that

$$\langle L,v\rangle = \int_\Omega f \cdot v \, dx + \int_{\partial\Omega\setminus\Gamma} g \cdot v \, d\sigma,$$

for some volume load $f \in [\mathbf{L}^2(\Omega)]^3$ and some surface stress $g \in [\mathbf{L}^2(\partial\Omega \setminus \Gamma)]^3$.

$$Y \cap \ker K \cap rc\, K \text{ is a subspace.} \tag{14}$$

■

Let us make a few remarks.

REMARK 1 Theorem 1, as well as the quoted results ([2], [3], [4], [5]), may be extended to variational inequalities with non-symmetric bilinear forms (hence, no longer equivalent to minimum problems): see [6], [7]. We also mention an analogous result for non-linear variational inequalities, proved in [8].

REMARK 2 The condition (13) is *necessary* (although not sufficient) in order that G be bounded from below (see [7]).

As an application of the abstract results, in [7] the equilibrium problem of a partially supported plate is discussed thoroughly. (3) In the following example, we briefly describe the problem and the main results.

EXAMPLE 2 Let Ω be the configuration at rest of a thin, horizontal plate, subject to a vertical, downward load. A portion E of Ω is constrained to lie on or above a rigid, plane horizontal obstacle, so that only non-negative vertical displacement is permitted to the points of E (we assume the displacement to be small).

Without precising the (well known) mathematical model which describes the preceding mechanical problem, we just quote the main results of existence. We distinguish two cases.

(a) The resultant of the applied load vanishes. Then an equilibrium configuration exists if and only if the resultant momentum vanishes too.

(b) The resultant of the applied load does not vanish. In this case, a sufficient condition for the existence of an equilibrium configuration is that the center of the applied forces (which is well defined) belong to the *algebraic interior* (4) of the convex hull of E.

For instance, in the case (b) the plate may be supported by a segment, provided that the center of forces belongs to the segment and does not coincide with either endpoint; even more, the plate may be supported by

(3) The case of the supported beam is also detailed in [7].

(4) Here, by *algebraic interior* (or *relative interior*) of a subset A of $\mathbf{R}^N$ we mean the interior of A in the topology of the smallest linear space containing A. For instance, the algebraic interior of a segment in $\mathbf{R}^2$ consists in the segment itself (endpoints not included), while the (topological) interior is clearly empty.

a point obstacle coinciding with the center of forces.

The previous results do not include anything for the case (b), when the center of forces belongs to the algebraic boundary of E. In this case, an appropriate investigation is needed. It may be shown, for instance, that *there is no* solution of the model problem when E is a circle and the center of forces belongs to its boundary. This situation may be depicted as follows. Because of the geometry of the problem, the constraint reaction must be concentrated at the center of forces, so the problem has a solution if and only if there exists a solution of an auxiliary equilibrium problem with point obstacle, and this solution is non-negative on E. Since a local study shows that this cannot happen, we see that there is no solution in the case at hand (see [7] for the details).

■

The abstract theorem 1 permits also to re-obtain the existence results for the Signorini problem (example 1), already proved by G. Fichera [2], [3]. Again, we find that a solution exists, provided that the system of forces has a resultant which pushes the elastic body against the rigid one and that the central axis of the forces meets Γ at its "interior". Without discussing in further detail the existence theorem for the Signorini problem, let us make some comments about the unilateral condition expressed by the convex (see (6))

$$\mathbf{K} \equiv \{v \in [\mathbf{H}^1(\Omega)]^3 : v \bullet \nu \leq 0 \ \ on \ \Gamma\}. \tag{15}$$

REMARK 3 The unilateral condition is prescribed *just* on (a part of) the boundary of Ω. On one hand, this is useful for the mathematical treatment of the problem. For instance, it is possible to prove the local $\mathbf{H}^2$-regularity for the solution, using essentially the fact that a system of *equations* is satisfied inside Ω (see [3], [9]).

On the other hand, the assumption that only a part of $\partial\Omega$ cannot penetrate the rigid obstacle is quite acceptable from the mechanical point of view. Indeed, deformations are assumed to be small and regular, so that no interior point of Ω is allowed to bubble out of the surface and pierce the rigid body. Yet, a pure mathematical investigation of the model might not exclude such situations. The main reason for this drawback is that the mathematical model loses the memory of the assumptions of the linear approach, hence cannot eliminate solutions corresponding to large or not invertible deformations.

A possible way to get rid of the previous drawbacks is to make a slight modification of the unilateral condition (we point out that the linearized approach for the energy functional of example 1 is still adopted). Precisely, we will consider the problem of equilibrium of an elastic body Ω subject to a system of applied forces and confined to lie *entirely* into a box Q with rigid contour. Of course, the interesting case is that of an *unbounded* box: this we will assume henceforth.

This framework is more general than the one considered in example 1, so it will be referred to as *Generalized Signorini-like Problem* (*GSP*). The corresponding set of admissible displacements is the following:

$$\mathbf{K}_Q \equiv \{v \in [\mathrm{H}^1(\Omega)]^3 : x + v(x) \in Q,\ x \text{ a.e. in } \Omega\}. \tag{16}$$

This set is obviously closed, as soon as Q is closed; furthermore, it is non-empty (a natural requirement is that $\Omega \subset Q$, hence $v \equiv 0$ belongs to $\mathbf{K}_Q$).

The troubles arise about convexity: clearly, $\mathbf{K}_Q$ is convex if and only if Q is convex. In one case, existence of a solution may be derived from theorem 1; in the other one, we are dealing with a non-convex problem, so we need a different abstract approach. Before attacking this argument, let us make some comments about (16).

REMARK 4 A set of admissible displacements, intermediate between (15) and (16), is the one considered in [10], say

$$C_Q \equiv \{v \in [\mathrm{H}^1(\Omega)]^3 : x + v(x) \in Q,\ x \in \partial\Omega\}, \tag{17}$$

when the box Q coincides with the upper half space. A condition of type (17) has the advantage of allowing the so called *advancing contact* (5), while (15) was suitable only for *receding contact*. C_Q is obviously a convex, hence theorem 1 applies. The compatibility conditions on the load read as follows: (*GSP*) has a solution if the system of applied forces (6) has non-zero, vertical downward resultant and the central axis meets the zone of *initial* contact (7).

REMARK 5 Consider the condition (15). Assume that Ω is a sphere and that Γ is the portion of the boundary which lies south of the polar circle. Then, a point of the Equator may move according to an *admissible* displacement if it points toward its diametric opposite, no matter how far it goes from its initial position. So, it eventually penetrates the rigid obstacle on the "opposite side" of the sphere. This is not prevented by the condition (15), although physically unacceptable. In this sense, the condition (15) is *local*, say reasonable only near a point where it is imposed. This is in

(5) Say, the region of $\partial\Omega$ which comes in contact with the rigid obstacle may be larger than it was in the rest configuration.

(6) In this framework, only volume forces are allowed.

agreement with the assumption of small deformations.

On the contrary, the conditions (16) or (17) are ***global***, hence suitable for a finite approach to the equilibrium problem. We do not detail the non-linear approach here: in the framework of the stored energy function considered by J. Ball [11], a unilateral problem was discussed for the first time in [12] and later considered in [13] in view of global injectivity properties. We finally quote [14], where existence is proved under less restrictive conditions.

The previous considerations carry attention to the following three features of concrete minimum problems:

(a) non-convexity due to the constraint is possible, even when the "dominant part" of the functional G is quadratic (e.g. when G satisfies (1));

(b) shape more general than (1) occurs in some cases, in particular G itself may not be convex;

(c) in any case, a mixture of non-coerciveness and of non-convexity must be handled to give existence results in practical problems.

As we already suspect, the lack of coerciveness will imply restrictions on the data of the problem. Non-convexity will cause formal complications. Furthermore, even when G satisfies (1) and the non convexity is due to $\mathbf{K}$, the equivalence with variational inequalities fails to be true, as it is easy to see even with finite dimensional examples. So, we need an existence theorem for problem (P) in the case of quite general functionals, in particular non-coercive and non-convex.

The statement of the theorem will involve the behavior at infinity of G. This is usually describe by means of the so called ***recession functional***, whenever G and $\mathbf{K}$ are convex. So we need a generalization of this functional, suitable in the non-convex case, which reduces to the usual recession functional in the convex framework.

For v in V, define the ***topological recession functional*** at v in the following way:

$$G_\infty(v) \equiv \inf \liminf_{n \to +\infty} \frac{1}{\lambda_n} G(\lambda_n v_n), \tag{18}$$

where the infimum is taken over all sequences $\{\lambda_n\}_{\mathbf{N}}$, $\{v_n\}_{\mathbf{N}}$ such that

(7) Note that the role played by the potential contact area, fixed in advance, in the classical Signorini problem, is now played by the subset of points of $\partial\Omega$ which are initially in contact with the rigid plane (and this is one of the data of the problem).

$\lambda_n \to +\infty$, v_n converges weakly to v and $v_0+\lambda_n v_n$ belongs to $\mathbf{K}$ for some v_0 and for n large enough.

THEOREM 2 *Assume that*

$$G \text{ is weakly lower semicontinuous;} \tag{19}$$

$$(\text{compactness}): \quad \left\{ \begin{array}{l} v_n \overset{w}{\to} v,\ \lambda_n \to +\infty,\ G(\lambda_n v_n) \le \text{constant and} \\ v_0+\lambda_n v_n \in \mathbf{K} \text{ for some } v_0 \text{ and } n \text{ large enough} \\ \Rightarrow\ v_n \to v \text{ strongly in } V: \end{array} \right. \tag{20}$$

$$(\text{compatibility}): \quad G_\infty(v) > 0, \quad \forall v \in V,\ v \ne 0. \tag{21}$$

Then (P) *is solvable.*

Sketch of the proof. Set

$$\mathbf{K}_R \equiv \mathbf{K} \cap \{v \in V: \| v \| \le R\}, \qquad R>0.$$

Let u_R be a minimizer of G on $\mathbf{K}_R$ (there is some, for R large enough). If $\|u_R\|$ were not bounded, then the normalized family $\frac{u_R}{\|u_R\|}$ would converge weakly (up to subsequences) to some w. The compactness assumption (20) would give the strong convergence. On the other hand, it should be $G_\infty(w) \le 0$, hence (21) would imply $w=0$, which contradicts the strong convergence.
So, $\|u_R\|$ is uniformly bounded, hence it converges weakly: it is immediate to see that the limit solves (P).

■

Theorem 2 applies to quite general functionals, although it is not "bilateral", in a sense we are going to specify.
Let us compare theorem 1 and theorem 2: of course, this may be done if G and $\mathbf{K}$ satisfy (1) and (7), respectively. In such case, the topological recession functional G_∞ is given by

$$G_\infty = \left\{ \begin{array}{ll} +\infty & \text{if } v \notin Y \cap rc\,\mathbf{K} \\ -\langle L,v\rangle & \text{if } v \in Y \cap rc\,\mathbf{K} \end{array} \right. \tag{22}$$

So, the assumption (21) implies the abstract necessary condition (13) of theorem 1. Moreover, it is obvious that

$$Y \cap \ker L \cap rc\,\mathbf{K} = \{0\}, \tag{23}$$

so that the sufficient condition (14) is fulfilled too. Since the compactness assumptions of both theorems (12) and (20) are equivalent, we get that in the particular case at hand theorem 2 is contained in theorem 1. The inclusion is *strict*, because (23) is much stronger than (14). For instance, theorem 2 does not apply ***directly*** to the generalized Signorini problem if the box Q is convex and has a bilateral recession direction. In this case, it is readily seen that no system of applied forces may satisfy (21).

REMARK 6 Referring to the literature, we say that the condition (21) is of Lions-Stampacchia type while the condition (14) is of Fichera type. Non-bilaterality is in some sense physiological in non-convex problems, as the following finite dimensional example shows. We mention here that an abstract bilateral condition may be obtained, though. Instead of detailing it, we refer the reader to [14].

EXAMPLE 3 Let V be the euclidean space $\mathbf{R}^2$ and set

$$\mathbf{K} \equiv \{(x,y) \in \mathbf{R}^2 : y \geq \exp[-x^2]\}.$$

Let us minimize $G(x,y) = y^2$ on $\mathbf{K}$. The functional is of type (1), with $L = 0$ and $a((x,y),(x,y)) = 2y^2$. $\mathbf{K}$ is not convex (it actually is star shaped) and the natural extension of $Y \cap \ker L \cap rc\, \mathbf{K}$ is the (entire) straight line $\{(x,y): y = 0\}$. This is a subspace, yet G has no minimum in $\mathbf{K}$.

■

As an application of theorem 2, we state an existence result for the generalized Signorini problem quoted above, which (we briefly summarize) consists in solving problem (P) with V, G, a and $\mathbf{K}$ given respectively in (2), (1), (4), (16).

THEOREM 3 GSP *has a solution if the following hypotheses hold:*

$$Q \text{ is contained in a cone } C \text{ of } \mathbf{R}^3; \tag{24}$$

$$\left\{ \begin{array}{c} \textit{the system of applied forces has a resultant} \\ R \neq 0 \textit{ and the resultant momentum with respect} \\ \textit{to any point of the central axis vanishes;} \end{array} \right. \tag{25}$$

$$\textit{the central axis meets the convex hull of } \Omega; \tag{26}$$

$$\langle L, \beta \rangle < 0, \qquad \forall \beta \in C. \tag{27}$$

■

That the hypotheses (24)-(27) imply existence of a solution is quite natural, from the physical point of view. Indeed, we are requiring that the system of forces pushes the body toward the vertex of the cone C, hence prevents it from going along an ***escape direction***. Note that (27) yields that the cone C contains no entire directions.

In [14] it is shown that the abstract conditions of theorem 2 are satisfied, if (24)-(27) hold. Here we just make some remarks.

REMARK 7 The compactness (20) holds, because the unbounded directions of V along which G is bounded from above belong to the kernel of the bilinear form $a(\bullet,\bullet)$, which is finite dimensional (rigid body motions). Analogously, in the finite approach of [11], bad directions have finite dimension.

REMARK 8 In (27), the strict inequality cannot be relaxed, in general. Indeed, consider as rigid box Q a cone, resting on a horizontal plane, parallel to a generatrix. An elastic body subject to gravity has no equilibrium position inside Q. In fact, it must touch the surface of Q outside the generatrix. Hence a reaction appears, pushing the body toward the open "end" of the cone.

We finally mention an extension of theorem 2 which covers minimum problems for functionals defined on Banach spaces, possibly non-reflexive. This case occurs in practice, for instance when dealing with unilateral problems for masonry-like materials. Again, the details of the abstract theorem and of the application to the concrete problem can be found in [14].

REFERENCES

[1] Signorini, A. *Questioni di elasticità non linearizzata e semi-linearizzata*. Rend. Mat., **18** (1959), 95-139.

[2] Fichera, G. *Problemi elastostatici con vincoli unilaterali: il problema di Signorini con ambigue condizioni al contorno*. Atti Acc. Naz. Lincei Mem. Sez. I, (8) **7** (1964), 71-140.

[3] Fichera, G. *Boundary value problems in elasticity with unilateral constraints*. Handbuch der Physik, Band VIa/2, Springer Verlag, Berlin (1972), 347-389.

[4] Lions, J.L. and Stampacchia, G. *Variational inequalities*. Comm. Pure Appl. Math., **20** (1967), 493-519.

[5] Schatzman, M. *Problèmes aux limites non linéaires, non coercifs*. Ann. Sc. Norm. Sup. Pisa, (3) **27** (1973), 641-686.

[6] Baiocchi, C., Gastaldi, F. and Tomarelli, F. *Inéquations variationnelles non coercives*. C.R. Ac. Sci. Paris., **299** (1984), 647-650.

[7] Baiocchi, C., Gastaldi, F. and Tomarelli, F. *Some existence results on non-coercive variational inequalities*. Ann. Scuola Norm. Sup. Pisa (to appear).

[8] Gastaldi, F. and Tomarelli, F. *Some remarks on non-linear and non-coercive variational inequalities*. Boll. Un. Mat. Ital. (to appear).

[9] Kinderlehrer, D. *Remarks about Signorini's problem in linear elasticity*. Ann. Scuola Norm. Sup. Pisa, IV, **8** (1981), 605-645.

[10] Kinderlehrer, D. *Estimates for the solution and its stability in Signorini's problem*. Appl. Math. Optim.,**8** (1982), 159-188.

[11] Ball, J. *Convexity conditions and existence theorems in nonlinear elasticity*. Arch. Rational Mech. Anal., **63** (1977), 337-406.

[12] Ciarlet, P.G. and Necas, J. *Unilateral problems in nonlinear, three dimensional elasticity*. Arch. Rational Mech. Anal., **87** (1985), 319-338.

[13] Ciarlet, P.G. and Necas, J. *Almost everywhere injectivity, self-contact and the non-interpenetration in non-linear, three-dimensional elasticity*. (To appear).

[14] Baiocchi, C., Buttazzo, G., Gastaldi, F. and Tomarelli, F. *General existence results for unilateral problems in Continuum Mechanics*. (To appear).

DYNAMICS IN THE PRESENCE OF UNILATERAL CONTACTS AND DRY FRICTION: A NUMERICAL APPROACH

M. Jean, J.J. Moreau
Laboratoire de Mécanique Générale des Milieux Continus
Université des Sciences et Techniques du Languedoc

1. INTRODUCTION

This paper is essentially devoted to mechanical systems with finite degree of freedom. However, as a test for adapting the proposed numerical methods to continuous systems, the representation of a one-dimensional medium by a finite chain of mass points is presented in fine.

The theoretical background has been defined by one of the authors in the Volume of the last meeting [1] and will only be recalled under some simplifying assumptions in the sequel. This background involves in particular that the possible shocks, occurring if two parts of the system suddenly enter into contact, are <u>inelastic</u>. The paper referred to was restricted to frictionless contacts ; in contrast <u>dry friction</u> is taken here into account, under the traditional law of Coulomb or some extension of it to anisotropic contacting surfaces. Mathematically, this law results in some non-smooth relation between the involved variables, i.e. not everywhere differentiable ; the same is true for the expression of unilateral constraints. Treating both non-smooth aspects jointly is made possible here by using simple concepts of Convex Analysis, widely known today. Of course, some smooth dissipative actions - viscosity - are easy to be introduced additionally.

Through the frictional effects as well as the inelasticity of possible shocks, irreversibility emerges as a dominant feature in the present approach.

No attention is paid in this paper to the existence and uniqueness study of the investigated motions. It has long been observed [2][3] that when bodies exhibit dry friction in their possible points of contact, time intervals of smooth motion may end with instants at which some velocity jump occurs necessarily ; this is called a tangential shock. Sufficient conditions imposed to the system constants and to initial data have recently been given [4] , in order that the motion begins with a time interval exempt from such an accident (and from contact breaking off as well) ; thereby a local existence and uniqueness theorem has been obtained.

Since the algorithms presented in the sequel are devised to treat shocks on an equal footing with regular motions, they prove able to include such events in the computation.

An example presented in Sect. 8 is that of the chattering motion of a solid body elastically driven along a plane wall (a piece of chalk pushed at an angle against a blackboard). Oscillations with intermittent breaking of contact are exhibited. The slip velocity may drop recurrently to zero ; contrary to common belief, this "stick-slip" phenomenon is not essentially connected with the possible dependence of the friction coefficient on slip velocity. Anyway, the proposed algorithm allows one to take such a dependence into account, even under its extreme form : distinguishing between the "static" (i.e. for zero slip velocity) and "dynamic" friction coefficients.

Another example presented in section 6 is that of the motion of a chain of material points in contact with a wall moving at constant velocity along one of the axes. The points are connected together by springs and dashpots and each of them is pushed against the wall by a single pair of springs and dashpots. The points are submitted to Coulomb friction from the wall. The example presented shows a "stick-slip" phenomenon.

2. ANALYTICAL SETTING

As usual in Analytical Dynamics, the totality of the possible positions of the system $\mathcal{S}$, regardless of the unilateral constraints which will be imposed further, constitutes a differential manifold $\mathcal{Q}$, with dimension ℓ equal to the freedom of $\mathcal{S}$. Let (q^i), $i = 1,2,\ldots,\ell$ denote (at least local) coordinates in $\mathcal{Q}$. For every differentiable motion $t \to q(t)$, the derivatives $\dot{q}^i(t)$ are the components of the velocity $\dot{q}(t)$, an element of the tangent space to $\mathcal{Q}$ at the point $q(t)$. The kinetic energy of $\mathcal{S}$ is supposed expressed as a time-independent positive definite quadratic form of $\dot{q}$, namely

$$T(q,\dot{q}) = \frac{1}{2} a_{ij}(q)\dot{q}^i\dot{q}^j \tag{2.1}$$

where $a_{ij} = a_{ji}$ are known functions of $q = (q^1,\ldots,q^\ell)$.

For simplicity we shall make in the whole paper the following assumptions.

ASSUMPTION 1. $\mathcal{S}$ is partitioned into a finite collection of subsystems $(\mathcal{S}_\alpha)$, $\alpha = 1,\ldots,\pi$, which may interact through configuration-dependent and/or velocity-dependent forces, but without any mutual hinge or contact. Therefore $\mathcal{Q}$ equals the product of the corresponding manifolds $\mathcal{Q}_\alpha$ and $T(q,\dot{q})$ splits into the sum of the respective kinetic energies, namely for subsystem S_α

$$T_\alpha(q_\alpha,\dot{q}_\alpha) = \frac{1}{2} a^\alpha_{mn}(q_\alpha)\dot{q}^m_\alpha \dot{q}^n_\alpha \quad . \tag{2.2}$$

Here $(q^1_\alpha, q^2_\alpha,\ldots)$ denote (at least local) coordinates of the point q_α in the manifold $\mathcal{Q}_\alpha$. Unlike Latin indices, the repetition of Greek indices will not be understood as implying summation.

ASSUMPTION 2 . For every α , it is supposed that, at least locally in manifold $\mathcal{Q}_\alpha$, the coordinate system $(q^1_\alpha, q^2_\alpha,\ldots)$ may be constructed in such a way that the coefficients a^α_{mn} in (2.2) do not depend on q_α . This means that the Riemann metric defined on manifold $\mathcal{Q}_\alpha$ by the following expression of arc-length ds

$$ds^2 = a^\alpha_{mn}(q_\alpha)dq^m_\alpha dq^n_\alpha$$

actually turns out to be Euclidean.

Such is trivially the case if subsystem S_α consists of a single mass-point moving through some region of the physical space.

In another usual case, S_α will consist in a rigid body, allowed to perform only motions parallel to some fixed plane. Locating each of its possible positions by the Cartesian (orthonormal) coordinates X,Y of the center of mass G and by the rotation angle Θ , one obtains

$$T_\alpha = \frac{1}{2} m(\dot{X}^2 + \dot{Y}^2) + \frac{1}{2} I \dot{\Theta}^2 \quad . \tag{2.3}$$

Here m denotes the mass of the rigid body and I its moment of inertia about the axis drawn through G , normal to the plane of motion. This makes $[a^\alpha_{mn}]$ a diagonal matrix, constant with regard to the triplet

$(q^1_\alpha, q^2_\alpha, q^3_\alpha) = (X, Y, \Theta)$.

As already said, the investigated unilateral constraints are not taken into account at the stage of defining the parametrization of S . Such superimposed constraints will consist in impenetrability effects about which we shall make now some simplifying hypotheses.

ASSUMPTION 3 . Unilateral contacts never occur between two different members of the previously defined partition ; they take place either inside some S_α or between some of these subsystems and some external body with prescribed motion. In addition, it is supposed that the geometric effect of impenetrability upon the concerned subsystem S_α is expressed by a single scalar inequality

$$f_\alpha(t, q^1_\alpha, q^2_\alpha, \ldots) \leqslant 0 \quad , \tag{2.4}$$

where the real function f_α is C^1 , with nonzero gradient (at least in a neighborhood of the hypersurface $f_\alpha = 0$) . This inequality defines for every t a closed subset L_α of manifold $\mathcal{Q}_\alpha$.

Under these assumptions, let us first write the Lagrange equations governing the motion of the system S during a time interval of smooth motion, i.e. an interval, if any, where the functions $t \to q^i(t)$ happen to be twice differentiable. Since the matrix a^α_{mn} has been supposed constant with regard to t and to $q_\alpha \simeq (q^1_\alpha, q^2_\alpha, \ldots)$, the Lagrange equations relative to subsystem S_α reduce to

$$a^\alpha \ddot{q}_\alpha = r^\alpha + Q^\alpha(t, q, \dot{q}) \ . \tag{2.5}$$

The writing on the left-hand side is that of product matrix, i.e. the covector with components $a^{\alpha}_{mn}\ddot{q}^{n}_{\alpha}$. The covector $r^{\alpha} \simeq (r^{\alpha}_1, r^{\alpha}_2, \ldots)$ represents the force of constraint arising from the possible unilateral contact experienced by S_{α} ; the function $Q^{\alpha} \simeq (Q^{\alpha}_1, Q^{\alpha}_2, \ldots)$ expresses all other forces, a priori depending on the position and velocity of the whole system S and possibly also on t as independent variable, that subsystem S_{α} undergoes from any part. It is only through such terms as Q^{α} that the Lagrange equations corresponding to the respective members S_{α} of the partition of S are coupled together.

Under the writing (2.5) is summarized a set of scalar equations whose number equals the freedom of subsystem S_{α} . There is to join with them the geometric inequality (2.4) and some mechanical information we shall state in the section to come, about the force of unilateral constraint r^{α} .

3. LAWS OF FRICTIONAL UNILATERAL CONTACT

The primary statement concerning the force of constraint $r^\alpha \simeq (r_1^\alpha, r_2^\alpha, \ldots)$, introduced in (2.5), is that it vanishes at every instant t where the corresponding contact is not in effect, i.e. when the position $q_\alpha(t) \simeq (q_\alpha^1(t), q_\alpha^2(t), \ldots)$ verifies (2.4) as a strict inequality.

On the contrary, every position q_α of S_α such that (2.4) holds as an equality corresponds to effective contact. The governing laws have to be made precise in the framework of the 3-dimensional physical space ; afterwards in sect. 4 we shall transcribe them in terms of analytical variables.

Let us first consider the case where the investigated contact takes place between some part B_1 of S_α and an external body B_o with prescribed motion. Such a contact will always be supposed located at a single point M and the definition of a plane is assumed, which will be declared a common tangent plane at the point M to both boundaries of B_o and B_1. This does not preclude bodies presenting edges or vertices ; one of them may even reduce to a single material point, provided the boundary of the other one is a regular surface. Anyway, let us denote by M_o and M_1 the respective particles of B_o and B_1 which happen to lie at point M at the instant in view.

The contact actions exerted by B_o upon B_1 are supposed represented by a single force R applied to particle M_1 ; hence any torque expressing "resistance to rolling" is neglected. Kinematically, for every imagined differentiable motion of the system, passing at time t by the considered position, the difference of the velocity vectors of M_o and M_1 ,

$$U = V(M_1) - V(M_o) \tag{3.1}$$

is referred to as the slip velocity of B_1 relative to B_o .

Let us denote by E_3 the linear space of the vectors of physical space. Let $N \in E_3$ be the unit vector normal to the common tangent plane at point M , directed toward B_1 .

This Section is devoted to motions smooth enough for Lagrange's equations to make sense. The existence of accelerations is then postulated ; therefore velocities are continuous functions of time. Elementary kinematics entails in that case that the slip velocity satisfies

$$N.U = 0 \ , \tag{3.2}$$

i.e. U belongs to the two-dimensional subspace T of E_3 , tangent to both contacting bodies.

The admitted law of dry friction stipulates a relation between the elements $U \in T$ and $R \in E_3$. Here are two equivalent formulations, generalizing to possibly anisotropic contacting surfaces the traditional law of Coulomb.

DECOMPOSED FORM (see more details in [5] [6]).

The reaction vector R lets itself be uniquely decomposed into

$$R = R_T + R_N N \quad , \quad R_T \in T \quad , \quad R_N \in \mathbb{R} \; . \tag{3.3}$$

It is first stated that the real number R_N satisfies

$$R_N \geqslant 0 \quad , \tag{3.4}$$

i.e. no adhesive effect takes place between the bodies in contact.

Secondly there is given a closed convex subset $\mathcal{D}_1$ of T, containing the zero of E_3. For every $R_N \geqslant 0$ put

$$\mathcal{D}(R_N) = R_N \mathcal{D}_1 \tag{3.5}$$

and denote by ψ the indicator function of the subset $\mathcal{D}(R_N)$ of T, i.e.

$$\psi(X) = \begin{cases} 0 & \text{if} \quad X \in \mathcal{D}(R_N) \\ +\infty & \text{if } X \in T \setminus \mathcal{D}(R_N) \; . \end{cases}$$

This is a lower semi-continuous convex function defined in T. Recall that for every $X \in T$ the subdifferential $\partial\psi(X)$, relative to the Euclidean scalar product in T, equals the set of the elements Y of T which are outward normal to the closed convex set $\mathcal{D}(R_N)$ at point X in the conventional sense of Convex Analysis. This subdifferential is empty if $X \notin \mathcal{D}(R_N)$; otherwise it contains at least the zero of T and reduces to this single element in case X is interior to $\mathcal{D}(R_N)$.

Then the expected relation writes down as

$$- U \in \partial\psi(R_T) \,. \tag{3.6}$$

For every fixed R_N this is a resistance law of the rate-independent sort, also familiar in Plasticity theory ; some equivalent forms are classically given to such laws, in terms of a "principle of maximal dissipation" or also in terms of a "dissipation function" (actually the support function of the set $-\mathcal{D}(R_N)$) ; see e.g. [6] [10] .

The traditional law of Coulomb for isotropic contact is simply obtained by taking as $\mathcal{D}_1$ the closed disk in T , with center at the origin and radius equal to the friction coefficient.

CONICAL FORM.

Standard arguments of Convex Analysis [7] permit to translate the above formulation into a statement avoiding the decomposition (3.3) .

Let us denote by C the closed convex cone in E_3 (with vertex at the origin) generated by the set $\mathcal{D}_1 + N$. In the special case of isotropic Coulomb friction, this set is the closed region delimited in E_3 by what is traditionally called the cone of friction : a conical surface of revolution about vector N.

Then the previous formulation is found equivalent to

$$- U \in \text{proj}_T \; \partial\psi_C(R) \,. \tag{3.7}$$

Here proj_T denotes the orthogonal projection mapping from E_3 onto T . In the sense of Convex Analysis in E_3 , the set $\partial\psi_C(R)$ constitutes the outward normal cone to C at the point R . Since this set is empty when $R \notin C$, one observes in particular that the no-adhesion assumption (3.4) is involved in such a writing.

What precedes has been presented for a part B_1 of S_α in contact with some external body. The formulation is quite similar in the case where equality $f_\alpha = 0$ expresses the contact, occuring at some point M of physical space, between two bodies B_1 and B_2, both parts of subsystem S_α. Again the definition of a common tangent plane at M to the respective boundaries is assumed ; M_1 and M_2 denote the respective particles of B_1 and B_2 which happen to lie at M, at the considered instant, and U the slip velocity of B_1 relative to B_2, i.e.

$$U = V(M_1) - V(M_2) . \tag{3.8}$$

The contact actions exerted by B_2 upon B_1 are supposed represented by a single force R applied to M_1 ; hence, in view of Newton's third law, the contact actions exerted by B_1 upon B_2 are represented by the single force $-R$ applied to M_2.

The law of dry friction stipulates a relation between R and U, exactly in the same terms as in the preceding case.

Let us finally take notice of the simplifying circumstance of <u>two-dimensional contact</u>. In fact, it may happen that, due to the primitive constraints of the system (i.e. the perfect constraints which have permitted its parametrization) the set of the values of $V(M_1) - V(M_2)$ (resp. $V(M_1)$ in case of external contact) which, regardless to contact permanence, are possible, is not the whole of E_3 but only a two-dimensional linear subspace E_2, a priori depending on q_α. Then the role of T is played by some one-dimensional subspace of E_2 ; the cone C simply reduces to an angular region whose boundary consists of two half-lines $\mathcal{D}_1$ and $\mathcal{D}_2$ (possibly non-symmetric with respect to the normal direction N : this reflects difference in the friction coefficients associated with either directions of sliding).

The active part of the reaction R consists now in the orthogonal projection of this vector on E_2 . Similar formulations as above then hold, with E_3 replaced by E_2 , but two-dimensionality brings the following simplification.

Let A_1 and A_2 denote generators of $\mathcal{D}_1$ and $\mathcal{D}_2$. The vector $B = A_2 - A_1$ cannot be parallel to N ; then, for every $U \in T$, (3.7) is readily found equivalent to

$$R \in C \tag{3.9}$$

$$R \in \mathcal{D}_1 \quad \text{if} \quad B.U > 0 \tag{3.10}$$

$$R \in \mathcal{D}_2 \quad \text{if} \quad B.U < 0 \tag{3.11}$$

4. EQUATIONS OF SMOOTH MOTIONS

The law of frictional unilateral contact formulated above has now to be transcribed into the analytical setting of Sect. 2 .

Let us first consider the case of the contact occurring between a part $\mathcal{B}_1$ of $\mathcal{S}_\alpha$ and some external body $\mathcal{B}_0$. The contacting particle M_1 of $\mathcal{B}_1$ a priori depends on the position $q_\alpha \in \mathcal{Q}_\alpha$ attained by the system $\mathcal{S}_\alpha$ at the considered instant and also, since $\mathcal{B}_0$ has been assumed to possess a certain prescribed motion, on the time t as independent variable. Supposing t fixed and given the position $q_\alpha(t) \in \mathcal{Q}_\alpha$, let us consider a chain of positions of $\mathcal{S}_\alpha$ depending in a differentiable way on some real variable τ . Using local coordinates $(q_\alpha^1, q_\alpha^2, \ldots)$ in $\mathcal{Q}_\alpha$, define it by

$$q_\alpha^i = \kappa^i(\tau) , \qquad (4.1)$$

with real functions $\tau \longmapsto \kappa^i(\tau)$, $i = 1,2,\ldots$ such that $\kappa^i(0) = q_\alpha^i(t)$; this is conventionally called a virtual displacement at fixed time t . If some particle P of $\mathcal{S}_\alpha$ has been identified, every value of q_α yields a position of P , relatively to the reference frame $\mathcal{F}$ adopted in physical space. When q_α is expressed by (4.1) as a function of τ , the said position of P becomes a τ-dependent point in $\mathcal{F}$ that we shall denote by $\tau \rightarrow P^F(\tau)$. Analysing the geometry of the system and the way parameters $(q_\alpha^1, q_\alpha^2, \ldots)$ have been defined, one may obtain for the vector derivative $\frac{\delta P^F}{\delta\tau}(0)$ an expression in which the real derivatives $\frac{\delta\kappa^i}{\delta\tau}(0)$ figures linearly (using here δ instead of d to denote differentials is traditional in order to prevent confusion between virtual displacements and proper motions). This is true in particular if P equals the particle M_1 identified above ; one obtains an expression of the form

$$\frac{\delta M_1^F}{\delta\tau} = G(\frac{\delta\kappa}{\delta\tau}) . \tag{4.2}$$

Here $\frac{\delta\kappa}{\delta\tau}$, with components $\frac{\delta\kappa^i}{\delta\tau}(0)$, constitutes an element of the <u>tangent space</u> E_α to the manifold $\mathcal{Q}_\alpha$ at point $q_\alpha(t)$ and G denotes a certain linear mapping of E_α into $\mathcal{E}_3$.

This calculation is first used to express the covariant components (or "generalized components") of the force R exerted on the particle M_1 of $\mathcal{B}$. By definition these components are the real numbers $r_1, r_2, \ldots$ such that, for every virtual displacement as above,

$$R.\frac{\delta M_1^F}{\delta\tau} = r_i \frac{\delta\kappa^i}{\delta\tau} . \tag{4.3}$$

The real numbers r_i may be viewed as the components of some element r of the dual of E_α , the linear space E_α^* <u>cotangent</u> to the manifold $\mathcal{Q}_\alpha$ at point $q_\alpha(t)$. Denoting by $\langle .,. \rangle$ the bilinear pairing between E_α and E_α^* and using (4.2) , we write equivalently

$$R.G(\frac{\delta\kappa}{\delta\tau}) = \langle r, \frac{\delta\kappa}{\delta\tau} \rangle .$$

Since the choice of functions $\tau \rightarrow \kappa^i(\tau)$ allows one to identify $\frac{\delta\kappa}{\delta\tau}$

with an arbitrary element of E_α , this is equivalent to

$$r = G^*(R) , \tag{4.4}$$

where $G^* : E_3 \to E_\alpha^*$ is the transpose of the linear mapping $G : E_\alpha \to E_3$ (the space E_3 of the vectors of the physical space is paired with itself through the Euclidean scalar product).

The same linear mapping G is involved in the calculation of the velocity of particle M_1 in the course of some motion of S_α . Such a motion is defined by giving the position $q_\alpha \in Q_\alpha$, i.e. the local coordinates $q_\alpha^1, q_\alpha^2, \ldots,$ as differentiable functions of time. Then, for every identified particle P of the system, the position of P , relative to the reference frame F becomes a t-dependent point of this frame, whose vector derivative constitutes the velocity vector $V(P)$. For simplicity, we restrict ourselves in the sequel to scleronomic systems, i.e. the position of P in F is determined by $q_\alpha \simeq (q_\alpha^1, q_\alpha^2, \ldots)$ independently of time ; this is in fact the only practical case where the expression of kinetic energy reduces to the form assumed in (2.1) . Then expressing $V(P)$ from the t-derivatives $\dot{q}_\alpha^1, \dot{q}_\alpha^2, \ldots$ requires exactly the same calculation as was made above for expressing $\delta \overset{F}{P}/\delta\tau$ in some virtual displacement, from the τ-derivatives $\frac{\delta\kappa^1}{\delta\tau}$, $\frac{\delta\kappa^2}{\delta\tau}$,... (in the absence of the scleronomy assumption, there would appear in $V(P)$ some additional term). Hence, by making P equal to the particle M_1 one obtains

$$V(M_1) = G(\dot{q}_\alpha) .$$

The contacting particle M_o of external body B_o a priori depends on the position $q_\alpha \in Q_\alpha$ of S_α and also on t as independent variable, since, by assumption, B_o exhibits some given motion. Hence the velocity $V(M_o)$ appears as a vector function U_e of $(t, q_\alpha^1, q_\alpha^2, \ldots)$ which has to be calculated from geometric inspection. This finally yields an expression of the slip velocity (3.1) under the form

$$U = G(\dot{q}_\alpha) + U_e(t, q_\alpha). \tag{4.5}$$

Of course the linear mapping $G : E_\alpha \to E_3$ depends on $q_\alpha \in Q_\alpha$ and also on time t as independent variable, since these elements are needed in order to identify the contact particle M_1 .

What precedes makes sense only for t and q_α such that the contact is effective, i.e. inequality (2.4) holds as an equality. In the sequel we shall need the definition of G and U_e to be extended in a smooth arbitrary way to every (t, q_α) , at least in some neighborhood of hyper-surface $f_\alpha(t, q_\alpha) = 0$.

The case of contact occurring between two parts B_1 and B_2 of S_α lets itself be treated in the same way. Then both opposite reactions R and $-R$ experienced by the contact particles M_1 and M_2 are forces acting on the system. Thus the analytical treatment will involve the co-variant components of this pair of forces, i.e. the real numbers $r_1, r_2, \ldots$ such that, instead of (4.3) one has

$$R.\frac{\delta M_1^F}{\delta\tau} - R.\frac{\delta M_2^F}{\delta\tau} = r_i \frac{\delta\kappa^i}{\delta\tau}$$

for every virtual displacement. Here a linear mapping $G : E_\alpha \to E_3$ may be found, such that

$$\frac{\delta M_1^F}{\delta\tau} - \frac{\delta M_2^F}{\delta\tau} = G\left(\frac{\delta\kappa}{\delta\tau}\right) .$$

This finally yields the same writing as in (4.4) . Besides, supposing again the system scleronomic, one finds as expression of the slip velocity $U = V(M_1) - V(M_2)$

$$U = G(\dot{q}_\alpha) \quad , \tag{4.6}$$

which is the same as (4.5) , with $U_e = 0$.

It is now easy to list a complete set of equations governing any possible smooth motion of the system. This consists in writing, for every subsystem S_α, the Lagrange equations (2.7) , the expression (4.5) or (4.6) of the slip velocity U , the expression (4.4) of r from the reaction R and, finally, the law of frictional contact as stated in Sect. 3 . Let us symbolize the latter as

$$R \in \text{cou}_\alpha(t, q_\alpha, U) \tag{4.7}$$

Here the right-hand member represents the subset of E_3 (resp. E_2 in the case of two-dimensional contact) consisting of the values of the contact reaction R which agree with the formulated dry friction law if, the position of S_α being q_α at time t , the slip velocity happens to assume the value U . It is an essential feature of this law that it cannot be "solved" in order to express R as a single-valued function of U (nor U as a single-valued function of R) . Instead, the symbol cou_α represents a multivalued mapping or multifunction that we propose to call the Coulomb multifunction associated with the possible contact affecting subsystem S_α .

The description of friction, in sect. 3 , was developed under such smoothness assumptions that U essentially belongs to the linear subspace

T of E_3, tangent to contacting bodies. Let us put explicitely :

CONVENTION 1 . In the case of contact, i.e. $f_\alpha(t,q_\alpha) = 0$, the value of the multifunction cou_α is defined equal to the empty set whenever $U \notin T$.

In other words, the existence of some R verifying (4.7) implies $U \in T$.

Furthermore, the no-contact case $f(t,q_\alpha) > 0$ may be embodied in the same writing by putting :

CONVENTION 2 . In the case $f_\alpha(t,q_\alpha) > 0$, the value of the multifunction cou_α is declared to be the singleton $\{0\}$ whichever is U in E_3.

Incidentally, the geometric impenetrability condition could also be included by agreeing that the multifunction assumes the value $\emptyset$ whenever it is not satisfied. We shall avoid this trick for it does not prove convenient when discussing numerical approximation.

In exploiting the above set of conditions, one may wish to eliminate the variables R and U. This finally yields

$$a^\alpha \ddot{q}_\alpha - Q^\alpha(t,q,\dot{q}) \in G^*(cou_\alpha(t,q_\alpha,G(\dot{q}_\alpha) + U_e(t,q_\alpha))) \tag{4.8}$$

a second order differential inclusion.

On the contrary one may keep the reactions as primary unknowns. One has the following set of equations :

- if α is the index of a subsystem with contact at time t

$$a^{\alpha}\ddot{q}_{\alpha} = Q_{\alpha}(t,q,\dot{q}) + G^{*}_{\alpha}(R_{\alpha}) \tag{4.9}$$

$$U_{\alpha} = G_{\alpha}(\dot{q}_{\alpha}) + U_{e\alpha} \tag{4.10}$$

$$R_{\alpha} = R_{T_{\alpha}} + R_{N_{\alpha}} N_{\alpha} \tag{4.11}$$

$$R_{N_{\alpha}} \geq 0 \tag{4.12}$$

$$- U_{\alpha} \in \partial\psi(R_{T_{\alpha}}) \tag{4.13}$$

where $\partial\psi$ is the subdifferential of the indicator function of the convex set $\mathcal{D}(R_{N_{\alpha}}) = R_{N_{\alpha}} \mathcal{D}_1$;

- if α is the index of a subsystem with no contact

$$a^{\alpha}\ddot{q}_{\alpha} = Q_{\alpha}(t,q,\dot{q}) .$$

The initial conditions $q(t_o),\dot{q}(t_o)$ being fixed, let us write as

$$U_{\alpha} = A_{\alpha}(R) \quad , \quad R = (R_1,\ldots,R_{\alpha},\ldots)$$

the result of the elimination of q_{α} between (4.9) and (4.10) . This defines the mapping A_{α} , under the usual assumptions ensuring existence and uniqueness for the solution to the differential equations (4.9) .

Furthermore, the inclusion (4.13) may equivalently be written as a variational inequality, expressing the "principle of maximal dissipation", namely

$$R_{T_{\alpha}} \in \mathcal{D}(R_{N_{\alpha}}) \text{ and } \forall\, \Phi \in \mathcal{D}(R_{N_{\alpha}}) \ , \ (\Phi - R_{T_{\alpha}}).U_{\alpha} \geq 0$$

where the dot refers to scalar product in E_3 .

Hence conditions (4.9) to (4.13) are finally equivalent to a system of quasi-variational inequalities.

In usual cases, due to the positiveness of inertia matrices and the convexity of elastic potential, the mapping $A \overset{\sim}{-} (A_1,\ldots,A_\alpha,\ldots)$ turns out to be monotone in the sense of Minty and the mapping $Id - \rho A$ is a contraction for every $\rho > 0$ in some interval.

Finally the problem may be formulated as the following fixed point condition

$$R_{T_\alpha} = \mathrm{proj}_{\mathcal{D}(R_{N_\alpha})}(R_{T_\alpha} - \rho\, A_\alpha(R))$$

together with (4.11),(4.12) .

5. ALGORITHM OF CONICAL TYPE

By introducing the additional unknown vector function of t

$$u_\alpha = \dot{q}_\alpha \quad , \tag{5.1}$$

one gives to the differential inclusion (4.8) the form

$$a^\alpha \dot{u}_\alpha - Q^\alpha(t,q,\dot{q}) \in G^\star(\mathrm{cou}_\alpha(t,q_\alpha,G(u_\alpha) + \mathcal{U}_e(t,q_\alpha))) \quad . \tag{5.2}$$

Let $t_o, t_1, \ldots, t_i, \ldots$ be a time discretization, with mesh length $t_{i+1} - t_i = h$ independent of i for simplicity. Dropping the subscript α which refers to the considered member of the partition of S, we shall denote by q_i, u_i some approximants of vectors $q_\alpha(t_i)$, $u_\alpha(t_i)$. Then (5.1) is discretized into

$$q_{i+1} = q_i + h\, u_i \quad . \tag{5.3}$$

The discretization of (5.2) has to be performed in such a way that each step, starting with known q_i and u_i, yields a computable value of u_{i+1}. In simple cases, this value is expected to be uniquely defined (anyway, it has long been recognized [2] that, in the dynamics of systems with dry friction, uniqueness of the motion consequent to initial data is not granted in general).

Recall that q , $\dot{q}$ in Q^{α} , on the left-hand side of (5.2) , refer to the whole system S and not only to the member S_{α} of the partition. When computing an approximant of Q^{α} , one may assign to t and q in this function the starting values of the considered step or, alternatively, some anticipated values such as q_{i+1} or $(q_i + q_{i+1})/2$; the influence of this choice upon numerical precision is discussed in [7] . As for $\dot{q}$ in Q^{α} , we decide to adopt the starting values of the step ; in that respect, our computation looks like an "implicit" scheme : this is justified by the assumed smoothness of the function Q^{α}. For the sequel, let us denote by $Q_i^{!}$ the approximant of vector Q^{α} constructed in that way.

Incidentally, in the absence of assumption 2 , made in sect. 2 for simplicity, there would appear in the left-hand members of Lagrange equations some quadratic terms in $\dot{q}$: the same would apply to them.

Similarly, the linear mapping G of E_{α} to E_3 and the velocity $U_e \in E_3$ of a possible external obstacle, which are smooth known functions of t and q , will be calculated from the starting values t_i , q_i or through some anticipation as above.

In contrast, the Coulomb multifunction behaves critically since, with regard to variables t and q_{α}, it is discontinuous on the boundary of the permitted region. Strictly speaking, cou_{α} is not defined if $f_{\alpha}(t,q_{\alpha}) > 0$; however, in the course of computation, a certain amount of violation of the unilateral constraint may be faced, so we agree to extend in a smooth arbitrary way the definition of cou_{α} , at least when condition $f_{\alpha}(t,q_{\alpha}) \lesssim 0$ is moderately violated. For instance, in the case of the traditional (isotropic) Coulomb law, a cone of revolution in E_3 will be considered, with axis and angle depending continuously on t and q_α, so as to reduce to the proper Coulomb cone when the contact condition $f(t,q_\alpha) = 0$ holds. The definition of cou_α derives from this cone in the way explained in sect. 3 .

Recall that $cou_\alpha(t,q_\alpha,U) = \{0\}$ if $f_\alpha(t,q_\alpha) < 0$. Hence every discretization step rests on the consideration of some *test point* (t_i', q_i') ; an expedient choice proves to be $t_i' = t_i + h/2$, $q_i' = q_i + h\, u_i/2$.

If $f_\alpha(t_i', q_i') < 0$ it will be admitted that $cou_\alpha = \{0\}$ for the considered step, so that (5.2) simply discretizes into

$$a(u_{i+1} - u_i)/h - Q_i' = 0 \qquad (5.4)$$

corresponding to the dynamics of the system in the absence of contact. Since a is an invertible matrix, this yields as end of the computation step

$$u_{i+1} = u_i + h\, a^{-1} Q_i' \quad . \qquad (5.5)$$

In any case, we shall denote by u_i' the left member of (5.5) . Furthermore let g be the gradient of f_α at the test point, i.e. the element of E_α^* with components $\partial f_\alpha / \partial q_j^\alpha$.

For simplicity, suppose the function f_α *constant in* t *and* $U_e = 0$.

When $f_\alpha(q_i') \geq 0$, computation goes on as follows.

If $\langle u_i', g \rangle \leq 0$ we estimate that motion takes place without contact action and adopt again (5.5) as step end.

If, on the contrary, $\langle u_i', g\rangle > 0$, the nontrivial multifunction $U \to cou_\alpha(t, q_\alpha, U)$ has to be considered, with (t, q_α) conveniently fixed at the same test point (t_i', q_i') as above, and shortly written $cou(U)$. In order to yield a precise characterization of u_{i+1} , the discretization has to involve this multifunction through some "implicit" procedure, i.e. the value substituted for u_α in the right-hand side of (5.2) will be the unknown u_{i+1} itself. Since the set $cou(U)$ is essentially conic, one may equivalently multiply both members by the positive number h , thus obtaining

$$a(u_{i+1} - u_i') \in G^*(cou(G(u_{i+1}))) \; . \tag{5.6}$$

We place ourselves in the regular case where the linear mapping $G : E_\alpha \to E_3$ (resp. E_2 in bidimensional cases) is surjective, i.e. every value $U = G(\dot{q}_\alpha)$ in E_3 (resp. E_2) of the relative velocity of contacting bodies at point M , without regard to contact permanence nor impenetrability, may be obtained by choosing suitably the time-derivative $\dot{q}_\alpha \in E_\alpha$. Then the transpose G^* is injective ; in addition, the following geometrical remark will be of use.

If contact holds in the considered position of the system, a value u of $\dot{q}_\alpha$ is compatible with further preservation of condition $f_\alpha(q_\alpha(t)) \leq 0$ if $\langle u, g\rangle \leq 0$. Now $f_\alpha \leq 0$ expresses that, in physical space, the subsequent imagined motion does not make the bodies interpenetrate. This proves the implication

$$\langle u, g\rangle \leq 0 \Rightarrow G(u).N \geq 0 \quad (\text{i.e. } \langle u, G^*(N)\rangle \geq 0)$$

which, in view of the unilateral version of Lagrange multiplier theorem, is equivalent to

$$\exists\ \lambda \geq 0 \ : \ G^{\star}(N) = -\lambda g \ . \tag{5.7}$$

Actually $\lambda > 0$, due to injectivity of $G^{\star}$. This shows in particular that the element $U = G(u)$ lies in the common tangent space T to contacting bodies if and only u belongs to T , the subspace of E_{α} orthogonal to the element g of $E_{\alpha}^{\star}$.

How inclusion (5.6) may determine u_{i+1} will only be demonstrated here by the simple example of *two-dimensional friction*, as formulated at the end of sect. 3 (another very clear instance, developed in [7] , is that of a single mass-point, unilaterally confined by some moving boundary).

The cone C in E_2 is generated by a couple of elements A_1 , A_2 . Thus its image $C = G^{\star}(C)$ equals the flat cone in E_{α} whose edges are the half-lines D_1 and D_2 respectively generated by $a_1 = G^{\star}(A_1)$ and $a_2 = G^{\star}(A_2)$. Due to (5.7) , C contains $-g$; put $b = a_2 - a_1$.

In view of the form (3.9) , (3.10) , (3.11) that relation $R \in \mathrm{cou}(U)$ takes in two-dimensional cases, the relation $s \in G^{\star}(\mathrm{cou}(G(u)))$, holds between some elements $s \in E^{\star}$ and $u \in T$ if and only if

$$s \in C \tag{5.8}$$

$$s \in D_1 \qquad \text{if} \qquad \langle b,u\rangle > 0 \tag{5.9}$$

$$s \in D_2 \qquad \text{if} \qquad \langle b,u\rangle < 0 \ . \tag{5.10}$$

The dimension of E_{α} and $E_{\alpha}^{\star}$ equals the freedom of S_{α} . For picture clarity let us imagine it to be 3 ; through the choice of adapted bases in E_α and E^*_α , the positive definite matrix a may be made unit ; this amounts to identify both spaces with a single copy of Euclidean $\mathbb{R}^3$ (at a more elaborate level it may be said that, due to the Riemannian structure of manifold $\mathcal{Q}_\alpha$, the tangent and cotangent spaces at each point q_α merge into a single Euclidean linear space ; this standpoint is systematically used in [1] , [7]) .

Recall that we are now in the case where $<u'_i,g>$, i.e. the scalar product $u'_i.g$ in the sense of Euclidean $\mathbb{R}^3$, is strictly positive. Therefore the (non homogeneous) cone $u'_i + C$ has nonempty intersection with the linear subspace T ;

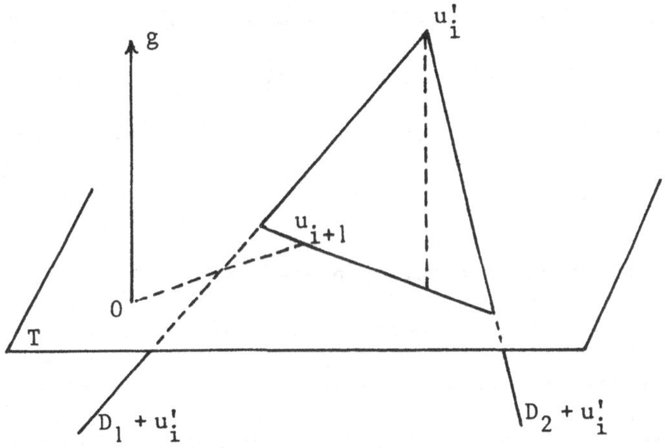

- Figure 1 -

this intersection is a line interval S , possibly unbounded. Clearly in (5.9) and (5.10) , if $u \in T$, the vector b may equivalently be replaced by its orthogonal projection on T ; this is precisely a directing vector of S .

One readily concludes in the existence of a single point $u_{i+1} \in T$ satisfying (5.6), namely *the nearest point in* S *to the origin of* $\mathbb{R}^3$; computing its coordinates is elementary.

For instance, in the case shown on Fig. 1, u_{i+1} happens to lie in the interior of S; hence this vector is orthogonal to the plane of the flat cone C, i.e. $G^{*}(E_2)$. This yields $G(u_{i+1}) = 0$, meaning that the corresponding slip velocity is zero in that case.

REMARK 1. Some elementary exercises in frictional Dynamics concern the motion of a mass-point on a surface. Exploiting Coulomb law in the traditional way requires an expression of the normal component of acceleration : this classically involves velocity and the curvature tensor of the surface. Then one may be surprized at observing that our algorithms do not make use of the second derivatives of f_α. Actually it turns out that curvature effects are completely accounted for by calculating the covector grad f_α at each step.

REMARK 2. The above discretization procedure may be qualified as mixed implicit-explicit. A strong argument in favor of such a policy is that, in particular, it yields exactly zero velocity in cases where the initial conditions are those of equilibrium. This contrasts with some traditional approaches of dry friction ; for instance, in one dimensional situations, it is classical to express the law of dry friction through the function sgn(u), conventionally zero for $u = 0$. Thereby, the multivalued character of Coulomb law is disregarded ; if explicit discretization schemes are derived from such formulations, they can only render equilibrium as a saw-tooth motion, with amplitude depending on the time-step.

Also observe that, at each step of our algorithm, there is no difficulty in making the Coulomb mapping depend on u_i ; this accounts in particular for the usual empirical distinction between the "static" and "dynamic" friction coefficients.

6. ALGORITHM BASED ON REACTIONS

In presenting the main features of this algorithm we shall restrict ourselves to the case where the subsystems S_α are points, connected with springs and dashpots moving in a plane Ox_1x_2 (see fig. 2). All of them are confined in the region $x_2 \geq 0$. The obstacle B_o has rectilinear boundary $x_2 = 0$, moving along the x_1 axis at velocity E. We consider small displacements so that the equation (2.5) governing the system may be written as

$$m^\alpha \ddot{q}_\alpha + V^\alpha(\dot{q}) + K^\alpha(q) = P_\alpha(t) + R_\alpha \quad , \quad q_\alpha = (q_\alpha^1, q_\alpha^2) \quad .$$

$V_\alpha(\dot{q})$ represents the viscosity forces from the dashpots ; $K_\alpha(q)$ represents the elastic forces from the springs ; $P_\alpha(t)$ is some force explicitely given as a function of time. R_α is written only if α is the index of a point in contact with the obstacle. In that case G_α^* is nothing but the identity mapping from E_2 onto itself. The above equations may also be written as

$$M\ddot{q} + V\dot{q} + Kq = P(t) + G^*R \quad , \tag{6.1}$$

where the inertia matrix M is diagonal ; V and K are symmetric non-negative matrices. Let us denote by n the number of points and χ the number of contact points, assuming that for some interval of time, these points remain in contact. Contact points are indexed by $\beta \in \{1,\dots,\chi\}$ and they have a corresponding index in the α list, that we denote α_β. The 2χ vector R is the column of the χ two-dimensional vectors R_α. G^* is thus a mapping from $\mathbb{R}^{2\chi}$ into $\mathbb{R}^{2n}$, with the identity matrix on the α_β,β block and zero elsewhere.

There is a wide range of methods to compute an approximation of the solution to an ordinary differential equation such as (6.1). In numerous cases, explicit or implicit methods provide two matrices $\mathbb{A}_h$, $\mathbb{B}_h$, (h denotes the step length $t_{i+1} - t_i$) deriving an approximate value q_{i+1} of $q(t_i)$ and $\dot q_{i+1}$ of $\dot q(t_{i+1})$ from an approximate value q_i of $q(t_i)$ and $\dot q_i$ of $\dot q(t_i)$ namely

$$\begin{bmatrix} q_{i+1} \\ \\ \dot q_{i+1} \end{bmatrix} = \mathbb{A}_h \begin{bmatrix} q_i \\ \\ \dot q_i \end{bmatrix} + \mathbb{B}_h h\bar\varphi \tag{6.2}$$

$$\bar\varphi = \bar P + G^* \bar R$$

$\bar P$ is an approximant of P on $[t_i,t_{i+1}]$ and $\bar R$ is an approximant of R on $[t_i,t_{i+1}]$. One has also to choose the values $\bar q$ and $\bar{\dot q}$ of approximants of q and $\dot q$ on $[t_i,t_{i+1}]$, in order to get an approximant of the slip velocity on $[t_i,t_{i+1}]$. A possible choice is

$$\bar q = (1-s)q_i + s\, q_{i+1}$$

$$\bar{\dot q} = (1-s)\dot q_i + s\, \dot q_{i+1} \qquad s \in [0,1]$$

The choice $s = 1$ proves the best one regarding stability or smoothing effects. The velocity $\dot{q}_{i+1}$ resulting from (6.2) may be written as

$$\dot{q}_{i+1} = \Gamma + B'h\, G^{*}\, \overline{R} \quad .$$

Here Γ contains terms with q_i, $\dot{q}_i$ and $\overline{P}$. The term $G^{*}\,\overline{R}$ can be decomposed into

$$G^{*}\,\overline{R} = G^{*}_{T}\,\overline{R}_{T} + G^{*}_{N}\,\overline{R}_{N}$$

where $\overline{R}_{T}$ is the χ column of tangential components of $\overline{R}$ and $\overline{R}_{N}$ the χ column of normal components of $\overline{R}$. Actually G_T and G_N, the transpose matrices of G^{*}_{T} and G^{*}_{N}, are projectors from $\mathbb{R}^{2\chi}$ on $T_{\alpha_1} \times T_{\alpha_2} \times \ldots \times T_{\alpha\chi}$ and from $\mathbb{R}^{2\chi}$ on $N_{\alpha_1} \times N_{\alpha_2} \times \ldots \times N_{\alpha\chi}$, respectively. Let us express that the velocity $(\dot{q}_{\alpha})_{i+1}$ lies in the tangent space T_{α} :

$$0 = G_N\dot{q}_{i+1} = G_N\Gamma + G_N B' G^{*}_{T}\, h\, \overline{R} + G_N B' G^{*}_{N}\, h\, \overline{R}_{N} \tag{6.3}$$

Since in most cases B' is an invertible matrix (in fact B is close to M), $G_N B' G^{*}_{N}$ is invertible and the above equation yields an expression of R_N as a function of R_T. The slip velocity expresses as

$$G_T\dot{q}_{i+1} - [E] = U = G_T\Gamma + G_T B' G^{*}_{T}\, h\,\overline{R}_{T} + G_T B' G^{*}_{N}\, h\,\overline{R}_{N} \tag{6.4}$$

By $[E]$ is denoted the χ column with components E, the velocity of the moving wall. The matrix $G_T B' G^{*}_{T}$ is symmetric positive definite. Inequalities

$$R_{N_\alpha} \geq 0 \qquad \alpha = \alpha_1, \ldots, \alpha_\chi \tag{6.5}$$

are tested. The inclusion

$$- \mathcal{U}_\alpha \in \partial\psi(\mathcal{R}_{N_\alpha})$$

may be equivalently transformed into the following fixed point condition

$$\mathcal{R}_{T_\alpha} = \mathrm{proj}_{\mathcal{D}_\alpha}(\mathcal{R}_{T_\alpha} - \rho(W_\alpha + G_T B' G_T^{*} h \bar{R}_T)) \tag{6.6}$$

with $\mathcal{D}_\alpha = R_{N_\alpha} \mathcal{D}_1$ and

$$W = G_T \Gamma + G_T B' G_N^{*} h \bar{R}_N ,$$

Observe that the impulse $h\bar{R}$ emerges as the natural variable. We shall develop in section 7 below how this allows for treating shocks as well as smooth motions. Of course the method applies to more general situations than those related to systems of points. The projectors have to be computed at each step. Nevertheless in the case of a plane obstacle the computation may be avoided in checking whether the set of contact points at some step is the same set as the one met at the previous step. If a motion with persistent contact is expected then the projectors may be computed once for all. The algorithms may be constructed in taking advantage of simplifications allowed by special forms of the matrix M and of the projectors. For instance M or $G_N B' G_T^{*}$, $G_N B' G_N^{*}$, $G_T B' G_T^{*}$ may be diagonal matrices (see for instance [12]).

The system of figure 3 is composed of ten material points. The points are connected together by viscoelastic dipoles (involving spring and dashpot) . Each of them is submitted to a constant vertical force (gravity) and is pushed against the obstacle by a viscoelastic dipole. The angle of rake is such that when a point moves to the right, the pressure against the obstacle increases. For this system one may observe a motion with persistent contact. The velocities and tangential components of the reaction as functions of time are plotted for points 1, 3, 5 on figure 3 . A stick-slip phenomenon is in evidence for points 3 and 5 (see arrows ▼) . The velocity settles at the value E of the obstacle velocity while a drop of the tangential component of the reaction occurs. Figure 4 shows a similar system except for the angle of rake of the dipoles pushing against the obstacle. When a point moves to the right, the pressure against the obstacle decreases. The x_1 and x_2 coordinates of the points 1, 2, 3, 4, 5 as functions of time are plotted on figure 4 . One sees (arrow ◄) that the point 3 leaves smoothly the obstacle. Then the point comes back into contact (arrow ►) with a shock after which it slides again. By plotting velocity one could see that stick-slip is also present in this case.

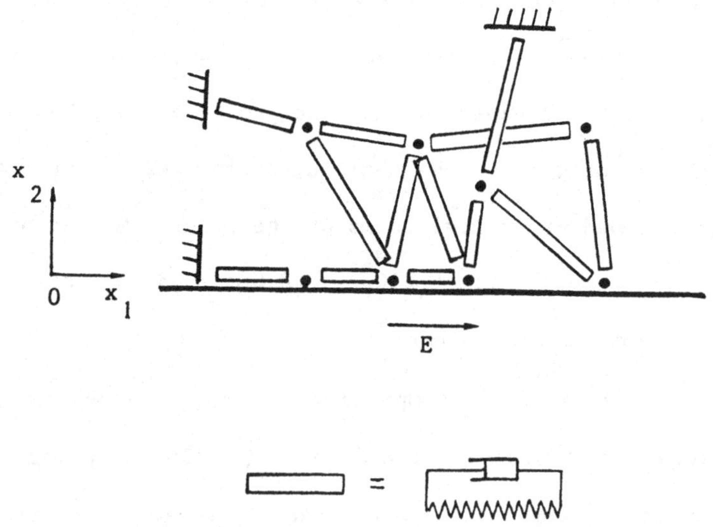

- Figure 2 -

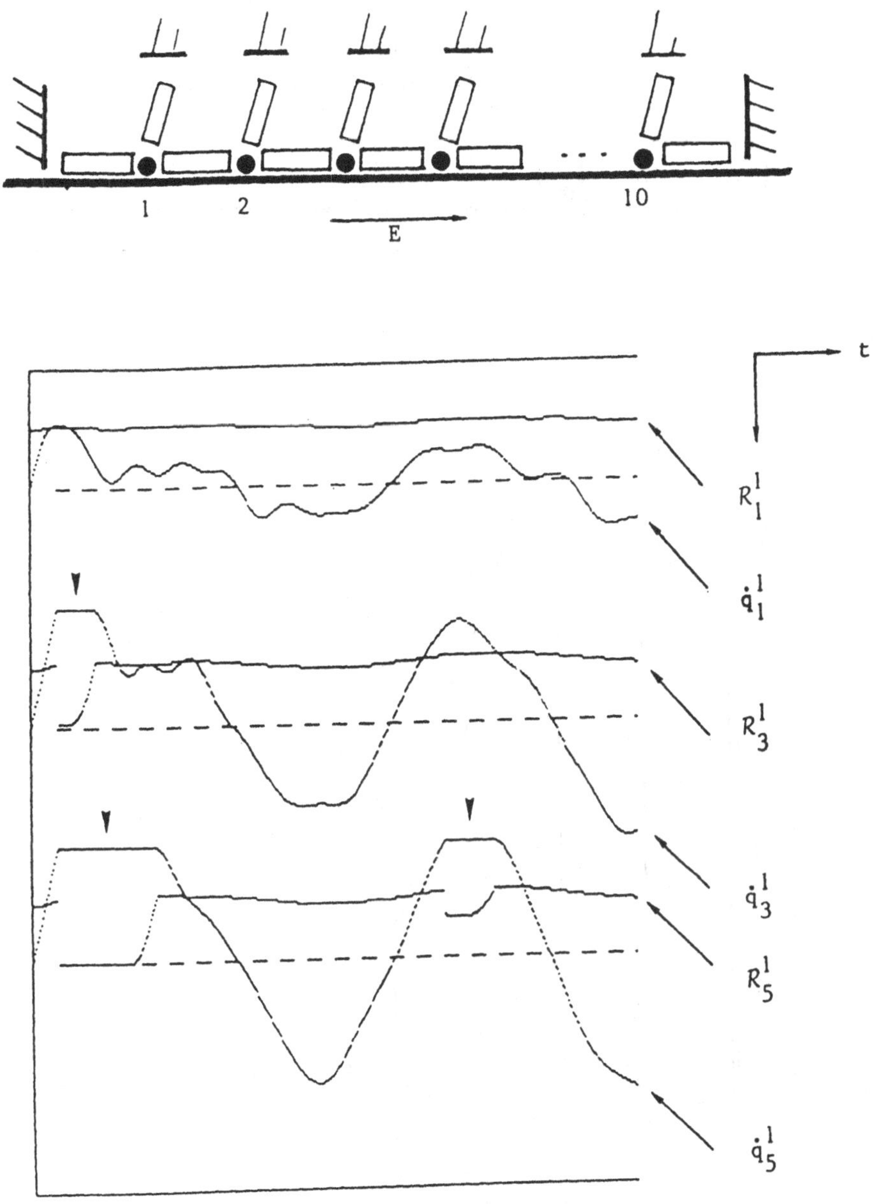

- Figure 3 -

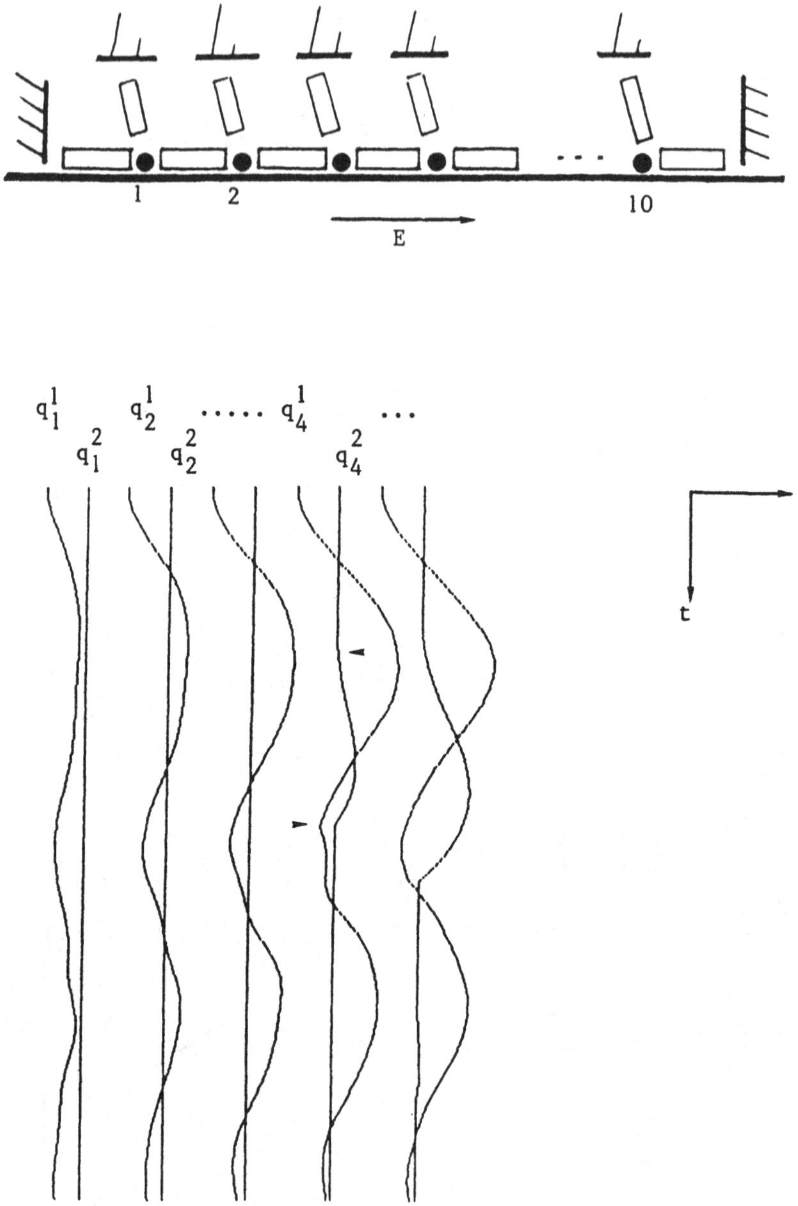

- Figure 4 -

7. SHOCK CALCULATION

All what precedes concerns motions smooth enough for the second derivative $\ddot{q}$ to exist. Such motions may verify $f_\alpha(t,q_\alpha) \leq 0$ either as a strict inequality or as an equality. In particular, some episode with persistent equality may be followed by contact break without any jump in velocity.

On the contrary, if an episode with $f_\alpha < 0$ ends at some instant t_s such that $f(t_s,q_\alpha(t_s)) = 0$, the continuity of $\dot{q}_\alpha$ at t_s cannot be expected in general. We shall assume the existence of the right-limit $u_\alpha^+ = \dot{q}_\alpha(t_s^+)$ and of the left-limit $u_\alpha^- = \dot{q}_\alpha(t_s^-)$.

This is a shock, the physical circumstances of which are inevitably intricate. Some insight can only be gained from relaxing, at least qualitatively, the idealization previously adopted. A certain amount of deformation will be accepted for the bodies in presence, so that the sudden velocity change develops through a little longer duration than the single instant t_s . By equalling the integrals of both members of the equations of Dynamics over this duration, one classically obtains an expression for the momentum change in terms of the percussion of constraint.

In the no-friction case, it is usually admitted that this percussion is normal to the colliding surfaces (this might be discussed ; see e.g.

[11] , sect. 9.7.c , Remarque). Even so the above balance of momentum does not provide enough information to determine u_α^+ . The complementary condition is commonly derived from assuming that the shock is elastic or, alternatively, inelastic.

Let us restrict ourselves here to the same simple case as in previous sections : the system is supposed scleronomic and the function f_α does not depend on time. Then the definition of an elastic shock simply reduces to asserting that the kinetic energy of the system is preserved, while inelastic shock is characterized by condition $\langle u_\alpha^+ , \text{grad}\, f_\alpha(q_\alpha)\rangle = 0$. One proves that the latter essentially involves energy dissipation (cf. [1] , where the concept of an inelastic shock is also extended to the case of constraints expressed by several inequalities).

In practice, physical arguments concerning the energy balance of a shock are generally uncertain. Even if the slighlty deformable material of which the system is built may safely be treated as elastic, one cannot a priori assert energy preservation . In fact the slight deformation resulting from the impact is not usually restricted to the small zone of contact ; it propagates to the whole of colliding parts and possibly also to other connected bodies, whether internal or external to the system. Therefore, after bounce, various parts may remain in a state of vibration which, at the macroscopic scale, amounts to energy loss.

In contrast, it seems easier to characterize some physical situations where the inelasticity of shocks is a pretty safe guess. A familiar example is that of a wood or metal ball landing on a floor of beaten earth.

Frictional effects between the colliding bodies still complicate the situation. A theory of frictional shock was proposed by G. Darboux and further improved by J. Pérès (see e.g. [3] , p. 316-328) . It consists in studying the evolution of the velocity as a function of some "microscopic time" ordering the very short duration of the shock, while position is treated as constant. This is interesting as an early example of a multiple scaling method but the physical relevance seems difficult to assess ; in particular the estimation of shock end proves critical.

We shall adopt a cruder approach of frictional shock by assuming that the percussion of contact relates to the final velocity u_α^+ , through the same generalized Coulomb law as does the force of contact in smooth motions.

The mathematical background of "nonsmooth Dynamics" has been formulated in [1] under the assumption that velocity u is a function of t with locally bounded variation, non differentiable in general. Then the role of the acceleration is played by the differential measure du , a vector measure on the considered time interval. Accordingly, forces are replaced by vector measures, comprising possible percussions ; in smooth cases these measures admit, relative to the Lebesgue measure dt , some density functions whose values are forces in the proper sense.

In that context, the differential inclusion (5.2) receives a precise generalized meaning. It turns out that the discretization algorithms of sections 5 and 6 yield approximate solutions of the corresponding problems through exactly the same computation procedures as in smooth cases. If some discretization interval (t_i,t_{i+1}) happens to include shock, it only comes that the difference $u_{i+1} - u_i$ has no more the same order of magnitude as $h = t_{i+1} - t_i$; this difference approximates the integral of the measure du over the considered time interval.

Thus the decision of relating frictional percussion to the right-limit of velocity appears to parallel exactly the choice previously made of an implicit discretization scheme in treating the differential inclusion, as well as the assumption of shock inelasticity.

In our opinion, this approach has the following methodological virtues. It standardizes a type of irreversible process, resulting in evolution problems with strong theoretical and numerical consistency. Some practical instances certainly exist where the assumptions made are in satisfactory agreement with reality. But, above all, this ideal situation provides a safe basis to which further empirical corrections could be added in rational order. This looks similar, for instance, to the concept of standard dissipative process in thermoplasticity [13] or to the use of Lax's entropy conditions when studying shockwaves in hyperbolic conservation laws (see e.g. [14]) .

8. AN EXAMPLE OF OSCILLATOR WITH FRICTIONAL UNILATERAL CONTACT

Figures 6, 7, 8, have been produced by a computer program using the algorithm of sect. 5.

A rigid body $\mathcal{B}$ performs a motion parallel to the (x,y)-plane. The plotter draws the profile of $\mathcal{B}$ in its initial position : the significant part of it consists of an arc of circle with center A and radius b.

Using as position parameters the coordinates X, Y of the center of mass G and some rotation angle Θ, one gives to the kinetic energy of $\mathcal{B}$ the form (2.3). In this computation $b = m = I = 1$; AG = 1,732.

The body is confined in the region $y \geq 0$; its possible contact with the fixed wall $y = 0$ involves dry friction with coefficient 0.5.

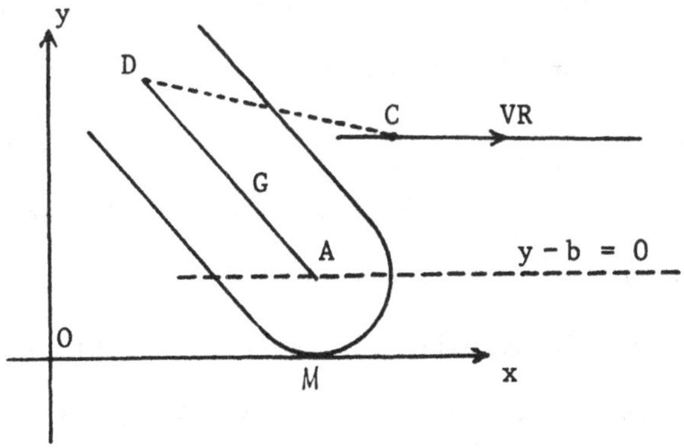

- Figure 5 -

Here are the driving forces :

1° Some particle D of $\mathcal{B}$ (located in this computation on the line AG, with GD = 1.6) experiences a linearly elastic pull-back, centered at some external point C with prescribed uniform motion : C is supposed to move along the line y = 2.7 with velocity 30 to the right. The pull-back force has its (x,y)-components proportional to the corresponding components of vector DC ; the respective stiffness coefficients are 200 for the x-component and 2000 for the y-component.

2° In addition, a constant torque C_r is exerted on $\mathcal{B}$. In the figured case, negative values of C_r tend to favor the contact of $\mathcal{B}$ with the wall.

Steady oscillations of the system, corresponding to three values of C_r are presented. The plotter draws, at constant time intervals, a line with origin A and directed to G. The unilateral constraint makes A lie in half-plane $y - b \geq 0$; this point belongs to the boundary line $y - b = 0$ whenever $\mathcal{B}$ keeps contact with the wall. Numbers refer to steps of the time-discretization algorithm.

Figure 6 is obtained for $C_r = -700$; discretization step length 0.002 ; position plotted every 4th step. Then $\mathcal{B}$ does not get loose from the wall. One observes angular oscillations ; also oscillations with longer period affect the space-lag from the pull-back center, as measured for instance by the variable $\ell = x(C) - x(G)$. The program might be used to draw additionally the two-dimensional trajectories concerning any couple among the 6 variables ℓ, Y, Θ, $\dot{\ell}$, $\dot{Y}$, $\dot{\Theta}$; this would display the typical pattern of a "doubly periodic" motion, revealing that the trajectories in $\mathbb{R}^6$ are coiled on some invariant torus. In fact, the present circumstances : persistent contact and slip velocity with constant direction make the motion verify as well the equations of smooth, bilateral dynamics. Therefore the system behaves like a classical smooth nonlinear oscillator.

Figure 7 corresponds to $C_r = -200$; step length 0.001 ; position plotted every 8th step. The motion, in variables ℓ , Y , Θ is now periodic. Each period consists of a contact-free episode, ending with a shock of $\mathcal{B}$ against the wall. The program is also able to display the slip velocity : its value following the shock is found zero ; the body gets loose immediately from the wall in a smooth way.

Figure 8 corresponds to $C_r = -100$; step-length 0.0005 ; position plotted every 15th step. The motion is again periodic but with some more complicated pattern. Every period exhibits two different episodes of contact-free motion. An episode of the first sort ends with a shock, followed by a nonzero time-interval with persistent contact and positive slip velocity. Afterwards the body $\mathcal{B}$ gets loose from the wall with zero normal velocity and performs the second sort of contact-free episode. This ends with a second shock, immediately followed by an episode of the first sort. The slip velocity resulting from the second shock is positive, but the normal velocity is zero, in agreement with our inelasticity assumption.

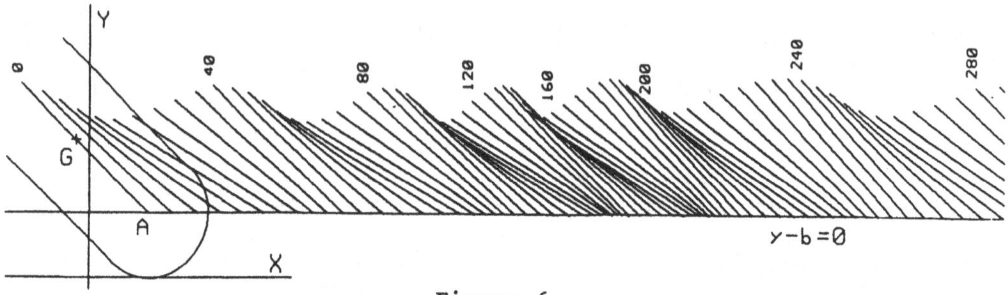

- Figure 6 -

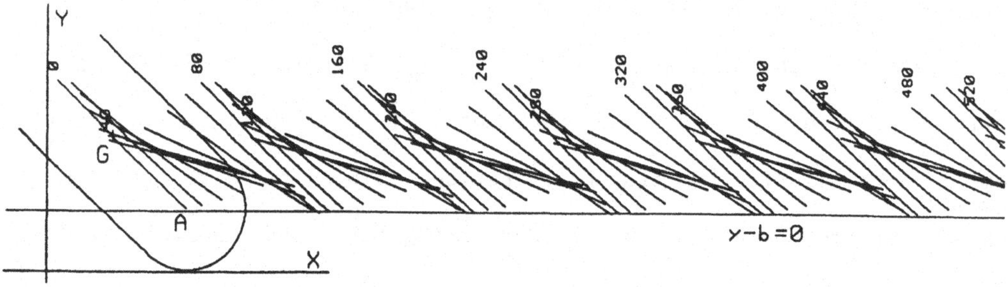

- Figure 7 -

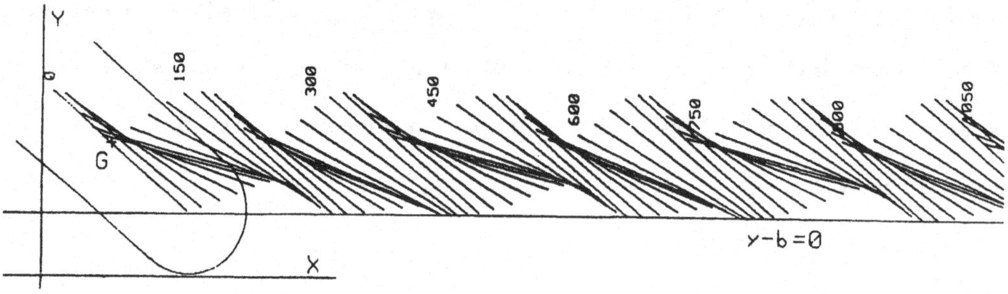

- Figure 8 -

REFERENCES

1. Moreau, J.J., Standard inelastic shocks and the dynamics of unilateral constraints, in : Unilateral Problems in Structural Analysis, CISM Courses and Lectures n° 288 (Ed. G. Del Piero and F. Maceri), Springer-Verlag, Wien-New York 1985, 173-221.
2. Delassus, E., Considérations sur le frottement de glissement, Nouv. Ann. de Mathématiques, 4ème Ser., 20 (1920), 485-496.
3. Pérès, J., Mécanique Générale, Masson, Paris, 1953.
4. Jean, M. and Pratt, E., A system of rigid bodies with dry friction, Int. J. Engng. Sci., 23 (1985), 497-513.
5. Moreau, J.J., Sur les lois de frottement, de plasticité et de viscosité, C.R. Acad. Sci. Paris, Sér. A, 271 (1970), 608-611.
6. Moreau, J.J., Application of convex analysis to some problems of dry friction, in : Trends in Applications of Pure Mathematics to Mechanics, vol. II (Ed. H. Zorski), Pitman, London-San Francisco-Melbourne 1979, 263-280.
7. Moreau, J.J., Dynamique de systèmes à liaisons unilatérales avec frottement sec éventuel ; essais numériques, Université des Sciences et Techniques du Languedoc, Montpellier, Lab. de Mécanique Générale des Milieux Continus, Note Technique 85-1, 1985.

8. Lötstedt, P., Coulomb friction in two-dimensional rigid body systems, Z. Angew. Math. u. Mech. 61 (1981), 605-615.

9. Lötstedt, P., Mechanical systems of rigid bodies subject to unilateral constraints, SIAM J. Appl. Math. 42(1982), 281-296.

10. Moreau, J.J., On unilateral constraints, friction and plasticity, in : New Variational Techniques in Mathematical Physics, CIME 2° ciclo 1973, (Ed. G. Capriz and G. Stampacchia), Edizioni Cremonese, Roma, 1974, 175-322.

11. Moreau, J.J., Mécanique classique, vol. 2 Masson, Paris, 1971.

12. Jean, M., Un algorithme numérique simple pour un système d'oscillateurs avec frottement de Coulomb sur un plan, Université des Sciences et Techniques du Languedoc, Lab. de Mécanique Générale des Milieux Continus, Note Technique 85-2, 1985.

13. Halphen, B. and Nguyen, Q.S., Sur les matériaux standards généralisés, J. de Mécanique, 14, 39, 1975.

14. Lax, P.D., Hyperbolic systems of conservation laws and the mathematical theory of shock waves, Regional Conf. Sér. in Appl. Math., n° 11, S.I.A.M., Philadelphia, 1973.

CONTACT PROBLEMS WITH FRICTION BY LINEAR COMPLEMENTARITY

A. Klarbring
Department of Mechanical Engineering
Institute of Technology of Linkoping

ABSTRACT

This study presents a new approach to the quasi-static contact problem with friction. It is assumed that the contacting bodies (structures) are linear elastic and that their displacement fields can be described by a finite number of variables. The approach is based on concepts from the discipline of Mathematical Programming, particularly those related to the Linear Complementarity Problem. It is pointed out that the problem studied may in some cases have multiple solutions, and that in other cases a solution may not even exist.

1. INTRODUCTION

The power of Mathematical Programming (MP) when dealing with certain types of discrete problems of structural mechanics has now been thoroughly established. Particularly, the knowledge of problems of elasto-plastic structural behaviour has considerably increased by their formulation as MP problems [1], [2]. Two major advantages of this possibility can be distinguished: firstly, MP solution algorithms with proven convergence and termination properties become available for the numerical treatment; secondly, MP plays an important theoretical role in that it unifies and provides a formalism through which important results on duality and on existence and uniqueness of solutions can be established.

Despite the fact that frictional systems often serve as models of plastic behaviour in elementary text-books and elsewhere, few attempts to treat contact problems with friction by MP methodology have been recorded in the literature [3] - [5]. These works have without exception treated problems where the constitutive model of surface friction is severely simplified. That is, the following assumptions are made: (1) the non-reversible character of frictional forces is assumed ignorable, which makes it possible to consider a static problem; (2) the distribution of normal contact forces is assumed to be a known function, which assures the existence of a potential for the governing equations. These assumptions are simply made in order to reduce the problem to a familiar one of constrained minimization; neither of them have any physical bearing. It has been suggested that assumption (2) can be removed by an iterative process where the above minimization problem is solved in each iteration [3]. However, the numerical realization of this procedure seems to be time consuming, and, at the same time, the physical insight gained in the theoretical treatment by fixed point theory is small.

In the present paper a novel MP approach to contact problems with friction, without assumptions (1) and (2), is introduced. It is a development of ideas previously explored by the author in refs. [6] - [8], where the restriction was made that the contact surface should be constant during the loading history. Here this restriction is removed and unilateral normal contact conditions as well as unilateral frictional conditions are present. However, the structure is still assumed to be linear elastic and a piecewise linear friction law, approximating Coulomb's law of friction, is used. In case of solid bodies in contact with each other, the finite dimensional structure assumed here can be obtained by a finite element discretization along the lines suggested in ref. [8].

In Section 2 of this paper the contact and friction conditions are introduced and, since a quasi-static problem is considered, their influence on incremental quantities is investigated. In Section 3 the contact boundary conditions are combined with a matrix stiffness equation which describes the linear elastic behaviour of the structure involved. The resulting mathematical system can be characterized as a new type of parametric Linear Complementarity Problem (LCP) [9], where complementarity conditions take a non-traditional form. That is, variables related to the normal contact displacements and forces should satisfy the usual complementarity condition while those related to the tangential ones should satisfy the type of complementarity condition present in the parametric LCP involving derivatives [10]. The piecewise linear structure of the problem implies that at least some of the solutions possible are piecewise linearly dependent on time (or some equivalent parameter). For every such linear part an incremental LCP is solved to determine the slope of the linear solution. In Section 4 this incremental LCP is investigated more closely and it is made clear that it is a problem for which we cannot always expect a unique solution or even the existence of solutions. This rather intriguing result is amplified in Section 5 where a very simple but still sufficiently rich example is investigated on this point.

2. CONTACT CONDITIONS

Consider a structure whose possible configurations can be described by a finite number of displacement variables. The "contact boundary" of the structure consists of k contact nodes, each one associated with a displacement vector $w^j \in R^3$. For each contact node we assume that an outward normal direction of the contact surface may be determined and that the contact displacement vector is referred to a local coordinate system such that normal and tangential subvectors, $w_N^j \in R^1$ and $w_T^j \in R^2$, define w^j. Dual vectors of forces P^j, P_N^j and P_T^j are similarly defined. Thus, if superscript t denotes transposition,

$$w^j = [w_T^j, w_N^j]^t, \quad P^j = [P_T^j, P_N^j]^t . \qquad (2.1),(2.2)$$

The unilateral normal conditions to be considered are, for each contact node

$$y^j = g_N^j - w_N^j, \quad y^j \geq 0, \quad P_N^j \leq 0, \quad y^j P_N^j = 0 , \qquad (2.3)$$

where g_N^j may be regarded as a normal gap between a unilateral rigid support and the contact node.

The relation between tangential force and displacement variables at a contact node will be given by a law of friction. On the basis of the developments of Moreau, Panagiotopoulos and Curnier, fairly general such laws were given in ref. [11]. If $(\dot{})$ denotes differentiation with respect to time, and ∂ subdifferentiation with respect to $-P_T^j$ we have

$$\dot{w}_T^j \in \partial J^j(-P_T^j; P_N^j). \qquad (2.4)$$

Here $J^j(\cdot; P_N^j)$ is, for each P_N^j, a lower-semicontinuous but generally non-convex and non-differentiable real valued function, which is termed a superpotential [11]. In this paper it will be assumed that $J^j(\cdot; P_N^j)$ is the indicator function of the closed set

$$C^j(P_N^j) = \{- P_T^j \in R^2 | \phi_\alpha^j(- P_T^j , P_N^j) \leq 0, \quad \alpha = 1, \ldots, n\} , \qquad (2.5)$$

where ϕ_α^j, $\alpha = 1, \ldots, n$, are affine functions of $- P_T^j$ and P_N^j. The subdifferential relation (2.4) then indicates that the rate or increment $\dot{w}_T^j$

belongs to a normal cone of $C^j(P_N^j)$ and it is easily seen, [11], that (2.4) is equivalent to

$$\dot{w}_{Ti}^j = \sum_{\alpha=1}^{n} \dot{\lambda}_\alpha^j \frac{\partial \phi_\alpha^j}{\partial(-P_{Ti}^j)}, \quad i = 1, 2, \tag{2.6}$$

$$\dot{\lambda}_\alpha^j \geq 0, \; \phi_\alpha^j(-P_T^j, P_N^j) \leq 0, \; \dot{\lambda}_\alpha^j \phi_\alpha^j(-P_T^j, P_N^j) = 0, \quad \alpha = 1, \ldots, n. \tag{2.7}$$

Here subindex i denotes components of a vector of R^2 and non-negative multipliers λ_α^j, associated with ϕ_α^j, have been introduced.

Furthermore, since ϕ_α^j is affine, the partial derivatives of ϕ_α^j are independent of $-P_T^j$ and P_N^j. Therefore, relation (2.6) can be integrated to yield

$$w_{Ti}^j = \sum_{\alpha=1}^{n} \lambda_\alpha^j \frac{\partial \phi_\alpha^j}{\partial(-P_{Ti}^j)} + g_{Ti}^j, \quad i = 1, 2, \tag{2.8}$$

where g_{Ti}^j, $i = 1, 2$, are constants depending on the initial conditions. Also, it must be

$$\phi_\alpha^j(-P_T^j, P_N^j) = -\sum_{i=1}^{2} \frac{\partial \phi_\alpha^j}{\partial(-P_{Ti}^j)} P_{Ti}^j + \frac{\partial \phi_\alpha^j}{\partial P_N^j} P_N^j - k_\alpha^j, \tag{2.9}$$

where k_α^j, $\alpha = 1, \ldots, n$ are constants which indicate the orthogonal distances between the origins $P^j = 0$ of R^3 spaces, and the affine manifolds $\phi_\alpha^j(-P_T^j, P_N^j) = 0$.

It is easily seen that, if we restrict ourselves to a twodimensional situation, the classical friction law of Coulomb is a special case of the general friction law described by (2.7)-(2.9); see Fig. 2.1a. In this case we have

$$\phi_1^j = P_T^j + \mu P_N^j, \quad \phi_2^j = -P_T^j + \mu P_N^j,$$

where μ is the coefficient of friction.

In a threedimensional situation, (2.4) represents Coulomb's law of friction if $J^j(\cdot;P_N^j)$ is the indicator function of the closed set

$$C^j(P_N^j) = \{-P_T^j \in R^2 \,\|\, P_T^j| + \mu P_N^j \leq 0\},$$

where $P_N^j \leq 0$. This set may be approximated by a polytope such as (2.5), as indicated in Fig 2.1b.

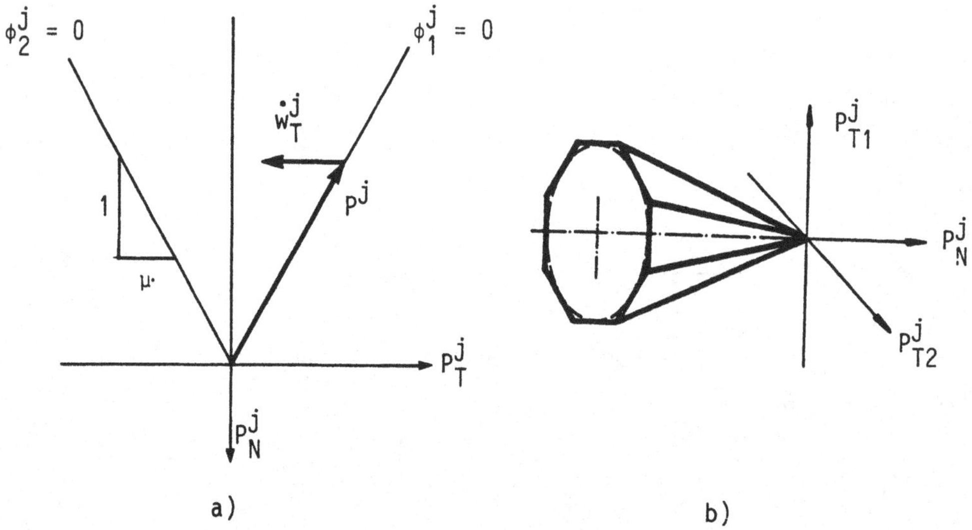

Fig. 2.1, Geometric illustration of piecewise linear Coulomb friction in two and three dimensions

A notable feature of the contact and frictional relations is given by the complementarity conditions (2.3) and (2.7). They indicate the choices between the physical states of contact and no contact, on one hand, and, on the other hand, slip and adhesion, i. e., nonzero or zero tangential displacement rate.

Of interest in the sequel will be the constraints that the complementarity conditions represent on solutions that are linearly dependent on time:

$$P_N^j(\tau) = P_{No}^j + \dot{P}_N^j\tau, \quad y^j(\tau) = y_o^j + \dot{y}^j\tau,$$

$$\phi_\alpha^j(\tau) = \phi_{\alpha o}^j + \dot{\phi}_\alpha^j\tau, \quad \lambda_\alpha^j(\tau) = \lambda_{\alpha o}^j + \dot{\lambda}_\alpha^j\tau,$$

with $\tau \in [0, T]$, $T > 0$ and where subindex "o" denotes prescribed values. We assume that T is sufficiently small so that an inequality valid at $\tau = 0$ will also be valid at $\tau = T$.

Consider first the complementarity conditions of (2.3). With respect to the above we can single out three different cases:

(i) if $y_o^j > 0$ and $P_{No}^j = 0$ then $\dot{y}^j \gtrless 0$, $\dot{P}_N^j = 0$;

(ii) if $P_{No}^j < 0$ and $y_o^j = 0$ then $\dot{y}^j = 0$, $\dot{P}_N^j \gtrless 0$;

(iii) if $P_{No}^j = 0$ and $\dot{y}_o^j = 0$ then $\dot{y}^j \geq 0$, $\dot{P}_N^j \leq 0$, $\dot{y}^j\dot{P}_N^j = 0$.

Concerning the complementarity conditions (2.7), we obtain the following conditions:

(iv) if $\phi_{\alpha o}^j < 0$ then $\dot{\lambda}^j = 0$, $\dot{\phi}^j \gtrless 0$;

(v) if $\phi_{\alpha o}^j = 0$ then $\dot{\lambda}_\alpha^j \geq 0$, $\dot{\phi}_\alpha^j \leq 0$, $\dot{\lambda}_\alpha^j\dot{\phi}_\alpha^j = 0$.

Note that these conditions are valid for the right-hand derivatives at $\tau = 0$ of more general smooth functions than the linear ones.

3. THE BASIC PROBLEM

In order to formulate our problem in a compact matrix form we assemble the vectors defined for each contact node as column vectors w_N, P_N, g_N, $y \in R^k$, w_T, P_T, $g_T \in R^{2k}$ and ϕ, λ, $k \in R^{nk}$, so that, e. g. the ith entry of w_N is $w_N^i \in R^1$.

Furthermore, for each contact node we define matrices of partial derivatives, $G_T^j \in R^{2\times n}$ and $G_N^j \in R^{1\times n}$, with elements

$$[G_T^j]_{i\alpha} = \frac{\partial \phi_\alpha^j}{\partial(-P_{Ti}^j)} \quad , [G_N^j]_\alpha = \frac{\partial \phi_\alpha^j}{\partial P_N^j} \quad .$$

These matrices are assembled in order to describe the contact conditions for all contact nodes simultaneously.

$$G_T = \text{diag}\,[G_T^1, \ldots, G_T^j, \ldots, G_T^k], \quad G_N = \text{diag}\,[G_N^1, \ldots, G_N^j, \ldots, G_T^k].$$

The normal contact conditions and the friction law are then written as

$$\begin{bmatrix} w_T \\ \hline w_N \end{bmatrix} = \left[\begin{array}{c|c} G_T & 0 \\ \hline 0 & -I \end{array}\right] \begin{bmatrix} \lambda \\ \hline y \end{bmatrix} + \begin{bmatrix} g_T \\ \hline g_N \end{bmatrix} , \tag{3.1}$$

$$\begin{bmatrix} \phi \\ \hline P_N \end{bmatrix} = \left[\begin{array}{c|c} -G_T^t & G_N^t \\ \hline 0 & I \end{array}\right] \begin{bmatrix} P_T \\ \hline P_N \end{bmatrix} - \begin{bmatrix} k \\ \hline 0 \end{bmatrix} , \tag{3.2}$$

$$y \leq 0, \; P_N \leq 0, \; y^t P_N = 0, \tag{3.3}$$

$$\dot\lambda \geq 0, \; \phi \leq 0, \; \dot\lambda^t \phi = 0. \tag{3.4}$$

Now turn to the linear elastic behaviour of the structure involved. In refs. [8], [11] and [12] the following relation was shown to be valid:

$$P = R + \kappa w, \tag{3.5}$$

where κ is a positive semi-definite stiffness matrix and R is a column vector, representing the actions of prescribed forces and displacements;

P and w are contact force and displacement vectors, which may be decomposed as $w = [w_T^t, w_N^t]^t$ and $P = [P_T^t, P_N^t]^t$. Equation (3.5) was derived in ref. [12] by using a force method. It was shown to hold if a fictitious structure, created by "welding" the contact surface so that w can be considered as prescribed, is in equilibrium under external loads. In refs. [8] and [11] it was shown that if the aforementioned fictitious structure is not a mechanism, (3.5) can be obtained by using a displacement method combined with a substructure technique.

Now introduce notations

$$N = \begin{bmatrix} -G_T & 0 \\ G_N & I \end{bmatrix}, \qquad \overset{o}{N} = \begin{bmatrix} G_T & 0 \\ 0 & -I \end{bmatrix},$$

$$G = \begin{bmatrix} g_T \\ g_N \end{bmatrix}, \qquad K = \begin{bmatrix} k \\ 0 \end{bmatrix}, \qquad \Phi = \begin{bmatrix} \phi \\ P_N \end{bmatrix}, \qquad \Lambda = \begin{bmatrix} \lambda \\ y \end{bmatrix}.$$

Combining (3.1), (3.2) and (3.5), one obtains

$$\Phi = N^t R - K + N^t \kappa G + N^t \kappa \overset{o}{N} \Lambda. \tag{3.6}$$

Furthermore, consider a proportional loading during a time interval [0, T]. At t = 0 the load, R, and multipliers, Λ, are known and equal to $\bar{R}$ and $\bar{\Lambda}$. The proportional loading is described by monotonically increasing multiplier ξ and a vector of additional loads R'. That is,

$$R(\tau) = \bar{R} + \xi(\tau)R', \quad \xi(0) = 0, \quad \dot{\xi}(\tau) > 0, \quad \forall\, \tau \in [0, T]. \tag{3.7}$$

The change of the multipliers during the loading path can be described by an incremental vector of multipliers $\Delta\Lambda$, so that

$$\Lambda(\tau) = \bar{\Lambda} + \Delta\Lambda(\tau), \ \tau \in [0, T]. \tag{3.8}$$

From (3.3), (3.4) and by substitution of (3.7) and (3.8) into (3.6) one obtains

$$\Phi = \bar{\Phi} + \xi N^t R' + N^t \kappa \overset{\circ}{N} \Delta\Lambda, \qquad (3.9)$$

$$y \geq 0, \; P_N \leq 0, \; y^t P_N = 0, \qquad (3.10)$$

$$\dot{\lambda} \geq 0, \; \phi \leq 0, \; \dot{\lambda}^t \phi = 0, \qquad (3.11)$$

where $\bar{\Phi} = N^t \bar{R} - K + N^t \kappa G + N^t \kappa \overset{\circ}{N} \bar{\Lambda}$.

Time dependent vectors Φ and $\Delta\Lambda$, satisfying (3.9)-(3.11) may be considered as a solution of the contact problem with friction. If only normal degrees of freedom are present, i. e. $P = P_N$, then (3.11) is omitted from the above problem, and (3.9) and (3.10) represent a Parametric Linear Complementarity Problem (PLCP) [9], which may be considered as a mathematical description of a frictionless contact problem. If, on the other hand, it is assumed that $P_N < 0$, and, consequently, that $y = 0$ during the whole loading path, then (3.10) can be omitted and (3.9) and (3.11) represent what has been called a PLCP involving derivatives. A similar problem arises in connection with the analysis of elasto-plastic structures and with this application in mind the PLCP involving derivatives was mathematically studied by Kaneko [10], who also proposed a solution algorithm for it. In the case when $P_N < 0$, this algorithm was applied to the present frictional problem in refs [7] and [8].

It should now be noted that for conditions (3.11) to be meaningful a certain amount of regularity of the solution has to be required. Before stating this more precisely we notice that, due to the requirement $\dot{\xi} > 0$, we may consider Φ and $\Delta\Lambda$ to be, instead of functions of τ, functions of $\xi \in [0, \Xi]$ for some $\Xi > 0$. $(\dot{\ })$ then denotes the derivative with respect to ξ.

A solution of (3.9)-(3.11) is now defined to be a map $(\Phi, \Delta\Lambda)$: $[0, \Xi] \to R^{(n+1)k} \times R^{(n+1)k}$, which satisfies the following conditions:

(i) $\Delta\Lambda(0) = 0$;

(ii) it is continuous everywhere on $[0, \Xi]$;

(iii) it is piecewise affine so that for a number (possibly ∞) of consecutive finite intervals (ξ_1, ξ_2), $\xi_2 > \xi_1$, it holds for each one that

$$\Phi(\xi) = \Phi_o + \dot{\Phi}(\xi - \xi_1), \tag{3.12}$$

$$\Delta\Lambda(\xi) = \Lambda_o + \dot{\Lambda}(\xi - \xi_1), \tag{3.13}$$

with Φ_o, $\dot{\Phi}$, Λ_o and $\dot{\Lambda}$ constant in the interval.

Notice that the number, length and location on the real line of the intervals are unknown and a part of the solution. Assume that the solution map is known up to a value $\xi_o \in [0, \Xi)$ which we know to be a critical point, i. e. a break point between two finite intervals. If (3.12) and (3.13) are valid for the successive finite interval, due to (ii), Φ_o and Λ_o obviously equal the values $\Phi(\xi_o)$ and $\Delta\Lambda(\xi_o)$. Regarding the rates $\dot{\Phi}$ and $\dot{\Lambda}$, substitution of (3.12) and (3.13) in (3.9) results in the rate equation

$$\dot{\Phi} = N^tR' + N^t\kappa\mathring{N}\dot{\Lambda}. \tag{3.14}$$

Besides this equation the rates should satisfy conditions (i)-(v) at the end of Section 2. To study this, subscripts A, B and C are introduced to denote the disjoint subsets of the index set $\{1, \ldots, k\}$, such that conditions (i), (ii) and (iii) of Section 2 are satisfied respectively. In the same way D and E denote the disjoint subsets of the set of products $\{(1, 1), \ldots, (1, \alpha), \ldots, (1, n), \ldots, (j, \alpha), \ldots, (k, n)\}$ such that conditions (iv) and (v) of Section 2 are satisfied.

If zeros are introduced in the appropriate places, and if $q = N^tR'$ and $M = N^t\kappa\mathring{N}$, equation (3.14) can be written as†

† If a vector x is of order n, $I \subset \{1, \ldots, n\}$, then x_I represents the subvector of x consisting of all elements x_i for $i \in I$; if a matrix A is of order n×m, $I \subset \{1, \ldots, n\}$ and $J \subset \{1, \ldots, m\}$, then $A_{I\cdot}$ represents the submatrix of A consisting of all the rows $A_{i\cdot}$ for $i \in I$, while $A_{\cdot J}$ represents the submatrix of A consisting of all the columns $A_{\cdot j}$ for $j \in I$; finally $A_{IJ} = (A_{I\cdot})_{\cdot J}$.

$$\begin{bmatrix} 0 \\ \dot{\Phi}_B \\ \dot{\Phi}_C \\ \dot{\Phi}_D \\ \dot{\Phi}_E \end{bmatrix} = \begin{bmatrix} q_A \\ q_B \\ q_C \\ q_D \\ q_E \end{bmatrix} + \begin{bmatrix} M_{AA} & M_{AB} & M_{AC} & M_{AD} & M_{AE} \\ M_{BA} & M_{BB} & M_{BC} & M_{BD} & M_{BE} \\ M_{CA} & M_{CB} & M_{CC} & M_{CD} & M_{CE} \\ M_{DA} & M_{DB} & M_{DC} & M_{DD} & M_{DE} \\ M_{EA} & M_{EB} & M_{EC} & M_{ED} & M_{EE} \end{bmatrix} \begin{bmatrix} \dot{\Lambda}_A \\ 0 \\ \dot{\Lambda}_C \\ 0 \\ \dot{\Lambda}_E \end{bmatrix} . \tag{3.15}$$

In order to identify a Linear Complementary Problem (LCP) for components C and E one should now transfer the zero components on the left hand side of (3.15) to the right hand side by means of a block pivot on $M_{AA} = - \kappa_{AA}$, which is non-singular for most practical situations. Then, rows and columns C and E of the pivot transformation, together with complementarity conditions (iii) and (v) of Section 2, constitute an LCP, whose solution (if it exists) determines the rates $\dot{\Phi}_C$, $\dot{\Phi}_E$, $\dot{\Lambda}_C$ and $\dot{\Lambda}_E$. This problem will be more thoroughly discussed in the next section. Here it should, however, be noted that it is a problem which in extreme cases may lack a solution and in other cases, when a solution exists, it need not be unique. Nevertheless, if it has a unique solution, all rates, $\dot{\Phi}$ and $\dot{\Lambda}$, are determined by (3.15).

Once (3.12) and (3.13) are determined for an interval, the end of the interval, ξ_2, may be obtained by direct substitution of these relations in the inequalities (3.10a), (3.10b) and (3.11b). When any component of $\Phi(\xi)$ and/or $\Delta\Lambda(\xi)$ violates a constraint a new interval and a new linear solution has to be considered.

If an LCP with a unique solution is defined for all critical points ξ_0, determined by considering inequalities as just indicated, it is likely that the solution map thus obtained is the unique one satisfying (i), (ii) and (iii). This is the case for the particular PLCPs involving derivatives considered in refs. [8] and [10]. However, regardless of the properties of the LCP, discontinuous solutions or non-piecewise linear solutions satisfying (3.9)-(3.11) may possibly exist besides a solution map as defined here. See Kaneko [10] for an example of this in the case of the LCP involving derivatives.

4. THE INCREMENTAL PROBLEM

As discussed in the previous section, a crucial part of the analysis of the contact-friction problem considered consists of the solution of the incremental LCP

$$\begin{bmatrix} \dot{\Phi}_C \\ \dot{\Phi}_E \end{bmatrix} = \begin{bmatrix} \bar{q}_C \\ \bar{q}_E \end{bmatrix} + \begin{bmatrix} \bar{M}_{CC} & \bar{M}_{CE} \\ \bar{M}_{EC} & \bar{M}_{EE} \end{bmatrix} \begin{bmatrix} \dot{\Lambda}_C \\ \dot{\Lambda}_E \end{bmatrix}, \tag{4.1}$$

$$\dot{\Phi}_C \leq 0, \ \dot{\Lambda}_C \geq 0, \ \dot{\Phi}_C^t \dot{\Lambda}_C = 0, \tag{4.2}$$

$$\dot{\Phi}_E \leq 0, \ \dot{\Lambda}_E \geq 0, \ \dot{\Phi}_E^t \dot{\Lambda}_E = 0, \tag{4.3}$$

where bars denote matrix blocks obtained when performing a block pivot on M_{AA} in (3.15). With obvious changes in notations we write (4.1)-(4.3) as

$$\dot{\bar{\Phi}} = \bar{q} + \bar{M}\dot{\bar{\Lambda}} \leq 0, \quad \dot{\bar{\Lambda}} \geq 0, \quad \dot{\bar{\Phi}}^t \dot{\bar{\Lambda}} = 0. \tag{4.4}$$

Since $\bar{M}$ is generally a non-symmetric matrix this LCP cannot be identified with the set of the Kuhn-Tucker conditions of a Quadratic Program, as had otherwise been possible. Nevertheless, the following property holds [13], [14]:

Minimum property: If (4.4) has a solution $(\dot{\bar{\Phi}}, \dot{\bar{\Lambda}})$, this solution also solves

$$\min\{- \dot{\bar{\Phi}}^t \dot{\bar{\Lambda}} \mid \dot{\bar{\Phi}} = \bar{q} + \bar{M}\dot{\bar{\Lambda}}, \ \dot{\bar{\Phi}} \leq 0, \ \dot{\bar{\Lambda}} \geq 0\}. \tag{4.5}$$

Conversely, if the minimum of problem (4.5) is zero, any optimal solution of this problem also solves (4.4). If the minimum (4.5) is not equal to zero, (4.4) has no solution.

For any $(\dot{\bar{\Phi}}, \dot{\bar{\Lambda}})$ satisfying the constraints of (4.5), all the relations of (4.4), except the orthogonality condition $\dot{\bar{\Phi}}^t\dot{\bar{\Lambda}} = 0$, are satisfied. Therefore, the optimization process is equivalent to explicitly imposing the orthogonality condition.

Furthermore, it is interesting to note that, at least in principle, the above miminum property is similar to a minimum principle of Coulomb's friction law, introduced and applied to rolling contact by Kalker [15].

It is easy to verify that $\bar{M}$ is generally an indefinite matrix and consequently (4.5) is a non-convex quadratic program. Therefore, it can be concluded that (4.4) needs not have a unique solution. It is also clear that even for a non-singular $\bar{M}$, (4.5) may lack a solution since the constraint set could be empty.

However, the following classical theorem, the proof of which can be found in ref. [14], makes it possible to decide if a unique solution exists.

Theorem: The LCP (4.4) has a unique solution $(\dot{\bar{\Phi}}, \dot{\bar{\Lambda}})$ for each $\bar{q}$ if and only if $-\bar{M}$ is a P-matrix.†

This theorem indicates that a solution $(\dot{\bar{\Phi}}, \dot{\bar{\Lambda}})$ of (4.4) is unique for any loading direction R' if $-\bar{M}$ is a P-matrix. The converse, however, does generally not hold since it is not certain that all $\bar{q}$ can be generated by some R'.

If some of the components of $\dot{\bar{\Lambda}}$ correspond to linearly dependent columns of $\overset{o}{N}$, -M cannot be a P-matrix since it is singular. Nevertheless, a meaningful partial uniqueness of solutions of the LCP (4.4) can be expected if $-\bar{M}^t$ is row adequate. Eaves [13] gives the following definition:

† A P-matrix is a matrix for which all principal minors are positive [14].

A square n by n matrix B is *row adequate* if it is a P_0-matrix and for every $I \subset \{1, \ldots, n\}$, $\det B_{II} = 0$ implies that the rows of $B_{I\cdot}$ are linearly dependent.

Here a P_0-matrix is a matrix for which all principal minors are non-negative.

To give a situation which implies that $-\bar{M}^t$ is row adequate we introduce the index set $I = C \cup E$. Some matrix algebra shows that

$$\bar{M} = (N_{\cdot I})^t(\kappa - \kappa_{\cdot A}\kappa_{AA}^{-1}\kappa_{A\cdot})\mathring{N}_{\cdot I} \ .$$

Assume that $-M_{JJ}$, $J \subset I$, is a P-matrix for all index sets J such that $\mathring{N}_{\cdot J}$ has full column rank, which is a sufficient condition for $-\bar{M}$ to be a P_0-matrix. In this case, singularity of $\bar{M}_{JJ}$ for some J implies that $\mathring{N}_{\cdot J}$ and, consequently, also $(N_{\cdot I})^t(\kappa - \kappa_{\cdot A}\kappa_{AA}^{-1}\kappa_{A\cdot})N_{\cdot J}$ has linearly dependent columns. According to the above definition $-\bar{M}^t$ is therefore row adequate.

Assume that (4.4) has two solutions, $(\dot{\bar{\Phi}}^1, \dot{\bar{\lambda}}^1)$ and $(\dot{\bar{\Phi}}^2, \dot{\bar{\lambda}}^2)$. It then holds that

$$\dot{\bar{\Phi}}^1 - \dot{\bar{\Phi}}^2 = \bar{M}(\dot{\bar{\Lambda}}^1 - \dot{\bar{\Lambda}}^2). \tag{4.6}$$

Furthermore, for components we have

$$(\dot{\bar{\Phi}}_i^1 - \dot{\bar{\Phi}}_i^2)(\dot{\bar{\Lambda}}_i^1 - \dot{\bar{\Lambda}}_i^2) = -\dot{\bar{\Phi}}_i^1\dot{\bar{\Lambda}}_i^2 - \dot{\bar{\Phi}}_i^2\dot{\bar{\Lambda}}_i^1 \geq 0, \quad \forall\ i \in I \ . \tag{4.7}$$

The following characterization of row adequate matrices is now useful [13]:

Theorem: A matrix B is row adequate if and only if for any vector x, $(B^t x)_i x_i \leq 0$ for $i = 1, \ldots, n$ implies $B^t x = 0$.

It follows from this theorem, (4.6), (4.7) and an assumption that $-\bar{M}^t$ is row adequate, that $\dot{\bar{\Phi}}^1 - \dot{\bar{\Phi}}^2 = 0$, which establishes the uniqueness of the

$\dot{\Phi}$-part of the solution. Regarding the $\dot{\Lambda}$-part, the theory of LCPs tells us that it is unique if it makes up part of a nondegenerate solution [13]. However, even if this is not the case, it may be shown that the contact displacement rate vector is uniquely defined. As a matter of fact,

$$(N_{\cdot I})^t \bar{\kappa} y = \dot{\Phi} - \bar{q} = z, \tag{4.8}$$

where z is known to be unique and where $\bar{\kappa} = (\kappa - \kappa_{\cdot A}\kappa_{AA}^{-1}\kappa_{A\cdot})$. Consequently, y is unique if $(N_{\cdot I})^t \bar{\kappa}$ has full column rank, which holds if $\bar{\kappa}$ is non-singular and $(N_{\cdot I})^t$ has full column rank. Furthermore, by the first row of (3.15) we have

$$0 = q_A - \kappa_{AA}\dot{\Lambda}_A + (N_{\cdot A})^t \kappa y,$$

which shows that $\dot{\Lambda}_A$ is also uniquely determinable. The contact displacement rate vector then follows from (3.1) as

$$\dot{w} = \overset{o}{N}_{\cdot A}\dot{\Lambda}_A + y. \tag{4.9}$$

Finally, we point out that, besides the result presented here, other conclusions concerning the qualitative behaviour of frictional problems can be shown to follow from the theory of LCPs. For instance, it can be decided whether the incremental LCP (4.4) has a finite or infinite number of solutions [14]. Such a conclusion is of physical interest since the situation of an infinite number of solutions might correspond to a limit state of gross slip of the frictional system.

5. COULOMB'S FRICTION LAW. AN ILLUSTRATIVE EXAMPLE

The friction law considered in the bulk of the development of this paper is more general than the two examples of Coulomb's friction given in Section 2. However, what should be especially noted about these special cases is that non-singularity of $\bar{M}$ cannot be expected when there is a contact node j for which $P_N^j = 0$, since columns of $\overset{o}{N}_{\cdot I}$ corresponding to $\phi_\alpha^j = 0$, are then linearly dependent. The discussion concerning row adequacy of Section 4 should therefore be applied when uniqueness of solutions is investigated.

In this context it may also be said that for sufficiently small friction coefficients uniqueness of solutions is likely to be expected. Indeed, if μ approaches zero, N approaches $\overset{o}{N}$ and, therefore, $-N^t\kappa\overset{o}{N}$ becomes a symmetric positive semi-definite matrix. The LCP (4.4) can then be considered to be the Kuhn-Tucker conditions of a quadratic program for which a solution, representing a unique displacement rate, always exists if κ is positive definite.

In order to illustrate the theory presented in the previous sections a very simple example will be considered. The aim is also to provide a counterexample that shows the possibility of multiple solutions in some cases and lack of solutions in other cases.

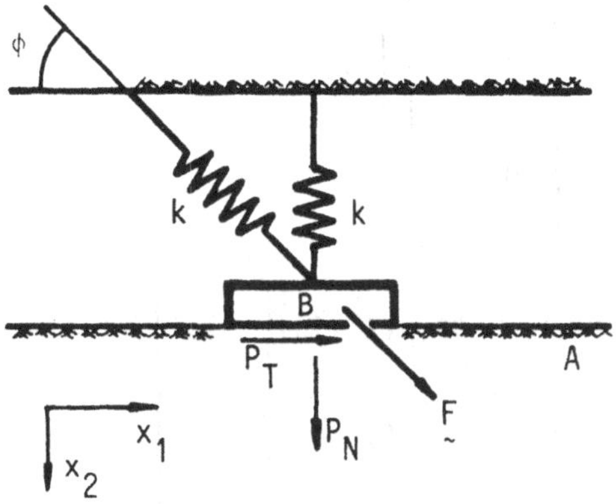

Fig. 5.1, A one-point contact problem

Consider a one-point contact problem as shown in Fig. 5.1. Introducing the notations $s = \sin\phi$ and $c = \cos\phi$ one verifies that, if k is the spring constant,

$$\kappa = k\begin{bmatrix} c^2 & cs \\ cs & s^2+1 \end{bmatrix}, \quad R = -\begin{bmatrix} F_1 \\ F_2 \end{bmatrix}.$$

Furthermore, if a twodimensional Coulomb's friction law is assumed we have

$$N = \left[\begin{array}{cc|c} 1 & -1 & 0 \\ \hline \mu & \mu & 1 \end{array}\right], \quad \overset{\circ}{N} = \left[\begin{array}{cc|c} -1 & 1 & 0 \\ \hline 0 & 0 & -1 \end{array}\right].$$

Since K and G are zero vectors in this case (3.9)-(3.11) take the following form:

$$\begin{bmatrix} \phi_1 \\ \phi_2 \\ P_N \end{bmatrix} = \begin{bmatrix} \bar{\Phi}_1 \\ \bar{\Phi}_2 \\ \bar{\Phi}_3 \end{bmatrix} + \xi \begin{bmatrix} 1 & \mu \\ -1 & \mu \\ 0 & 1 \end{bmatrix} \begin{bmatrix} R_1' \\ R_2' \end{bmatrix} -$$

$$k\begin{bmatrix} c^2+\mu cs & -c^2-\mu cs & cs+\mu(s^2+1) \\ -c^2+\mu cs & c^2-\mu cs & -cs+\mu(s^2+1) \\ cs & -cs & s^2+1 \end{bmatrix} \begin{bmatrix} \lambda_1 \\ \lambda_2 \\ -w_N \end{bmatrix} \tag{5.1}$$

$$w_N \leq 0, \quad P_N \leq 0, \quad w_N P_N = 0, \tag{5.2}$$

$$\dot{\lambda}_1 \geq 0, \quad \phi_1 \leq 0, \quad \dot{\lambda}_1 \phi_1 = 0, \tag{5.3}$$

$$\dot{\lambda}_2 \geq 0, \quad \phi_2 \leq 0, \quad \dot{\lambda}_2 \phi_2 = 0. \tag{5.4}$$

The incremental problem that arises if we consider a situation where the point B is resting on the foundation A, in a way such that the contact force is zero, will now be considered. This equilibrium state is obtained when spring forces and external forces balance each other. Since M_{AA} vanishes in this case the LCP (4.4) becomes identical with (5.1)-(5.4) when $\bar{\Phi}$ and ξ are removed and all variables are dotted. It is obvious that $-\bar{M}$ of this problem is not a P-matrix. However, it is easily verified that its transpose is row adequate if $c^2 + \mu cs$ and $c^2 - \mu cs$ are both positive, and uniqueness according to the discussion at the end of Section 4 can therefore be expected. To investigate a special case when these conditions are not satisfied we assume $\phi = 45^0$, $R_1' = -4F$ and $R_2' = -3F$. We then have the incremental LCP

$$\begin{bmatrix} \dot{\phi}_1 \\ \dot{\phi}_2 \\ \dot{P}_N \end{bmatrix} = -F \begin{bmatrix} 3\mu+4 \\ 3\mu-4 \\ 3 \end{bmatrix} - \frac{k}{2} \begin{bmatrix} 1+\mu & -1-\mu & 1+3\mu \\ -1+\mu & 1-\mu & -1+3\mu \\ 1 & -1 & 3 \end{bmatrix} \begin{bmatrix} \dot{\lambda}_1 \\ \dot{\lambda}_2 \\ -\dot{w}_N \end{bmatrix}, \qquad (5.5)$$

$$\dot{w}_N \leq 0, \; \dot{P}_N \leq 0, \; \dot{w}_N \dot{P}_N = 0, \qquad (5.6)$$

$$\dot{\lambda}_1 \geq 0, \; \dot{\phi}_1 \leq 0, \; \dot{\lambda}_1 \dot{\phi}_1 = 0, \qquad (5.7)$$

$$\dot{\lambda}_2 \geq 0, \; \dot{\phi}_2 \leq 0, \; \dot{\lambda}_2 \dot{\phi}_2 = 0. \qquad (5.8)$$

If μ is larger than 4/3 it may be seen that this problem has essentially three different solutions representing the physical states of adhesion, separation and slip:

I. Solution representing adhesion

$$\dot{\bar{\Phi}}^t = -F[3\mu + 4,\; 3\mu - 4,\; 3], \quad \dot{\bar{\Lambda}}^t = [0,\; 0,\; 0].$$

II. Solution representing separation

$$\dot{\Phi}^t = [0, 0, 0], \quad \dot{\Lambda}^t = [t, \frac{F9}{k} + t, \frac{F}{k}], \quad t > 0.$$

III. Solution representing slip

$$\dot{\Phi}^t = F[\frac{2\mu}{(1-\mu)}, 0, \frac{1}{(1-\mu)}], \quad \dot{\Lambda}^t = \frac{F}{k}[0, \frac{8-6\mu}{(1-\mu)}, 0].$$

The ray of solutions according to II represents the same displacement rate and should therefore be regarded as one single solution.

Let us now investigate the question of existence of a solution. Consider a situation where $\phi_1 < 0$, $\phi_2 = 0$ and $P_N < 0$; i. e. referring to Fig. 5.1, right hand slip is just about to take place. The incremental LCP (4.4) then becomes

$$\dot{\phi}_2 = - R_1' + \mu R_2' - (c^2 - \mu cs)\dot{\lambda}_2, \tag{5.9}$$

$$\dot{\phi}_2 \leq 0, \; \dot{\lambda}_2 \geq 0, \; \dot{\phi}_2 \dot{\lambda}_2 = 0. \tag{5.10}$$

If $c^2 - \mu cs$ is positive $\bar{M}$ is a P-matrix and this problem has a unique solution for all R_1' and R_2'. However, if μ is sufficiently large to make $c^2 - \mu cs$ negative it is obvious that if $- R_1' + \mu R_2'$ is positive no scalars $\dot{\phi}_2$ and $\dot{\lambda}_2$ exist which satisfy (5.9) and (5.10). If $- R_1' + \mu R_2'$ is negative there exist multiple solutions of the LCP (5.9) and (5.10).

Finally, it should be mentioned that J. Barber [16] has pointed out to the author that if dynamic terms are incorporated in the analysis, non-unique solutions do not occur for the problem considered here. A paper by Lötstedt [17] on Coulomb friction in twodimensional rigid body systems seems to indicate that this might be a general conclusion concerning finite dimensional problems of linear elasticity.

6. DISCUSSION

Contact problems with friction have in this paper been investigated in a finite dimensional setting. Naturally, such formulations are closely tied to numerical treatments of the problem. For the PLCP involving derivatives a solution algorithm based on Grave's principal pivoting algorithm was proposed by Kaneko [9]. This algorithm was in ref. [8] shown to apply to contact problems with friction where the contact surface is constant during the loading history. It seems that this method can apply to the present more general problem by simply using a different strategy for the pivoting. However, the situation with linearly dependent columns of $N._I$ in case of Coulomb's friction law, as mentioned in Section 5, has to be dealt with by special considerations.

The fact that problems for which no solution exists were detected in this study deserves attention. In an earlier study of finite dimensional contact problems with friction Janovský [18] stated that all such plane problems have a solution. However, in that study the nonreversible character of frictional forces was ignored and a static problem was considered. Heuristically, it seems that the discrepancy between Janovsky's study and the present one can be explained by the lack of a separation mode of deformation in an incremental problem for which the normal contact force is strictly less than zero at some contact node.

Furthermore, it is likely that discontinuous solutions exist in cases where no continuous ones do. Whether such solutions have any physical bearing or not is an interesting question for which the author has no answer.

REFERENCES

1. Grierson, D.E. et al: Mathematical programming and nonlinear finite element analysis, Comp. Meth. in Appl. Mech. and Eng., 17/18(1979), 497-518.

2. Cohn, M.Z., G. Maier and D Grierson (eds.): Engineering Plasticity by Mathematical Programming, Pergamon Press, New York 1979.

3. Panagiotopoulos, P.D.: A nonlinear programming approach to the unilateral contact- and friction-boundary value problem in the theory of elasticity, Ingenieur-Archiv, 44(1975),421-432.

4. Fredriksson, B., G. Rydholm and P. Sjöblom: Variational Inequalities in Structural Mechanics with Emphasis on Contact Problems, in: Proc. International Conference on Finite Elements in Nonlinear Solid and Structural Mechanics, Geilo 1977.

5. Oden, J.T. and E.B. Pires: Numerical analysis of certain contact problems in elasticity with non-classical friction laws, Comp. & Struct., 16,1-4(1983),481-485.

6. Klarbring, A.: Contact Problems With Friction - Using a Finite Dimensional Description and the Theory of Linear Complementarity, Linköping Studies in Science and Technology, Thesis No. 20, Linköping Institute of Technology, Linköping, Sweden 1984.

7. Klarbring, A.: The Influence of Slip Hardening and Interface Compliance on Contact Stress Distributions. A Mathematical Programming Approach, to appear in: Mechanics of Material Interfaces (Eds. A.P.S. Selvadurai, G. Voyiadjis), Elsevier 1985.

8. Klarbring, A.: A mathematical programming approach to threedimensional contact problems with friction, to appear in Comp. Meth. in App. Mech. and Eng.

9. Cottle, R.W.: Monotone solutions of the parametric linear complementary problem, Mathem. Program.,3(1972),210-224.

10. Kaneko, I.: A parametric linear complementarity problem involving derivatives, Mathematical Programming, 15(1978),146-154.

11. Klarbring, A.: General contact boundary conditions and the analysis of frictional systems, to appear in Int. J. of Solids and Struct.

12. Klarbring, A.: Quadratic programming in frictionless contact problem, to appear in Int. J. of Eng. Sci.

13. Eaves, B.C.: The linear complementarity problem, Management Sci, 17(1971),612-634.

14. Murty, K.G.: On the number of solutions of the complementarity problem and spanning properties of complementary cones, Linear Algebra and Its Appl.,5(1972),65-108.

15. Kalker,J.J.: A minimum principle for the law of dry friction, with application to elastic cylinders in rolling contact, J. of Appl. Mech., Dec. 1971, 875-887.

16. Barber,J.: Private Communication, October 12, 1984.

17. Lötstedt,P.: Coulomb friction in two-dimensional rigid body systems, ZAMM, 61(1981),605-615.

18. Janovský V.: Catastrophic Features of Coulomb Friction Model, in: Proc. Mathematics of Finite Elements and Applications, Brunel University 1981.

GENERALIZED VARIATIONAL PRINCIPLES AND UNILATERAL CONSTRAINTS IN ANALYTICAL MECHANICS

H.O. May
Institute for Technical Mechanics, RWTH Aachen

1. INTRODUCTION

The aim of this paper is to examine unilateral constraints in analytical mechanics. Examples of mechanical systems subject to unilateral constraints are moving rigid bodies which may enter into contact and detach from each other. But unilateral constraints also occur when two satellites are connected by a rope.

The text-books on analytical mechanics do not usually discuss the problems of unilateral constraints. Examples of works where remarks are made on this subject (usually Fourier's principle is mentioned) are the books of Appell |1|, Hamel |2| and Lanczos |3|, and for a more thorough investigation Pérès |4| and Bouligand |5|. Papers wherein the mechanical principles of one-sided constraints in the finite-dimensional case are discussed have been given by Delassus |6|, Stavrakova |7|, Moreau |8-10|, Lötstedt |11|.

Variational principles for unilateral constraints depending on the velocity will be investigated here. Some special aspects which differ from bilateral constraints have to be considered: for example, a constraint can sometimes be ignored and sometimes not, and in general there is no reduction in the degree of freedom for unilateral constraints. Moreover sudden changes (discontinuities) may occur, e.g. jumps in the value of the velocities for rigid unilateral constraints. We will study the dynamics of a system under the hypothesis that the constraints react without friction.

In order to describe bilateral smooth constraints, Jourdain's principle, which is a principle in equality form, can be used. Due to the unilateral character of one-sided constraints Jourdain's principle becomes an inequality in this case. Because we do not want to restrict our unilateral constraints too much, we will use Clarke and Rockafellar's mathematical notions, by means of which it is possible to give a very general description of unilateral constraints, see |12-15|. In this case Jourdain's principle is formulated as "hemivariational inequality", a term which was first introduced into mechanics by Panagiotopoulos in 1981. Corresponding to this the differential equations of motion must be replaced by differential inclusions expressing the unilateral character.

Finally, we describe a canonical formulation of nonholonomic constraints, which is different from the theory first given.

2. THE PARAMETRIC FORMALISM

In order to avoid having irregularities in the theory, as for example in the case of relativistic applications, we shall use a very general form of analytical mechanics (cf. |16,17|).

The space-time configuration of a mechanical system is described by n+1 variables $x = (x^o, x^1, \ldots, x^n)$, where x^o is added to the spatial variables in order to take the time into consideration. Now we can introduce a time t along a curve $\tau \to x(\tau)$, which is an absolutely continuous function with derivative $\dot{x} = (\dot{x}^o, \dot{x}^1, \ldots, \dot{x}^n)$ almost everywhere in $[0,T]$ using

$$\frac{dt}{d\tau} = \beta(x,\dot{x}) > 0 \tag{1}$$

where β is a function such that

$$\beta(x,\lambda\dot{x}) = \lambda\beta(x,\dot{x}) \qquad \forall \lambda > 0 . \tag{2}$$

Let us define the space-time-velocities by $v^i = dx^i/dt = \dot{x}^i/\beta$, i=0,1,...,n, which cannot be considered as independent variables, because the identity

$$\beta(x,v) = \beta(x, \frac{\dot{x}}{\beta}) = \frac{\beta(x,\dot{x})}{\beta(x,\dot{x})} = 1 \tag{3}$$

is valid. The most simple example for β is

$$\beta = \dot{x}^o/c \ , \ c = \text{const.} \ , \text{ i.e. } v^o = c \ .$$

3. d'ALEMBERT's PRINCIPLE OF UNILATERAL CONSTRAINTS

In analytical mechanics where constraints are expressed as equations, we can start with a generalized form of d'Alembert's principle in space and time |3,16|:

$$(\frac{dp_\lambda}{dt} - K_\lambda) \ \delta x^\lambda = 0 \tag{4}$$

From now on the summation convention from 0 to n with respect to the Greek indices will be valid, i.e., a further component is added to all space-components of the momentum $p = (p_0, p_1, \ldots, p_n)$, the force $K = (K_0, K_1, \ldots, K_n)$ and the variation $\delta x = (\delta x^0, \delta x^1, \ldots, \delta x^n)$ so that the equation of motion and the law of energy are included. $R_\lambda \delta x^\lambda = 0$ is valid for the reaction of bilateral ideal constraints, i.e. the virtual work of the reacting forces vanishes. (4) cannot be used for more general constraints, so that we have to start with the following form of the principle of viitual displacements:

$$(K_\lambda + R_\lambda - \frac{dp_\lambda}{dt}) \delta x^\lambda = 0 \quad \forall \delta x . \tag{5.I}$$

We can also study the sudden change of motion that results from the action of impulses on a dynamical system. In this case we can give an analogous variational equality (cf. |24|). Let $\tau \to \Phi(\tau)$ be a function and define

$$\Delta\Phi(\tau) = \Phi(\tau+0) - \Phi(\tau-0) = \Phi|_+ - \Phi|_- , \tag{6}$$

so that we get

$$(Q_\lambda + \overset{\circ}{R}_\lambda - \Delta p_\lambda) \delta x^\lambda = 0. \tag{5.II}$$

We have to consider this additionally as jumps in the velocities may occur in unilateral constraints. Q is an impulse acting on the system, which may arise when the force K becomes singular. Then Q_0 describes a jump in energy.

In order to avoid a contradiction between the relations I and II, we must establish the principle of constraints, which shall be valid from now on, see |18|:

"Constraints shall be maintained by forces so long as this is possible; otherwise, and only otherwise, by impulses."

(5.II) shall always be valid when (5.I) is fulfilled. If we suppose that the impressed quantities K and Q are independent of the reaction of constraints, then, if (5.I) is fulfilled, we must obtain

$$\overset{\circ}{R} = 0. \tag{7}$$

In order to investigate the case of unilateral constraints, consider a material point subjected to an impressed force K and a reaction R |19|. According to the principle of virtual work we may write

$$K_\lambda \, \delta x^\lambda + R_\lambda \, \delta x^\lambda = 0. \tag{8}$$

Suppose that the material point is constrained to remain on a smooth rigid surface $\Psi(x)=0$ in R^n. Then only bilateral variations are admissible and R is normal to the surface if no friction is considered, and the reaction R performs no work if a virtual displacement is imposed:

$$R_\lambda \, \delta x^\lambda = 0. \tag{9}$$

In the case of one-sided constraints the reaction is normal to the surface again, but now the virtual displacements may lead outside of the surface and therefore the variational inequality

$$R_\lambda \, \delta x^\lambda \geq 0 \qquad \forall \, \delta x \text{ admissible} \tag{10}$$

holds which means that the virtual work of the impressed forces cannot be positive (Fourier's principle, |2|). We assume further that the surface is not smooth and it presents first a convex corner (fig. 1a) and secondly a re-entrant corner (fig. 1b), where C denotes the set to which the system is restricted. In the first case the inequality (10) holds, but in the case of the re-entrant corner (10) cannot be valid

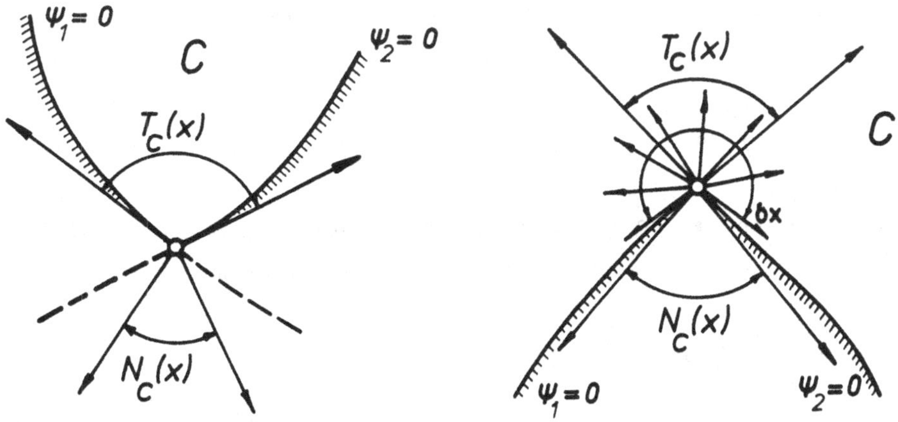

Fig. 1

(see fig. 1). In both cases the reaction at the corner is a linear combination of the normals to the surface, and the direction of R is not uniquely defined. From this we make the assumption that we may write

$$- R \in N_C(x) \tag{11}$$

where $N_C(x)$ denotes the normal cone at point $x \in C$. Let I_C be the indicator function of the set C, i.e.

$$I_C(x) = \begin{cases} o & \text{if } x \in C \\ +\infty & \text{otherwise} \end{cases} \tag{12}$$

For any set M and $x \in M$ we define its upper derivative $I_M^{\uparrow}$

$$I_M^{\uparrow}(x,\delta x) = \begin{cases} o & \text{if } \delta x \in T_M(x) \\ +\infty & \text{otherwise} \end{cases} \tag{13}$$

whereby T_M denotes the tangent cone at x, see |13|. From (11) we obtain the hemivariational inequality, see |12,13,15|,

$$I_C^{\uparrow}(x,\delta x) \geq -R_\lambda \ \delta x^\lambda \quad \forall \delta x \ , \tag{14}$$

which is an extension of (10) as is easy to see. Indeed, if $\delta x \in T_C(x)$, then (14) implies (10), if not, as in the case of a re-entrant corner, $R_\lambda \cdot \delta x^\lambda$ can achieve any value in R. A very general form of Fourier's principle then is

$$I_C^{\uparrow}(x,\delta x) \geq K_\lambda \ \delta x^\lambda . \tag{15}$$

The last relation can be extended to the dynamical case, where (14) is taken to describe the reaction force. Similarly we assume

$$I_C^{\uparrow}(x,\delta x) \geq -R_\lambda^{\circ} \ \delta x^\lambda \ . \tag{16}$$

If, for example, x is an interior point of C, then $I_C^{\uparrow}(x,\delta x) = 0$ for all

δx, and the reaction force vanishes. From (5) we get the hemivariational inequalities

$$I_C^{\uparrow}(x,\delta x) \geq (K_\lambda - \frac{dp_\lambda}{dt}) \cdot \delta x^\lambda \quad \forall \delta x, \; x \in C \quad \text{if} \quad \Delta v = 0 \tag{17.I}$$

$$I_C^{\uparrow}(x,\delta x) \geq (Q_\lambda - \Delta p_\lambda) \; \delta x^\lambda \quad \forall \delta x, \; x \in C \; . \tag{17.II}$$

For both quantities, derivative of momentum and jump of momentum, a similar relation is valid. (17.II) is the basic hemivariational inequality for the impulse of rigid constraints. The foregoing is always valid: if Q=0 and (17.I) is fulfilled, then the inequality (17.II) is insignificant. The hemivariational inequalities are an extension of Fourier's principle for the dynamical case. Due to the fact that they must be valid for all δx, we obtain the differential inclusion of motion in the case of one-sided constraints

$$K - \frac{dp}{dt} \in \partial I_C(x) \; , \quad x \in C \quad \text{if} \quad \Delta v = 0 \tag{18.I}$$

and the jump-condition

$$Q - \Delta p \in \partial I_C(x) \; , \quad x \in C \; , \tag{18.II}$$

where ∂I_C denotes the generalized gradient of I_C.

4. JOURDAIN's PRINCIPLE FOR UNILATERAL CONSTRAINTS

Jourdain's principle is often used to describe bilateral constraints

$$\Phi(x,\dot{x}) = 0 \tag{19}$$

depending on the velocities where only the velocities are varied. In a parametric form of mechanics the physical velocities will be replaced by $\dot{x}$ so that $\dot{x}$ has to be varied. The idea that only the velocities will be varied, results from the fact that no different reactions in the system shall be caused by Jourdain's and d'Alembert's principles. If, for example, there is a rigid constraint $\Gamma(x)=0$ where Γ is considered to be continuously differentiable, then the differential equation

$$\Phi(x,\dot{x}) = \dot{x}^{\lambda} \frac{\partial \Gamma}{\partial x^{\lambda}}(x) = 0 \ , \ \Gamma(x(0)) = 0 \qquad (20)$$

is valid for $\dot{x}$, i.e. the restriction of coordinates implies a restriction on the velocities. The system will react in the same way if the velocities only are varied in (20) (see |20|).

In the case of unilateral constraints these considerations are no longer as simple, because, for example, $\dot{x}^{\lambda}\ \partial f/\partial x^{\lambda}(x) \leq 0$ does not result from $f(x) \leq 0$ (see the later comment of fig. 2).

Let us consider first a unilateral rigid constraint, i.e. the motion is restricted to a closed set $C \subset R^{n+1}$. We first have to explain when a velocity is admissible. Consider a point x belonging to C. Intuitively a velocity v will be considered to be admissible if it leads from x into C. This concept, however, is still too limited for our purposes if, for example, we want to move along the boundary of C. More generally we can say that a velocity v will be called admissible if $v \in K_C(x)$. $K_C(x)$ denotes the cotangent cone of C at point x, with, |12, 14|,

$$K_C(x) = \{v \mid \exists \{\lambda_k\} \downarrow 0+,\ \{\overset{k}{v}\} \to 0 \ \text{with} \ x + \lambda_k \overset{k}{v} \in C\} \ . \qquad (21)$$

Obviously $K_C(x) = R^{n+1}$ if $x \in \text{int}(C)$, i.e. x is an interior point of C and any velocity is admissible. But if x belongs to the boundary of C certain velocities are no longer admissible. In consequence of this fact it is not possible to construct a set D of admissible coordinates and velocities which is closed and which results from the set C. Consider as

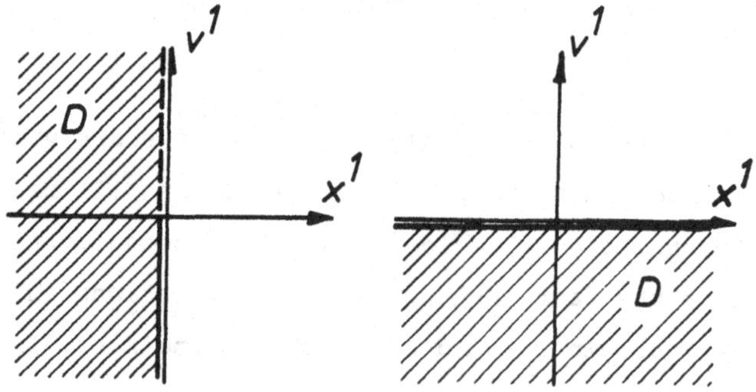

Fig. 2

an example a one-dimensional motion (n=1) subject to the constraint $x^1 \leq 0$. Then we have

$$v^1 \in R \qquad \text{if } x^1<0$$
$$v^1 \in [0,-\infty] \quad \text{if } x^1=0$$

(see fig. 2a). Obviously this set is not closed, and it does not make sense to replace D by its closure, because the physical states, characterized by the interrupted line, are not admissible. (In contrast to this the set $D=\{(x,v)\,|\,v^1 \leq 0\}$ is closed, see fig. 2b). A reaction of the system, caused by the constraints, is only possible if the boundary is reached, therefore a description of the reaction by the set D is not possible in this case. It is obvious that the reaction of the system consists of a jump in the velocities as is given by (17.II). It is easy to show that these considerations are true in general.

Let us consider now a motion being restricted to a set D of coordinates and velocities. We replace (x,v) by $(x,\dot{x})$, where the $\dot{x}^i$'s are independent of each other. We cannot in general require our set D to be closed as we have seen previously (moreover, the elements of D have to fulfill $dt/d\tau=\beta(x,\dot{x})>0$), but D has the property that for any point of D there is another point belonging to D and a line in D by which these points are connected. Jourdain's principle shall only be applicable to a point $(x,\dot{x})$ if this belongs to D. Here is taken into consideration that it is not possible to obtain an analogous description for unilateral rigid constraints by d'Alembert's and Jourdain's principles: if $(x,\dot{x}) \in \mathrm{int}(D)$ no reaction is caused by the constraint, a reaction only occurs if $(x,\dot{x}) \in b(D)$.

We cannot choose D to be admissible as we have seen from the example in fig. 2a. The interrupted line was senseless there. When reaching the boundary the system can only react by a jump in the velocities. Consider the set D of fig. 3: Reaching the point P of the set D the system can only react by a jump in the coordinates. We do not want to admit such systems.

Surely a point $(x,\dot{x}) \in D$ is admissible if there is an acceleration x which leads into D, i.e. there is a positive number δ, such that

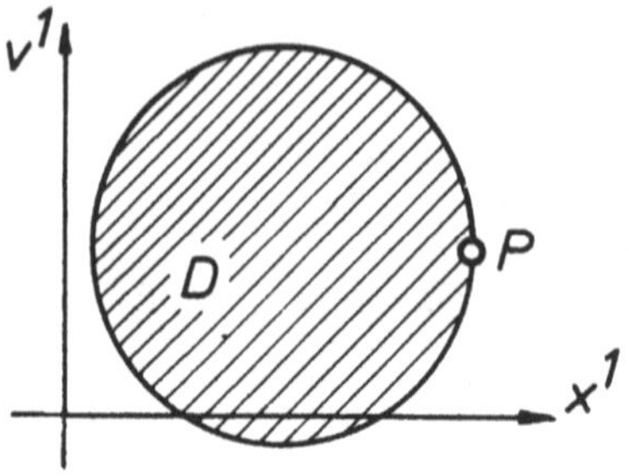

Fig. 3

$$(x,\dot{x}) + \tau\,(\dot{x},\ddot{x}) \in D \qquad \forall \tau \in [0,\delta]\ . \tag{22}$$

An interior point of D is always admissible, and obviously we are able to characterize admissible points of D with the help of the cotangent cone. Let $D \subset R^{2(n+1)}$ be a set and $(x,\dot{x}) \in D$. $(x,\dot{x})$ is called admissible if there is an $\ddot{x}$ such that $(\dot{x},\ddot{x}) \in K_D(x,\dot{x})$. From now on we only consider sets, consisting of points which are admissible.

In Jourdain's principle for bilateral constraints only the velocities are varied. If we want to give a general formulation of unilateral constraints, we find that there are two different possibilities of variational inequalities with which to describe the reaction forces:

$$1)\quad I^{\dagger}_{C(x)}(\dot{x},\delta\dot{x}) \geq -R_\lambda\ \delta\dot{x}^\lambda \qquad \forall\, \delta\dot{x}\ ,\ (x,\dot{x}) \in D \tag{23}$$

where $C(x) = \{\dot{x} \mid (x,\dot{x}) \in D\}$, and

$$2)\quad I^{\dagger}_{D}(x,\dot{x},0,\delta\dot{x}) \geq -R_\lambda\ \delta\dot{x}^\lambda \qquad \forall\, \delta\dot{x}\ ,\ (x,\dot{x}) \in D \tag{24}$$

The first inequality is connected with the calculation of "partial subgradients". Both possibilities are equivalent if the boundaries are smooth. But this is not true in general. Consider, for example, an unilateral constraint as is given in fig. 4. At the corner point P we have $C(x_P) = R$ and

$$I^{\dagger}_{C(x_P)}(x_P,\delta\dot{x}) = 0 \qquad \forall\, \delta\dot{x} \tag{25}$$

and therefore $R=0$, i.e. no reaction; a reaction at point P is only possible if the second form is taken. Because $(0,\delta\dot{x}) \notin T_D(P)\ \ \forall\, \delta\dot{x}$ we

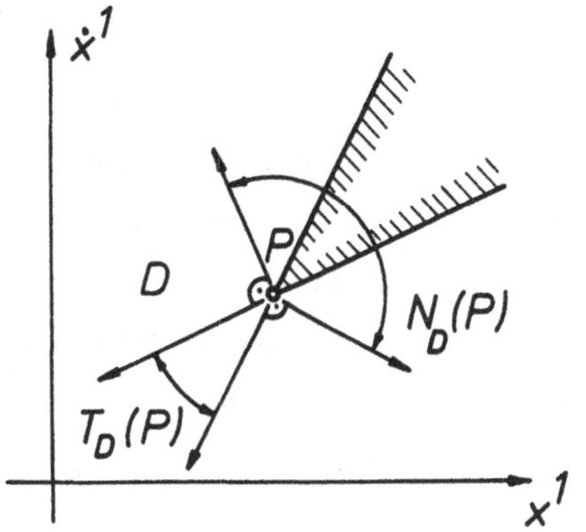

Fig. 4

obtain

$$I_D^{\dagger}(x_P,\dot{x}_P,0,\delta\dot{x}) = +\infty \qquad \forall\,\delta\dot{x} \tag{26}$$

so that we will give the priority to (24). Let us define now a "reduced normal cone"

$$\hat{N}_D(x,\dot{x}) = \{u \mid I_D^{\dagger}(x,\dot{x},0,\delta\dot{x}) \geq u_\lambda \delta\dot{x}^\lambda \quad \forall\,\delta\dot{x}\} \ . \tag{27}$$

According to the variational equality

$$\left(\frac{dp_\lambda}{dt} - K_\lambda - R_\lambda\right) \delta\dot{x}^\lambda = 0$$

and together with the condition (26) for the reaction force R, we receive Jourdain's principle in the form of a hemivariational inequality

$$I_D^{\dagger}(x,\dot{x},0,\delta\dot{x}) \geq \left(K_\lambda - \frac{dp_\lambda}{dt}\right) \delta\dot{x}^\lambda \quad \forall\,\delta\dot{x} \ , \ (x,\dot{x}) \in D \tag{28}$$

and the differential inclusion

$$\frac{dp}{dt} - K \in \hat{N}_D(x,\dot{x}) \tag{29}$$

concerning unilateral constraints which depend on the velocities. This description is valid for a set D of coordinates and velocities. If the constraints are rigid, d'Alemberts's principle applies.

Jumps in the velocities may also occur in the case of rigid constraints. Corresponding to this, jumps in the accelerations may occur if the constraints depend on the velocities. The influence of the boundary shall be as small as possible, which means that jumps in the acceleration (as well as jumps in the velocities) which are caused by restrictions, shall only occur if the motion is not possible otherwise. In order to explain this, we take as an example a restriction in the form of an inequality $g(x,\dot{x})\leq 0$, where g is assumed to be continuously differentiable. Now let $g(x,\dot{x})=0$ (something similar is valid for a constraint $f(x)\leq 0$, where $f(x)=0$ and $(\partial f/\partial \dot{x}^\lambda)\dot{x}^\lambda=0$). We consider

$$\frac{dg}{dt} = \frac{\partial g}{\partial x^\lambda}\dot{x}^\lambda + \frac{\partial g}{\partial \dot{x}^\lambda}\ddot{x}^\lambda \tag{30}$$

and assume $dg/dt|_->0$, then $dg/dt|_+\leq 0$. If $dg/dt|_+<0$, a particle fulfilling $g(x,\dot{x})=0$ would move away from the boundary, but the next moment it would turn towards it again if the force K depended continuously on the coordinates and the velocities, and a jump of the acceleration would again be necessary. This is prevented if $dg/dt|_+=0$.

In the case of non-smooth boundaries, $(x,\dot{x})$ must belong to the cotangent cone. Such an element always exists if $(x,\dot{x})\in D$ is admissible; so that we may assume in general

$$(\dot{x},\ddot{x})|_+ \in b(K_C(x,\dot{x})) \quad \text{if } \left(\frac{dp}{dt}+k\right)\Big|_+ \neq 0. \tag{31}$$

The second condition states that a reaction is only necessary to fulfill the restriction as, for example, when a particle leaves an L-shaped corner. In analogy this is valid for unilateral rigid constraints, because the case $\Delta\dot{x}=0$ (i.e. no jump in the velocities is necessary) is only possible at an admissible point of D.

5. LAGRANGIAN FORMULATION

Here we wish to restrict ourselves to the case where the set D of coordinates and velocities only consists of admissible points. We suppose the momenta p_i to be functions of the configuration and the velocities. Because of the relationship of the velocities $v^i=dx^i/dt=\dot{x}^i/\beta(x,\dot{x})$ the

p_i's could be considered as positively homogeneous functions of degree zero with respect to $\dot{x}$, so that $(\partial p_i/\partial \dot{x}^\lambda)\dot{x}^\lambda = 0$ must be valid. From now on, we shall discuss such systems for which the variational form, see |16,21|,

$$K_\lambda \ \delta x^\lambda = \delta p_\lambda \ v^\lambda \tag{32}$$

is valid. This relation is connected to the formulation of a generalized kinetic energy and the total energy of the system. Due to the fact that the δx^i's and the $\delta \dot{x}^i$'s could be chosen independently, we find $(\partial p_\lambda/\partial \dot{x}^i)\dot{x}^\lambda = 0$. Now we introduce a Lagrangian function L by

$$L(x,\dot{x}) = \dot{x}^\lambda \ p_\lambda(x,\dot{x}) \tag{33}$$

and we get the classical relation for the momenta, but now in space and time (see |16|, |21|)

$$p_i = \frac{\partial L}{\partial \dot{x}^i} \ . \tag{33}$$

With the principle of virtual displacements and the previously discussed hemivariational inequality for the reaction force, we receive

$$I_D^{\uparrow}(x,\dot{x},0,\delta\dot{x}) \geq \left(\frac{\partial L}{\partial x^\lambda} - \frac{d}{d\tau}\frac{\partial L}{\partial \dot{x}^\lambda}\right) \delta\dot{x}^\lambda \quad \forall \delta\dot{x} \ , \ (x,\dot{x}) \in D \ , \ \Delta\dot{x}=0 \tag{34}$$

or equivalently the Euler-Lagrange inclusion

$$\frac{\partial L}{\partial x} - \frac{d}{d\tau}\frac{\partial L}{\partial \dot{x}} \in \hat{N}_D(x,\dot{x}) \ . \tag{35}$$

As we have seen, there are no fundamental problems when passing to the Lagrangian picture of mechanics. It is less disturbing that the velocities only are varied and not the coordinates as well, because only transformations in the physical coordinates are considered.

6. THE HAMILTONIAN FORMALISM

We described our reaction forces by a variation of the velocities only, as we did not want to receive different reactions from Jourdain's and d'Alembert's principles for bilateral constraints. If we would pass to the canonical theory as before, we would totally destroy the symmetry between coordinates and impulses, and the invariance with respect to the canonical transformation would be lost. Of course, one can object to this in that is not necessary to require a canonical theory for nonholonomic constraints as well. Nevertheless, such theories for bilateral constraints exist (see e.g. |21, 22|) and, therefore, we want to discuss another possibility of handling nonholonomic constraints, where the above-described disadvantage does not appear. On the other hand one still has the problem with d'Alembert's principle which was discussed earlier, and this problem is only eliminated if one admits solely the Hamiltonian form of mechanics as being essential, as some people do.

First we construct a Hamiltonian function. We have assumed, that the momenta are functions of the configuration and the velocities. We shall assume in addition, that this is solvable for p:

$$p = h(x,v) \leftrightarrow v = g(x,p) \tag{36}$$

In order to do this, the v^i's must be considered as independent variables. But then in addition, because $1=f(x,v)$ there must exist a function, which we shall call Hamiltonian, and for which we have the identity

$$H(x,p) = 0 \, . \tag{37}$$

Now we must forget the relation between momenta and velocities and take the standpoint that the relation (37) and not (36) is given by the mechanical system. Hence we have to consider (37) as a constraint for (x,p), and this constraint shall also be fulfilled for all admissible variations.

In order to handle holonomic constraints we started with the principle of virtual displacements. If we stay in the picture of coordinates and momenta we can add a differential form in δp

$$\frac{dp_\lambda}{dt}\,\delta x^\lambda - K_\lambda\,\delta x^\lambda - P_\lambda\,\delta x^\lambda - Q^\lambda\,\delta p_\lambda = 0 \tag{38}$$

where $(dp_\lambda/dt)\delta x^\lambda$ belongs to the initial, $K_\lambda \delta x^\lambda$ to the impressed forces as before, and $P_\lambda \delta x^\lambda$, $Q^\lambda \delta p_\lambda$ are caused by the reactions. For the impressed forces we shall use again (32). In the case of the reaction forces a general inclusion $-(P,Q) \in J(x,p)$ depending on the coordinates and the momenta must be valid in order to preserve a canonical transformation scheme. Now we want to consider a motion being restricted on a set ¶ of coordinates and momenta, where we assume ¶ to be physically admissible. Moreover, we only discuss the case $\Delta p=0$. The most obvious relation for (P,Q) then is the hemivariational inequality

$$I_{\P}^{\dagger}(x,p,\delta x,\delta p) \geq -P_{\lambda}\ \delta x^{\lambda} - Q^{\lambda}\ \delta p_{\lambda} \qquad (39)$$

or in the case of (38)

$$I_{\P}^{\dagger}(x,p,\delta x,\delta p) \geq v^{\lambda}\ \delta p_{\lambda} - \frac{dp_{\lambda}}{dt}\delta x^{\lambda} \quad \text{if} \quad \Delta p=0 \ , \ (x,p) \in \P \ . \qquad (40)$$

Due to (37) which should be valid for all variations we cannot choose these variations independently, hence with a Lagrangian multiplier γ we get

$$I_{\P}^{\dagger}(x,p,\delta x,\delta p) \geq (\dot{x}^{\lambda} -\gamma \frac{\partial H}{\partial p_{\lambda}})\delta p_{\lambda} - (\dot{p}_{\lambda} +\gamma \frac{\partial H}{\partial x^{\lambda}})\delta x^{\lambda} \quad \text{if} \quad \Delta p=0, \ (x,p) \in \P . \qquad (41)$$

Here we have γ at our disposal, so that the variations could be chosen independently. Hence, we get the canonical inclusion for one-sided constraints

$$(-\dot{p} -\gamma \frac{\partial H}{\partial x} \ , \ \dot{x} -\gamma \frac{\partial H}{\partial p}) \in \partial I_{\P}(x,p) \quad \text{if} \quad \Delta p=0 \ . \qquad (42)$$

The inclusion in this form is invariant to canonical transformation (in a generalized sense, compare |23|). Moreover the first group reduces to the case of rigid restrictions

$$-\dot{p} -\gamma \frac{\partial H}{\partial x} \in \partial I_{C}(x) \quad \text{if} \quad \Delta p=0 . \qquad (43)$$

But if, in the case of bilateral holonomic restrictions, one might have the idea of determining the derivative with respect to time and then replacing the velocities with the help of the second group of canonical equations, one will receive different results.

REFERENCES

1. Appell, P.: Traité de Mécanique Rationelle, Vol. 1,2. Gauthier-Villars, Paris 1904

2. Hamel, G.: Theoretische Mechanik, Springer, Berlin, Heidelberg 1978

3. Lanczos, C.: The Variational Principles of Mechanics, University of Toronto Press, Toronto 1966

4. Peres, J.: Mécanique générale, Masson, Paris 1953

5. Bouligand, G.: Compléments et exercises sur la mécanique des solides, Vuibert, Paris

6. Delassus, E.: Théorie des liaisons finies unilatérales. Ann. Sci. Ec. Norm., (3), 34(1917), 95-179

7. Stavrakova, N.E.: The Principle of Hamilton-Ostrogradskii for Systems with One-Sided Constraints. PMM 29(1965), 738-741

8. Moreau, J.-J.: Les liaisons unilatérales et le principe de Gauss, Comptes rendus 256(1963), 871-874

9. Moreau, J.-J.: Liaisons unilatérales sans frottement et chocs inélastiques, C.R.Acad.Sc.,Série II, 296(1983), 1473-1476

10. Moreau, J.-J.: Standard Inelastic Shocks and the Dynamics of Unilateral Constraints, in: Unilateral Problems in Structural Analysis (Ed. G. Del Piero/F. Maceri), Springer, Wien, New York 1985

11. Lötstedt, P.: Mechanical Systems of Rigid Bodies Subject to Unilateral Constraints, SIAM J. Appl. Math. 42(1982), 281-296

12. Rockafellar, R.T.: The Theory of Subgradients and its Applications to Problems of Optimization. Convex and Nonconvex Functions, Heldermann, Berlin 1981

13. Clarke, F.H.: Optimization and Nonsmooth Analysis, John Wiley 1983

14. Hestenes, M.R.: Optimization Theory: The Finite Dimensional Case, Wiley, New York 1975

15. Panagiotopoulos, P.D.: Inequality Problems in Mechanics, Convex and Nonconvex Energy Functions, Birkhäuser, Basel, Boston 1985

16. Heinz, C.: Vorlesungen über analytische Mechanik, lectures given at RWTH Aachen, not yet published

17. Heinz, C.: Neue Aspekte in der Feldtheorie, Acta Mechanica 41(1981), 23-33

18. Kilmister, C.W. and J.E. Reeve, Rational Mechanics, Longmans, London 1966

19. May, H.-O. and P.D. Panagiotopoulos: F.H. Clarke's Generalized Gradient and Fourier's Principle, ZAMM 65(1985), 125-126

20. Päsler, M.: Prinzipe der Mechanik, de Gruyter, Berlin 1968

21. Dirac, P.A.M.: Homogeneous Variables in Classical Dynamics, Proc. Camb. Phil. Soc. 29(1933), 389-400

22. Weber, R.W.: Kanonische Theorie nichtholonomer Systeme, Thesis, ETH Zürich 1981

23. May, H.-O.: Variational Principles and Differential Inclusions for Unilateral Constraints in Analytical Mechanics, Meccanica 19(1984), 315-319

24. Pars, L.A.: A Treatise on Analytical Mechanics, Heinemann 1968

IOFFE'S FANS AND UNILATERAL PROBLEMS: A NEW CONJECTURE

P.D. Panagiotopoulos
Department of Civil Engineering, Aristotle University, Thessaloniki

ABSTRACT

The present paper deals with a general theory of variational "principles". First we consider laws expressed by convex and nonconvex superpotentials. Then a new type of superpotentials, the F-superpotentials, is proposed by using the notion of "fans" introduced recently by Ioffe. Then a conjecture is made which permits, first, the formulation of general variational principles in inequality form for systems with or without dissipation, and secondly, the derivation of new generalizations of the classical derivative. To this end we define the V-superpotentials whose definition is based on the notion of virtual work (or power). By means of this conjecture a general class of "generalized standard materials" is defined. To achieve it a new "Hypothesis of Dissipation" is introduced. The respective variational expressions are derived and one type of them, a variational-hemivariational inequality, is studied concerning the existence and the approximation of the solution. First the variational-hemivariational inequality is regularized and, simultaneously, by means of Galerkin's method, its formulation in a finite dimensional space is given. Using appropriate a priori estimates, a compactness argument, and the well-known monotonicity argument, we prove that the solution of the finite dimensional regularized problem converges to the solution of the continuous problem.

1. INTRODUCTION

In Mechanics there is a variety of variational inequality "principles"; they arise when material laws and/or boundary conditions are derived by convex, generally nondifferentiable and non everywhere finite potentials called convex superpotentials or superpotentials in the sense of Moreau [1]. Such relations derived by a convex superpotential are of monotone nature. We can mention here the Signorini boundary condition, the friction boundary condition, the Hencky material, the holonomic elasto-plastic materials, the locking materials etc. [2][3]. The necessity for the derivation of variational principles for material laws and boundary conditions of nonmonotone nature has led the author to the theory of hemivariational inequalities [4] (for a complete list of references see [3] Ch.4) at the beginning of the present decade. Such variational expressions arise from nonconvex and nondifferentiable potentials, which are non everywhere finite, and which we call nonconvex superpotentials or superpotentials in the sense of F.H. Clarke [3][4]. Both the variational and the hemivariational inequalities are expressions of the principle of virtual work (or power) in its inequality form due to the unilateral (or irreversible) character of the variations of certain quantities.

The "differential" operator acting on the superpotentials is the sub-differential ∂ in the convex case and the generalized gradient $\bar{\partial}$ of F.H. Clarke and R.T. Rockafellar [5] in the nonconvex case. Both these operators constitute generalizations of the classical "derivative" and are multivalued operators; this fact explains why the operator formulations of all unilateral problems, i.e. of problems giving rise to variational or hemivariational inequalities, are multivalued differential or integral inclusions.

2. CONVEX AND NONCONVEX SUPERPOTENTIALS. IOFFE'S FANS AND F-HEMIVARIATIONAL INEQUALITIES

Let us consider a mechanical system Σ characterized by the triplet $\{U,\langle u,f\rangle,F\}$. Here U is the vector space of all fields of "generalized velocities" and F the vector space of all fields of "generalized forces" acting on Σ. U and F are placed in separating duality by the bilinear form-work (or power)- $\langle u,f\rangle u\in U, f\in F$. To this end we assume that for each $u\neq 0$ (resp. $f\neq 0$) there exists $f\in F$ (resp. $u\in U$) such that $\langle u,f\rangle\neq 0$.

Let us consider that on U a convex, l.s.c. and proper functional Φ is defined and let us consider a convex superpotential law on Σ of the form (Fig. 1a)

$$-f\in \partial\Phi(u). \tag{2.1}$$

Analogously we may consider on U a nonconvex functional $\tilde{\Phi}$ taking values in [$-\infty,+\infty$] and a nonconvex superpotential law in the form (Fig. 1b)

$$-f \in \bar{\partial}\tilde{\Phi}(u) \tag{2.2}$$

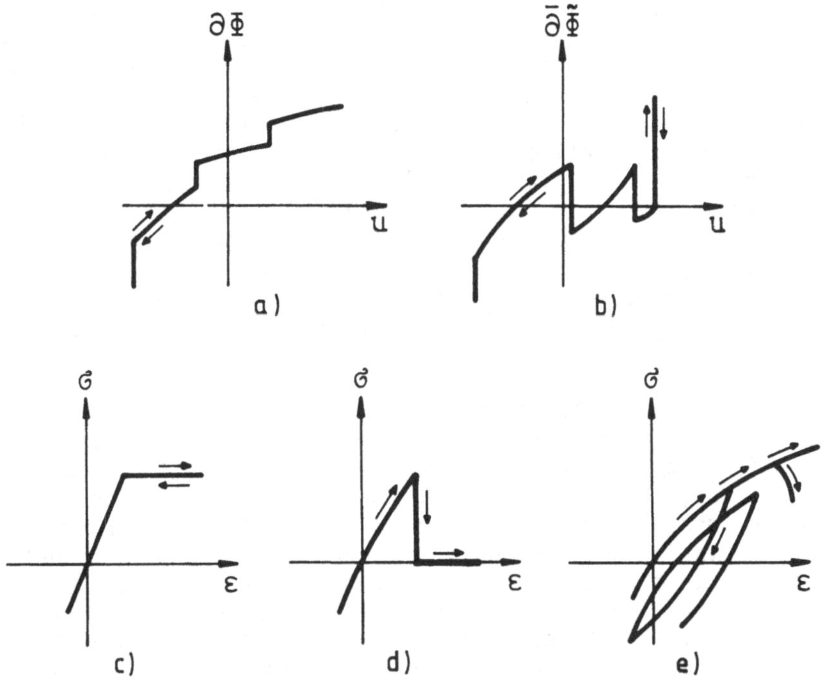

Fig. 1, On the Superpotential Laws (σ= stress, ε= strain)

For examples of materials and boundary conditions giving rise to (2.1) or to (2.2) we refer to [3]. It is important to note that both (2.1) and (2.2) cannot describe, as given here, dissipative phenomena. Thus, for instance, the stress-strain diagrams of Fig. 1c and Fig. 1d can be put in the form (2.1) and (2.2) if no unloading with dissipation occurs. Let us consider, e.g., the adhesive contact boundary condition described in Fig. 1d in the case of loading. The same diagram does not hold in the case of unloading as it is obvious. This situation recalls the comparison between the flow theory of plasticity and the deformation theory of plasticity. In order to be able to describe dissipative phenomena by means of the superpotential law an analogous extension of them as in the case of the flow theory of plasticity is necessary. An attempt towards this direction of rather technical nature was given in [6]: the stress-strain diagram was splitted into monotone increasing and decreasing parts "the macro-increments" and for every part a variational inequality could be written.

Another attempt could be the following: a functional $\tilde{\Phi}$ can be defined with the property that the graph $\{\overline{\partial}\tilde{\Phi}(\xi),\xi\}$ is the one given in Fig. 1e. In this case we would have difficulties with the analytic description of the loading and unloading. A kind of path labelling is necessary and this fact leads naturally to the need for consideration of another variable in order to describe the dissipative phenomenon. The hypothesis of normal dissipation [7][8] may be considered in this respect. A basic "product" of this hypothesis are the generalized standard materials. The whole theory formulated initially in the context of convexity has been generalized ([3] p.156) for the case of nonconvexity with surprising results: roughly speaking the lack of the assumption of convexity of the superpotential did not influence the form of the laws derived. In this case the hidden variables play the role of the additional variables, mentioned before with respect to the onedimensional examples. At this point it should be noted that in some cases the "labelling of the path" can be achieved by the consideration of observable variables, as it is the case of adhesive joints where the intensity of the adhesion (see e.g. [9] and the papers of M. Frémond cited there in the literature) plays this role.

Every generalization of the notion of derivative leads to a new variational problem and permits variational formulations for problems in mechanics which cannot be achieved by means of the classical calculus of variations. Thus, the subdifferential leads to the variational inequalities, and the generalized gradient in the sense of Clarke (resp. Rockafellar) to hemivariational inequalities containing the directional derivative of Clarke $\Phi^{\circ}(.,.)$ (resp. the upper subderivative $\Phi^{\uparrow}(.,.)$) [3]. The resulting variational problems express the principle of virtual work (or power). Thus, e.g. $\Phi^{\uparrow}(\varepsilon,\varepsilon^{*}-\varepsilon)$, $\Phi^{\uparrow}(u,u^{*}-u)$ etc. express a virtual work (or power) for the virtual generalized displacement $\varepsilon^{*}-\varepsilon$, $u^{*}-u$ etc. Thus we could describe a (σ,ε) or a $(-f,u)$ relationship simply by specifying the expressions of the virtual work $A(\varepsilon,\varepsilon^{*}-\varepsilon)$ and $B(u,u^{*}-u)$ and we would obtain a variational expression similar to a hemivariational inequality but involving instead of $\Phi^{\circ}(u,u^{*}-u)$ etc the virtual work (or power) expressions $A(\varepsilon,\varepsilon^{*}-\varepsilon)$ and $B(u,u^{*}-u)$.

Very close to the above is the notion of fans introduced in 1981 by Ioffe [10], which constitutes a special case of the aforementioned general thoughts. Let $g:U\times U\rightarrow R$ be an extended real valued functional which is bi-sublinear, i.e. convex and positively homogeneous of degree one in every one of the variables. We assume that U is a locally convex Hausdorff topological vector space, D is its dual space and $\langle u,f\rangle$ is the duality pairing.

Then the multivalued mapping $u\rightarrow D(u)$ with

$$D(u) = \{f\in F \mid g(u,u^{*}-u) \geq \langle f,u^{*}-u\rangle \quad \forall u^{*}\in U\} \qquad (2.3)$$

defines a "fan" from U into F. We assume that if $f\neq\emptyset$, $g(u,u^{*})>-\infty$ $\forall u^{*}\in U$

and $u^*\to g(u,u^*)$ is weakly lower semicontinuous. If $f\neq\emptyset$ we may pose $g(u,u^*)=-\infty$. It can be shown that $0\in D(0)$, that $D(\lambda x) = \lambda D(x)\ \forall x$ and $\lambda>0$, that all sets $D(x)$ are convex and finally that $D(x_1+x_2)\subset\overline{D(x_1)+D(x_2)}$. We may thus write a fan-superpotential (F-superpotential) law in the form

$$-f\in D(u). \tag{2.4}$$

Due to (2.3), if in (2.1) and (2.2) $\partial\Phi$ and $\overline{\partial}\tilde{\Phi}$ are replaced by fans, a fan-hemivariational inequality can be formulated involving instead of $\tilde{\Phi}^{\circ}(.,.)$ or $\tilde{\Phi}^{\uparrow}$ the bisublinear functional $g(.,.)$ defining the fan. We call it F-hemivariational inequality.

If $u\mapsto D(u)$ is given then we find the function $g(u,u^*)$ by the relation

$$g(u,u^*) = \sup_{u^*}\{\langle f,u^*\rangle / f\in D(u)\} . \tag{2.5}$$

Let G be a convex, weakly closed set of linear operators $A:U\to F$. Then the multivalued mapping (Fig. 2a)

$$u\mapsto D(u) = \{f/f= Au \text{ for some } A\in G\}$$

is a fan generated by G.

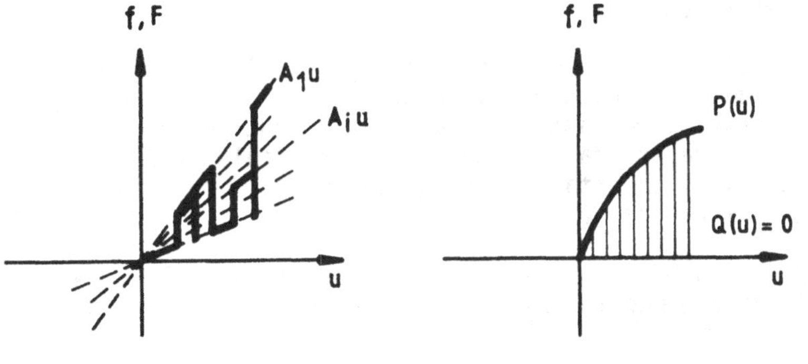

a) b)

Fig. 2, Graphs of Certain Fans

Let K be a convex closed cone in F and let us consider the partial ordering $<$ in F introduced by K. Further we consider a mapping $P:U\mapsto F$ such that $P(\lambda u)=\lambda P(u)$ for $\lambda>0$ and $P(u_1+u_2)< P(u_1)+P(u_2)$.

Then the mappings

$$u \to D(u) = \{f \in F / f < P(u)\} \quad (2.6)$$

$$u \to D(u) = \{f \in F / -P(-u) < f < P(u)\} \quad (2.7)$$

$$u \to D(u) = \{f \in F / Q(u) < f < P(u)\} \quad (2.8)$$

are fans provided that $-Q$ has the same properties as P above and $Q(u) < P(u)\ \forall u \in U$. Accordingly (cf. Fig. 2b) this kind of fans can be used for the description of unloading phenomena, for instance, if such a relation holds between ε and σ.

3. THE CONJECTURE AND ITS IMPLICATIONS IN NONSMOOTH MECHANICS

As pointed out in the previous section, in the variational, hemivariational and F-hemivariational inequalities the expressions $\Phi(u^*)-\Phi(u)$, $\Phi^{\uparrow}(u,u^*-u)$ or $\Phi^{\circ}(u,u^*-u)$, and $g(u,u^*-u)$ express the virtual work (or power) for the virtual "generalized velocity" u^*-u at the position of the system characterized by the actual "value" u of the "generalized velocity". These expressions are resulting from the definitions of the subdifferential $u \to \partial\Phi(u)$, the generalized gradient $u \to \overline{\partial}\Phi(u)$ and Ioffe's fan $u \to D(u)$. Accordingly every such generalization of the notion of classical derivative leads to a new type of variational expression and permits the derivation of variational formulations for more complicated mechanical problems. Thus the continuous development of new generalizations of the notion of classical derivative extends the frontiers of "nonsmooth mechanics" [3].

But as it is well-known in mechanics the physical meaning of a mechanical expression permits the mechanical theory to free itself from the parallel mathematical development and in some cases to outrun the mathematical theory. In the present case of the superpotentials (convex, nonconvex, etc) we have a very important "mechanical" tool in our hands: the principle of virtual work (or power).

Accordingly we propose the following: On the mechanical system Σ we consider a multivalued mechanical law (or constraint) of the form

$$-f \in W(u) \quad (3.1)$$

equivalent to the expression

$$W(u) = \{f \in F / G(u,u^*-u,\{B\}\{\beta\},\text{path},\text{history},\ldots) \geq \langle f,u^*-u\rangle\ \forall u^* \in U\}. \quad (3.2)$$

Here G is an appropriately defined function generally of a nonlocal nature which expresses the virtual work (or power) of the system under consideration. It may depend on several observable variables $\{B\}$ and on several hidden variables $\{\beta\}$, and may depend on the "path" of the variation u^*-u on the history of the mechanical system, etc.

Obviously (3.1) is much more general than (2.1), (2.2) and (2.4) since it goes beyond the geometry of the configuration under consideration, and enters into the "material behaviour" of the system. This last remark could be of some help in the further development of new generalizations of the notion of derivative in nonsmooth analysis (cf. [3], p.374). For instance, "normal" and "tangent" cones should be defined not only in terms of purely geometrical variables but also in terms of other parameters of physical origin (e.g. reactions with or without dry friction on a "deformable" reentrant corner). By definition (cf. (3.2)) the law (3.1) - which we shall call V-superpotential (virtual work-superpotential) law - may serve for the description of dissipative phenomena.

The foregoing considerations permit the introduction of the following "Hypothesis of Dissipation" which constitutes a generalization of the well-known "Hypothesis of Normal Dissipation" (cf. [7][8] and [3] p.156).

(H) For every real thermodynamic process a V-superpotential W exists such that the "flux" $u \in U = \mathbb{R}^n$ associated with the "force" $f \in F = \mathbb{R}^n$ satisfies the relation (3.1).

By the second principle of thermodynamics the inequality

$$f_i u_i \geq 0 \tag{3.3}$$

must always be fulfilled (cf. [3] App. III). By definition W is defined by an inequality analogous to the one in (3.2), and as obvious if

$$G(u,-u,\{B\},\{\beta\},\text{path, history}) \leq 0 \tag{3.4}$$

then (3.3) is satisfied. We call "V-substationarity point" a point u_0 such that

$$0 \in W(u_0) \Longleftrightarrow G(u_0,u^*-u_0,\{B\},\{\beta\},\text{path},\text{history},\ldots) \geq 0 \quad \forall u^* \in U. \tag{3.5}$$

We can prove easily the following proposition:

Proposition 3.1: To any flux u there corresponds at least one force f, if and only if at u

$$G(u,0,\{B\},\{\beta\},\text{path},\text{history}) \geq 0. \tag{3.6}$$

Proof: Indeed (3.6) implies that $W(u) \neq \emptyset$. Conversely, if $W(u) \neq \emptyset$ there exists f such as to satisfy (3.2) and thus for $u^* = u$ we obtain (3.6), q.e.d.

Analogously to (3.1) we may consider dissipation mechanisms given by a relation of the form

$$u \in \tilde{W}(f) \tag{3.7}$$

where $\tilde{W}$ is a given V-superpotential.

With the notations of [8] and of [3]p.157 we define a V-generalized standard material (i.e. compatible to a given expression of the virtual work) as a material for which (3.7) holds with

$$u = ((\tilde{d}F_p/dt)F_p^{-1}, -\tilde{d}\{\alpha_j\}/dt),\ f = \{R,\{A_j\}\}. \tag{3.8}$$

Obviously if $\tilde{W}$ is generated by $\tilde{G}$, (cf.(3.2)), and there exists $\tilde{\Phi}_1$ such that $\tilde{G}(f,f^*-f,\{\tilde{B}\}\dots) = \tilde{\Phi}_1^{\uparrow}(f,f^*-f)$ then we get a generalized standard material in the sense of F.H. Clarke, whereas if $\tilde{\Phi}_1$ is convex l.s.c. and proper then a generalized standard material, as defined in [8], is obtained.

In order to illustrate better the importance of a V-superpotential we shall show that this notion is inherent in the study of elastoplastic materials with hardening whose elastic properties change with plastic deformation. We shall derive for this situation a material law of the form

$$d\varepsilon^p \in W(\sigma) \tag{3.9}$$

where $d\varepsilon^p$ denotes a plastic strain increment.

To describe a situation in which (3.9) may result, let us consider a subset K of the stress space defined by the yield criterion $F(\sigma) \leq k^2$ such that $0 \in K$. We shall slightly modify Drucker's "proof" with which he justified the convexity of the yield surface. Let us consider a stress state $\sigma^* = \{\sigma^*_{ij}\}$ on or inside the yield surface, and an external agency

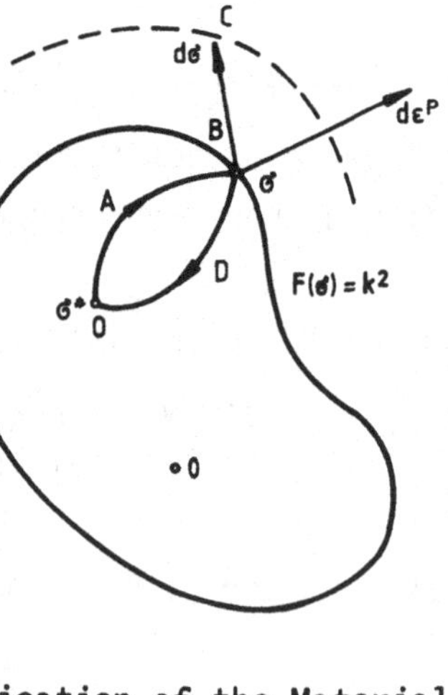

Fig. 3, Justification of the Material Law (3.9)

causing a change of the stresses until a stress state $\sigma=\{\sigma_{ij}\}$ is reached at the boundary of the yield surface. Suppose further that the agency produces a plastic-strain increment $d\varepsilon^p$ and a corresponding stress increment $d\sigma$(BC in Fig. 3) directed outwards from K. Further, the loading releases $d\sigma$ and returns the stresses to the initial stress state σ^* along the elastic loading path BDO. If the loading path OABDO lies entirely in K or on the boundary of K, then, due to the reversibility of elastic deformation, the work done over the cycle OABDO is zero, if the elastic properties are constant. However, it is assumed here that due to the plastic deformation the elastic properties do change, and we denote by G the nonrecoverable elastic work. We assume that G depends on σ, on $\sigma^*-\sigma$ and on the path followed and that for $\sigma=\sigma^*$, $G=0$. The work produced over the cycle OABCDO is positive and is given by

$$(\sigma_{ij}-\sigma^*_{ij})\,d\varepsilon^p_{ij}+d\varepsilon^p_{ij}\,d\sigma_{ij}+G(\sigma,\sigma^*-\sigma,\text{path})>0\,. \tag{3.10}$$

This inequality holds for every $\sigma^*\in K$ and for every path OABDO which does not pass out of K. (3.10) is also valid for $\sigma^*=\sigma$. Accordingly, $d\varepsilon^p_{ij}\,d\sigma_{ij}>0$. But this expression can be made as small as possible and therefore from (3.10) we get that

$$(\sigma_{ij}-\sigma^*_{ij})d\varepsilon^p_{ij}+G(\sigma,\sigma^*-\sigma,\text{path})\geq 0 \qquad \forall\sigma^*\in K. \tag{3.11}$$

Now let us assume that by definition $G=\infty$ if $\sigma,\sigma^*\notin K$. (3.11) thereby yields for $\sigma\in K$ the relation

$$d\varepsilon^p\in W(\sigma) \tag{3.12}$$

where W is the V-superpotential generated by G.

In the next section we shall study an important class of the V-hemivariational inequalities: the variational-hemivariational inequalities. They are obtained if the V-superpotential W is given by

$$W(\cdot)=\bar{\partial}\Phi(\cdot)+\text{grad}\Psi(\cdot) \tag{3.13}$$

or

$$W(\cdot)=\bar{\partial}\Phi(\cdot)+\partial\Psi(\cdot) \tag{3.14}$$

where Φ is a Lipschitz functional and Ψ is a convex, l.s.c. proper functional.

4. FORMULATION OF VARIATIONAL-HEMIVARIATIONAL INEQUALITIES

We shall study this type of variational problem with respect to a concrete mechanical problem: the interior nonmonotone-semipermeability problem when Fourier's heat conduction law is replaced by a nonlinear law derived by a convex superpotential.

If $q=\{q_i\}$ i=1,2,3 denotes the heat flux vector, and θ the absolute temperature, we can write such a law in the form ([3]p.267)

$$-q \in \partial\psi(\{\theta_{,i}\}) \tag{4.1}$$

where the comma denotes the differentiation and ψ is a convex, l.s.c, and proper functional on $\mathbb{R}^3$. Moreover we assume that the radiant heating per unit volume Q is

$$Q = \bar{Q} + \bar{\bar{Q}} \tag{4.2}$$

where $\bar{\bar{Q}}$ is given and $\bar{Q}$ is derived by a nonconvex superpotential. For the physical meaning of the interior nonmonotone semipermeability the reader is referred to [11]. We consider further an open, bounded, connected subset Ω of $\mathbb{R}^3$ referred to a fixed Cartesian coordinate system $Ox_1x_2x_3$ and we formulate the following static problem:
Find q and θ such as to satisfy

$$q_{i,i} = Q \text{ in } \Omega, \tag{4.3}$$

(4.1),(4.2), the interior nonmonotone semipermeability conditions, and the boundary condition

$$\theta = 0 \tag{4.4}$$

on the boundary Γ of Ω, which is assumed to be appropriately regular.

According to [11] and to [3](p.271) we define the dependence of $\bar{Q}$ on θ as follows: Let $b:\mathbb{R}\to\mathbb{R}$ be a locally bounded measurable function i.e. $b\in L^\infty_{loc}(\mathbb{R})$. Then we "fill in" the gaps of b and we obtain the multivalued function $\hat{b}$ in the following way: For $\delta > 0$ and $\xi\in\mathbb{R}$ let

$$\bar{b}_\delta(\xi) = \underset{|\xi-\xi_1|<\delta}{\operatorname{esssup}} b(\xi_1) \quad \text{and} \quad \underline{b}_\delta(\xi) = \underset{|\xi-\xi_1|<\delta}{\operatorname{essinf}} b(\xi_1) \tag{4.5}$$

which are increasing and decreasing functions of δ respectively and thus the limes as $\delta\to 0$ exist. We note them by $\bar{b}(\xi)$ and $\underline{b}(\xi)$ respectively, and we introduce the multivalued function

$$\hat{b}(\xi) = [\underline{b}(\xi), \bar{b}(\xi)] \tag{4.6}$$

where [.,.] denotes a closed interval in $\mathbb{R}$. Then if $b(\xi_{\pm 0})$ exists at every $\xi\in\mathbb{R}$, a locally Lipschitz-continuous function $\varphi:\mathbb{R}\to\mathbb{R}$ can be determined such that [12]

$$\hat{b}(\xi) = \bar{\partial}\varphi(\xi). \tag{4.7}$$

We can write, up to a constant, the equality $\varphi(\xi) = \int_0^\xi b(\xi_1)d\xi_1$. However, only the virtual work expression $\varphi^\circ(\xi,z)$ defined by

$$\varphi^\circ(\xi,z) = \limsup_{\substack{h\to 0\\ \lambda\to 0_+}} \frac{1}{\lambda}\int_{\xi+h}^{\xi+h+\lambda z} b(\xi_1)d\xi_1 \tag{4.8}$$

is needed. The relation between $\bar{Q}$ and θ is assumed to be

$$-Q\in \bar{\partial}\varphi(\theta) = \hat{b}(\theta) \text{ in } \Omega. \tag{4.9}$$

Further we introduce for $\theta\in \mathring{H}^1_p(\Omega)=\mathring{W}^{1,p}(\Omega)$, $p=n>3$ the functional

$$\Psi(\text{grad}\theta) = \begin{cases} \int_\Omega \psi(\text{grad}\theta)d\Omega & \text{if } \psi(\text{grad}\theta)\in L^1(\Omega) \\ \infty & \text{otherwise} \end{cases} \tag{4.10}$$

and assuming that $\theta,\phi\in \mathring{H}^1_p(\Omega)$ we are led to the following problem (P): Find $\theta\in \mathring{H}^1_p(\Omega)$ such as to satisfy

$$\Psi(\text{grad}\phi) - \Psi(\text{grad}\theta) + \int_\Omega \varphi^\circ(\theta,\phi-\theta)d\Omega \geq (\bar{\bar{Q}},\phi-\theta) \quad \forall\phi\in \mathring{H}^1_p(\Omega). \tag{4.11}$$

Here (.,.) denotes the duality pairing in $L^p(\Omega)$. (4.11) "results" from (4.3) by multiplying it by $\phi-\theta$, applying the Green-Gauss theorem formally, and using the definitions of the subdifferential in (4.1) and the generalized gradient in (4.9).

Further we shall distinguish two cases: in the first, called "differentiable case" we assume that gradψ exists, in the second ψ may take the value $+\infty$, $\psi\not\equiv\infty$ and is generally l.s.c and non everywhere differentiable ("nondifferentiable case").

5. STUDY OF THE DIFFERENTIABLE CASE

Let us consider a mollifier p, i.e., $p\in C_c^\infty(-1,+1)$ with $p\geq 0$ and $\int_{-\infty}^{+\infty} p(\xi)d\xi = 1$. Let

$$p_\varepsilon(\xi) = \frac{1}{\varepsilon}\, p\left(\frac{\xi}{\varepsilon}\right). \tag{5.1}$$

We introduce the convolution

$$b_\varepsilon = p_\varepsilon * b \quad \varepsilon>0 \tag{5.2}$$

and we call b_ε the "regularized form of b". We consider also a Galerkin basis $\{w_i\}$ of $V=\mathring{H}^1_p(\Omega)=\mathring{W}^{1,p}(\Omega)$ and let V_n be the corresponding n-dimensional subspace of V. Then we formulate the finite-dimensional problem $(P_{\varepsilon n})$ which is a finite dimension formulation of the regularized form (P_ε) of the initial problem (P).

$(P_{\varepsilon n})$: Find $\theta_{\varepsilon n} \in V_n$ such that

$$(\mathrm{grad}\psi(\theta_{\varepsilon n,i}),\mathrm{grad}\phi) + \int_\Omega b_\varepsilon(\theta_{\varepsilon n})\phi d\Omega = (\bar{\bar{Q}},\phi) \quad \forall \phi \in V_n \tag{5.3}$$

We assume for some ξ that

$$\operatorname*{ess\,sup}_{(-\infty,-\xi)} b(\xi) \le 0 \le \operatorname*{ess\,inf}_{(\xi,+\infty)} b(\xi). \tag{5.4}$$

Moreover we assume that the energy function ψ has the property:
For every $\phi \in \overset{\circ}{H}{}^1_p(\Omega)$ there exist positive constants c_1, and c_2 such that

$$(\mathrm{grad}\psi(\mathrm{grad}\phi),\mathrm{grad}\phi) \ge c(\sum_{\alpha=1} ||D^\alpha \phi||_{L^p})^p \tag{5.5}$$

and

$$|(\mathrm{grad}\psi(\mathrm{grad}\phi),\mathrm{grad}\phi)| \le c_2 ||\phi||^2 \tag{5.6}$$

where $\|\cdot\|$ denotes the $\overset{\circ}{H}{}^1_p$-norm.

It can be easily verified that (5.5) is fulfilled for the threedimensional generalizations of the polygonal $\{-q,\theta,_i\}$ laws (cf. Bingham's fluid and the projection operator).

Lemma 5.1: Suppose that (5.4), (5.5) hold. Then $(P_{\varepsilon n})$ has a solution.

Proof: We write (5.3) in the form

$$(T(\theta_{\varepsilon n}),\phi) = 0 \quad \forall \phi \in V_n, \tag{5.7}$$

where $T: V_n \to V_n'$. We apply Brouwer's fixed point theorem (cf. [13] p.53): if $f: \mathbb{R}^n \to \mathbb{R}^n$ is continuous such that for $r > 0$ $\sum_i (f(a))_i a_i \ge 0$ $\forall$ $a = \{a_i\}$ with $|a| = r$, then a_0 exists with $|a_0| \le r$ such that $f(a_0) = 0$. Indeed (5.4) implies that we can determine $\rho_1 > 0$ and $\rho_2 > 0$ such that

$$b_\varepsilon(\xi) \ge 0 \text{ if } \xi > \rho_1, b_\varepsilon(\xi) \le 0 \text{ if } \xi < -\rho_1 \text{ and } |b_\varepsilon(\xi)| \le \rho_2 \text{ if } |\xi| < \rho_1. \tag{5.8}$$

Moreover due to (5.5) and the inequality of Poincaré we can write the relation

$$(\mathrm{grad}\psi(\phi,_i),\mathrm{grad}\phi) \ge c\,\|\phi\|^p \quad \forall \phi \in \overset{\circ}{H}{}^1_p(\Omega) \quad c \text{ const} > 0. \tag{5.9}$$

Thus we find that

$$(T(\theta_{\varepsilon n}),\theta_{\varepsilon n}) \ge c_1 ||\theta_{\varepsilon n}||^p + \int_\Omega \theta_{\varepsilon n} b_\varepsilon(\theta_{\varepsilon n}) d\Omega - \|\bar{\bar{Q}}\|_* \|\theta_{\varepsilon n}\| \ge$$
$$\ge c_1 \|\theta_{\varepsilon n}\|^p - c_2 \|\theta_{\varepsilon n}\| - c_3 \tag{5.10}$$

since

$$\int_\Omega b_\varepsilon(\theta_{\varepsilon n})\theta_{\varepsilon n} d\Omega = \int_{|\theta_{\varepsilon n}(x)|>\rho_1} \ldots + \int_{|\theta_{\varepsilon n}(x)|\leq\rho_1} \ldots \geq 0 - \rho_1\rho_2 \text{mes}\Omega \qquad (5.11)$$

Here $\|\cdot\|_*$ denotes the $[\overset{\circ}{H}{}^1_p(\Omega)]'$-norm and c_1, c_2, c_3 are positive constants. (5.10) implies setting in the foregoing version of Brouwer's fixed point theorem $\theta_{\varepsilon n} = \sum_{i=1}^{n} a_i w_i$, and $f(a)_i = (T(\theta_{\varepsilon n}), w_i)$ that (5.7) admits a solution such that $\|\theta_{\varepsilon n}\| < c$. q.e.d.

Further we shall study the behaviour of the solution $u_{\varepsilon n}$ of the finite dimensional problem $(P_{\varepsilon n})$ as $\varepsilon \to 0$ and $n \to \infty$. Due to the fact that $\{\theta_{\varepsilon n}\}$ is bounded in $\overset{\circ}{H}{}^1_p$ we may extract a subsequence again denoted by $\{\theta_{\varepsilon n}\}$ such that

$$\theta_{\varepsilon n} \to \theta \qquad \text{weakly in } \overset{\circ}{H}{}^1_p(\Omega). \qquad (5.12)$$

But $H^1_p(\Omega)$ is compactly imbedded into $L^2(\Omega)$; thus we may write that

$$\theta_{\varepsilon n} \to \theta \qquad \text{strongly in } L^2(\Omega) \qquad (5.13)$$

and accordingly

$$\theta_{\varepsilon n}(x) \to \theta(x) \qquad \text{a.e. on } \Omega . \qquad (5.14)$$

Further we shall investigate the behaviour of $b_\varepsilon(\theta_{\varepsilon n})$ as $\varepsilon \to 0$ and $n \to \infty$.

Lemma 5.2: On the assumptions (5.4), (5.5) and (5.6), $\{b_\varepsilon(\theta_{\varepsilon n})\}$ is weakly precompact in $L^1(\Omega)$.

Proof: we show that by the Dunford-Pettis theorem (see [14], p. 239) for each $\alpha > 0$ a $\gamma > 0$ can be determined such that for $\omega \subset \Omega$ with mes $\omega < \gamma$

$$\int_\omega |b_\varepsilon(\theta_{\varepsilon n})| d\Omega < \alpha \qquad (5.15)$$

holds. Let us consider the inequality

$$|b_\varepsilon(\xi)\xi| + \xi_0 \sup_{|\xi|\leq\xi_0} |b_\varepsilon(\xi)| \geq \xi_0 |b_\varepsilon(\xi)| \qquad (5.16)$$

which implies that

$$\int_\omega |b_\varepsilon(\theta_{\varepsilon n})| d\Omega \leq \frac{1}{\xi_0} \int_\Omega |b_\varepsilon(\theta_{\varepsilon n})\theta_{\varepsilon n}| d\Omega + \int_\omega \sup_{|\theta_{\varepsilon n}(x)|\leq\xi_0^\varepsilon} |b_\varepsilon(\theta_{\varepsilon n})| d\Omega . \qquad (5.17)$$

But (5.3) and (5.6) together with (5.17) imply that

$$\int_\Omega |b_\varepsilon(\theta_{\varepsilon n})\theta_{\varepsilon n}|d\Omega = \int_{|\theta_{\varepsilon n}(x)|>\rho_1} b_\varepsilon(\theta_{\varepsilon n})\theta_{\varepsilon n}d\Omega + \int_{|\theta_{\varepsilon n}(x)|\leq\rho_1} |b_\varepsilon(\theta_{\varepsilon n})\theta_{\varepsilon n}|d\Omega \leq$$

$$\leq \int_\Omega b_\varepsilon(\theta_{\varepsilon n})\theta_{\varepsilon n}d\Omega + 2\int_{|\theta_{\varepsilon n}(x)|\leq\rho_1} |b_\varepsilon(\theta_{\varepsilon n})\theta_{\varepsilon n}|d\Omega = -(\mathrm{grad}\psi(\theta_{\varepsilon n,i}),\mathrm{grad}\theta_{\varepsilon n}) +$$

$$+ (\bar{\bar{Q}},\theta_{\varepsilon n}) + 2\int_{|\theta_{\varepsilon n}(x)|\leq\rho_1} |b_\varepsilon(\theta_{\varepsilon n})\theta_{\varepsilon n}|d\Omega \leq c + 2\,\rho_1\rho_2 \ \mathrm{mes}\Omega. \qquad (5.18)$$

We now choose ξ_0 such that

$$\frac{1}{\xi_0}\int_\Omega |b_\varepsilon(\theta_{\varepsilon n})\theta_{\varepsilon n}|d\Omega < \frac{\alpha}{2} \qquad (5.19)$$

for all n and ε and let γ be such that for mes$\omega < \gamma$ (cf. also (5.2))

$$\int_\omega \sup_{|\theta_{\varepsilon n}(x)|\leq\xi_0} |b_\varepsilon(\theta_{\varepsilon n})|d\Omega \leq \gamma \operatorname*{esssup}_{|\xi|\leq\xi_0+1} |b(\xi)| < \frac{\alpha}{2}\,. \qquad (5.20)$$

From (5.17),(5.18),(5.19) and (5.20) we obtain (5.15) q.e.d.

Lemma 5.2 has as a consequence that as $\varepsilon\to 0$ and $n\to\infty$

$$b_\varepsilon(\theta_{\varepsilon n}) \to \chi \qquad \text{weakly in } L^1(\Omega). \qquad (5.21)$$

Proposition 5.1: θ is solution of problem (P).

Proof: We recall that $\bigcup_n V_n$ is dense in V. In order to pass to the limit $\varepsilon\to 0$ $n\to\infty$ in (5.3) we have to pay special attention to the term $(\mathrm{grad}\psi(\mathrm{grad}\theta_{\varepsilon n}),\mathrm{grad}\phi)$. The following method is a combination of the well-known monotonicity argument (Minty's method) which is often used in the theory of variational inequalities with a compactness argument.

a) We show first that for $\varepsilon\to 0$ and $n\to\infty$ (5.3) and (5.21) imply the inequality

$$(\mathrm{grad}\psi(\theta,_i),\mathrm{grad}\phi) + \int_\Omega \chi\phi d\Omega = (\bar{\bar{Q}},\phi) \qquad \forall\phi\in V_n\,. \qquad (5.22)$$

Indeed $\phi\in V_n \subset \overset{\circ}{H}{}^1_p(\Omega) \subset L^\infty(\Omega)$. Accordingly $\int b_\varepsilon\phi \to \int\chi\phi$. Let us formulate the nonnegative expression

$$X_n = (\mathrm{grad}\psi(\theta_{\varepsilon n,i}) - \mathrm{grad}\psi(\phi,_i),\mathrm{grad}\theta_{\varepsilon n} - \mathrm{grad}\phi) \geq 0 \quad \forall\phi\in \overset{\circ}{H}{}^1_p(\Omega) \quad (5.23)$$

which by means of (5.3) becomes

$$X_n = (\bar{\bar{Q}},\theta_{\varepsilon n}) - \int_\Omega b_\varepsilon(\theta_{\varepsilon n})\theta_{\varepsilon n}d\Omega - (\mathrm{grad}\psi(\theta_{\varepsilon n,i}),\mathrm{grad}\phi) -$$

$$- (\mathrm{grad}\psi(\phi,_i),\mathrm{grad}(\theta_{\varepsilon n} - \phi)) \geq 0 \quad \forall\phi\in \overset{\circ}{H}{}^1_p(\Omega)\,. \qquad (5.24)$$

From (5.3) we obtain easily that $\| \mathrm{grad}\psi(\theta_{\varepsilon n,i}) \|_{[L^{p'}(\Omega)]^3} \leq c$. Therefore as $\varepsilon \to 0$ $n \to \infty$

$$\mathrm{grad}\psi(\theta_{\varepsilon n,i}) \to Y \quad \text{weakly in } [L^{p'}(\Omega)]^3. \tag{5.25}$$

Thus from (5.3),(5.21) and (5.25) we get the equality

$$(Y,\mathrm{grad}\phi) + \int_\Omega \chi\phi d\Omega = (\bar{\bar{Q}},\phi) \quad \forall\phi \in V_n \tag{5.26}$$

and its extension by density to V, (i.e. $\forall\phi \in V$). We denote this extension by (5.26a).

We assume at the present that as $\varepsilon \to 0, n \to \infty$

$$\lim \int_\Omega b_\varepsilon(\theta_{\varepsilon n})\theta_{\varepsilon n} d\Omega = \int_\Omega \chi\theta d\Omega . \tag{5.27}$$

This last relation will be shown further in Lemma 5.3. From (5.27) and (5.24),(5.25),(5.26a) we find that

$$\lim X_n = (Y,\mathrm{grad}(\theta-\phi)) - (\mathrm{grad}\psi(\phi,_i),\mathrm{grad}(\theta-\phi)) \geq 0 \quad \forall\phi \in \overset{\circ}{H}{}^1_p(\Omega) \tag{5.28}$$

Let us set in (5.28) $\theta-\phi = \lambda h$, $\lambda > 0$. We get the relation

$$(Y,\mathrm{grad}h) - (\mathrm{grad}\psi((\theta-\lambda h),_i),\mathrm{grad}h) \geq 0 \quad \forall h \in \overset{\circ}{H}{}^1_p(\Omega) . \tag{5.29}$$

Due to the monotonicity of $\lambda \to (\mathrm{grad}\psi((\theta-\lambda h),_i),\mathrm{grad}h)$ we may take $\lambda \to 0_+$ in (5.29) and we find for $\pm h$ that

$$Y = \mathrm{grad}\psi(\theta,_i) . \tag{5.30}$$

(5.30) and (5.26) imply (5.22).

b) It remains to show that $\chi \in \bar{\partial}\varphi(\theta)$ a.e. on Ω. This part of the proof is the same as in [3] (see eqs. (8.4.35÷44)) and is based on Egoroff's theorem.

In order to complete the proof we have to show (5.27).

Lemma 5.3: The relation (5.27) holds.

Proof: We recall (5.12),(5.13) and (5.21) and we form the difference

$$\int_\Omega [b_\varepsilon(\theta_{\varepsilon n})\theta_{\varepsilon n} - \chi\theta] d\Omega = \int_\Omega b_\varepsilon(\theta_{\varepsilon n})(\theta_{\varepsilon n} - \theta) d\Omega +$$

$$+ \int_\Omega \theta(b_\varepsilon(\theta_{\varepsilon n}) - \chi) d\Omega = A + B . \tag{5.31}$$

Note that due to (5.14) we may choose for $\gamma > 0$ a $\omega \subset \Omega$ with mes $\omega < \gamma$ such that as $\varepsilon\to 0, n\to\infty$ (Egoroff's theorem)

$$\theta_{\varepsilon n}\to\theta \quad \text{uniformly on } \Omega-\omega, \quad \theta\in L^{\infty}(\Omega-\omega). \tag{5.32}$$

We split A and B into a term over $\Omega-\omega$ and into a term over ω where ω is chosen as in Egoroff's theorem. Then using the Dunford-Pettis theorem for $\{b_\varepsilon(\theta_{\varepsilon n})\}$ (cf. (5.15)) and (5.12÷14),(5.21) and (5.32) we obtain the result. q.e.d.

6. STUDY OF THE NONDIFFERENTIABLE CASE

If in problem (P) the internal energy function $\psi(\cdot)$ is nondifferentiable, then we introduce a sequence of convex functions ψ_ρ depending on a parameter ρ such that

(i) As $\rho\to 0$

$$\int_\Omega \psi_\rho(\mathrm{grad}\phi)d\Omega\to\int_\Omega\psi(\mathrm{grad}\phi)d\Omega \quad \forall\phi\in \overset{\circ}{H}{}^1(\Omega). \tag{6.1}$$

(ii) If $\phi_\rho\to\phi$ weakly in $\overset{\circ}{H}{}^1_p(\Omega)$ for $\rho\to 0$ and $\int_\Omega\psi_\rho(\mathrm{grad}\phi_\rho)d\Omega< c$, then

$$\liminf_{\rho\to 0}\int_\Omega \psi_\rho(\mathrm{grad}\phi_\rho)d\Omega \geq \int_\Omega \psi(\mathrm{grad}\phi)d\Omega . \tag{6.2}$$

iii) Relations (5.5) and (5.6) hold for every ψ_ρ with c independent of ρ.

Now we define the following regularized problem $(P_{\varepsilon\rho})$

$(P_{\varepsilon\rho})$: Find $\theta_{\varepsilon\rho}\in \overset{\circ}{H}{}^1_p(\Omega)$ such that

$$(\mathrm{grad}\psi_\rho(\theta_{\varepsilon\rho,i}),\mathrm{grad}\phi)+\int_\Omega b_\varepsilon(\theta_{\varepsilon\rho})\phi d\Omega = (\bar{\bar{Q}},\phi) \quad \forall\phi\in \overset{\circ}{H}{}^1_p(\Omega) \tag{6.3}$$

and by means of a Galerkin basis the problem $(P_{\varepsilon\rho n})$

$(P_{\varepsilon\rho n})$: Find $\theta_{\varepsilon\rho n}\in V_n$ such that

$$(\mathrm{grad}\psi_\rho(\theta_{\varepsilon\rho n,i}),\mathrm{grad}\phi)+\int_\Omega b_\varepsilon(\theta_{\varepsilon\rho n})\phi d\Omega = (\bar{\bar{Q}},\phi) \quad \forall\phi\in V_n . \tag{6.4}$$

The following proposition holds:

Proposition 6.1: Suppose that (5.4) holds and that ψ_ρ satisfies (i),(ii), (iii). Then problem (P) has a solution.

Proof: For problem $(P_{\varepsilon\rho n})$ we can prove as in Lemma 5.1 that a solution exists and that

$$\|\theta_{\varepsilon\rho n}\| < c \tag{6.5}$$

where c is independent of ε,ρ and n. Thus as $\varepsilon\to 0$ $n\to\infty$

$$\theta_{\varepsilon\rho n}\to\theta_\rho \quad \text{weakly in } \overset{\circ}{H}{}^1_p(\Omega) \tag{6.6}$$

and thus strongly in $L^p(\Omega)$ and a.e. on Ω. Moreover Lemma 5.2 holds for $\{b_\varepsilon(\theta_{\varepsilon\rho n})\}$, as it is obvious, with α in (5.15) independent of ε, n and ρ. Thus

$$b_\varepsilon(\theta_{\varepsilon\rho n})\to\chi_\rho \quad \text{weakly in } L^1(\Omega)\,. \tag{6.7}$$

Note that (6.5) implies that as $\rho\to 0$

$$\theta_\rho\to\theta \quad \text{weakly in } \overset{\circ}{H}{}^1_p(\Omega). \tag{6.8}$$

On the other hand (5.15) implies that

$$\chi_\rho\to\chi \quad \text{weakly in } L^1(\Omega). \tag{6.9}$$

From (6.4) we obtain that

$$\|\mathrm{grad}\psi_\rho(\mathrm{grad}\theta_{\varepsilon\rho n})\|_{[L^{p'}(\Omega)]^3} < c \tag{6.10}$$

where c is independent of ε,n,ρ and thus as $\varepsilon\to 0, n\to\infty$

$$\mathrm{grad}\psi\,(\theta_{\varepsilon\rho n,i})\to Y_\rho \text{ weakly in } [L^{p'}(\Omega)]^3. \tag{6.11}$$

From (6.4) we obtain by passing to the limit $\varepsilon\to 0, n\to\infty$ and by density the variational equality

$$(Y_\rho,\mathrm{grad}\phi)+\int_\Omega\chi_\rho\phi\,d\Omega = (\bar{\bar{Q}},\phi) \quad \forall\phi\in\overset{\circ}{H}{}^1_p(\Omega). \tag{6.12}$$

As in Prop. 5.1 the monotonicity argument implies that

$$Y_\rho = \mathrm{grad}\psi_\rho(\theta_{\rho,i}). \tag{6.13}$$

To this end we prove as in Lemma 5.3 that

$$\lim_{\substack{\varepsilon\to 0\\ n\to\infty}}\int_\Omega b_\varepsilon(\theta_{\varepsilon\rho n})\theta_{\varepsilon\rho n}d\Omega = \int_\Omega\chi_\rho\theta_\rho d\Omega\,. \tag{6.14}$$

Further we pass to the limit with respect to ρ. From (6.12) and (6.13) we obtain, due to the convexity of ψ_ρ, that

$$\int_\Omega[\psi_\rho(\mathrm{grad}\phi)-\psi_\rho(\mathrm{grad}\theta_\rho)]d\Omega+\int_\Omega\chi_\rho(\phi-\theta_\rho)d\Omega \ge (\bar{\bar{Q}},\phi-\theta_\rho) \quad \forall\phi\in\overset{\circ}{H}{}^1_p(\Omega). \tag{6.15}$$

Let us take in (6.15) ϕ such that $\int\psi(\mathrm{grad}\phi)d\Omega<\infty$. Then, due to (6.1) $\int\psi_\rho(\mathrm{grad}\phi)d\Omega< c$ and from (6.15) we get that

$$\int_\Omega\psi_\rho(\mathrm{grad}\theta_\rho)d\Omega< c \tag{6.16}$$

and therefore due to (6.8), (6.2) holds. From (6.15) we obtain for $\rho\to 0$

$$\liminf[\int_\Omega\psi_\rho(\mathrm{grad}\phi)d\Omega+\int_\Omega\chi_\rho\phi d\Omega] \geq \liminf[\int_\Omega\psi_\rho(\mathrm{grad}\theta_\rho)d\Omega + \\ + \int_\Omega\chi_\rho\theta_\rho d\Omega + (\bar{\bar{Q}},\phi-\theta_\rho)] \tag{6.17}$$

for every $\phi\in \overset{\circ}{H}{}^1_p(\Omega)$ with $\int\psi(\mathrm{grad}\phi)d\Omega<\infty$. But from (6.1),(6.2),(6.9) and the fact that (same proof as in Lemma 5.3)

$$\lim_{\rho\to 0\,\Omega}\int\chi_\rho\theta_\rho d\Omega = \int_\Omega\chi\theta d\Omega \tag{6.18}$$

we conclude that the inequality

$$\int_\Omega\psi(\mathrm{grad}\phi)d\Omega-\int_\Omega\psi(\mathrm{grad}\theta)d\Omega+\int_\Omega\chi(\phi-\theta)d\Omega \geq (\bar{\bar{Q}},\phi-\theta) \\ \forall\phi\in \overset{\circ}{H}{}^1_p(\Omega) \text{ with } \int\psi(\mathrm{grad}\phi)d\Omega<\infty \tag{6.19}$$

is satisfied by $\theta\in \overset{\circ}{H}{}^1_p(\Omega)$ with $\int\psi(\mathrm{grad}\theta)d\Omega<\infty$.

Finally we show that $\chi\in \bar{\partial}\phi(\theta)$ a.e. on Ω and thus θ is a solution of problem (P), q.e.d.

REFERENCES:

1. Moreau, J.J.: La notion de sur-potentiel et les liaisons unilatérales en élastostatique, C.R.Acad.Sc., Paris 267A (1968),954-957.
2. Duvaut, G. and J.L. Lions: Les inéquations en Mécanique et en Physique, Dunod, Paris 1972.
3. Panagiotopoulos, P.D.: Inequality Problems in Mechanics. Convex and Nonconvex Energy Functions, Birkhäuser Verlag, Basel/Boston 1985.
4. Panagiotopoulos, P.D.: Nonconvex Energy Functions. Hemivariational Inequalities and Substationarity Principles, Acta Mechanica, 42 (1983), 160-183.
5. Rockafellar, R.T.: Generalized Directional Derivatives and Subgradients of Non-convex Functions, Can.J.Math., XXXII (1980), 257-280.
6. Panagiotopoulos, P.D.: Dynamic and Incremental Variational Inequality Principles, Differential Inclusions and their Applications to Co-existent Phases Problems, Acta Mechanica, 40 (1981), 85-107.
7. Germain, J.P.: Cours de Mécanique des Milieux Continus, I, Masson, Paris 1973.

8. Halphen, B. and N.Q. Son: Sur les Matériaux Standards Généralisés, J.de Mécanique, 14 (1975), 39-63.
9. Frémond, M.: Dissipation dans l'adhérence des solides, C.R.Ac.Sc. Paris, 300 II (1985), 709-714.
10. Ioffe, A.D.: Nonsmooth Analysis: Differential Calculus of Nondifferentiable Mappings, Trans.Am.Math.Soc., 266 (1981), 1-56.
11. Panagiotopoulos, P.D.: Nonconvex Problems of Semipermeable Media and Related Topics, ZAMM, 65 (1985), 29-36.
12. Chang, K.C.: Variational Methods for Non-differentiable Functionals and their Applications to Partial Differential Equations, J.Math. Anal.Appl., 80 (1981), 102-129.
13. Lions, J.L.: Quelques méthodes de résolution des problèmes aux limites non linéaires, Dunod/Gauthier-Villars, Paris 1969.
14. Ekeland, I. and R. Temam: Convex Analysis and Variational Problems, North-Holland, Amsterdam and American Elsevier, New York 1976.

CODES D'ELEMENTS FINIS POUR DES PROBLEMES DE CONTACTS UNILATERAUX AVEC FROTTEMENT FORMULES IN TERMES D'INEQUATIONS VARIATIONNELLES

M. Raous, J.C. Latil
CNRS-Laboratoires de Mécanique et d'Acoustique, Marseille

ABSTRACT

Faced with the growing importance taken by contact phenomena with friction in many problems issued from structural mechanics (metal forming, structure assembling, cracking, ...) a great number of big finite element codes have developed modules treating these very special limit conditions. Modelisation of friction together with an appropriate numerical treatment is a difficult problem which remains largely open. Future developments in this domain will have to rest on a strong concertation between structural mechanicians, tribologists, mathematicians, numerical analysts and physicists from industry.

However during this past decade a mathematical formulation together with numerical methods appropriate to Coulomb friction models with unilateral contact has been developed. Corresponding numerical procedures are or may be introduced into Finite Element codes and are to be compared to methods more directly inspired by the physics of the problem such as contact finite elements (evanescent third body) or iterative procedures on the boundary conditions.

A formulation in terms of a variational inequality may in turn be solved in different ways. We present here one of these, underlining its advantages and disadvantages with regard to other approaches. It consists of a diagonal fixed point algorithm based on a "Gauss-Seidel projected overrelaxed method" together with several adjustments which enable to gain in performance. This method is developed in the code PROTIS at the L.M.A. and also introduced in different large scale codes (MODULEF, EVPCYCL).

The performances of a numerical method associated to a given formulation have to be appreciated under different angles. There is not only the precision (in the phenomenon's description) or the cost of the computation (a factor wich tends to be not essential) but also the facility with which the procedure can be implemented into an existing code. We examine here the proposed method under all these aspects.

1. INTRODUCTION

Devant l'importance de la prise en compte des phénomènes de contact avec frottement dans beaucoup de problèmes en mécanique des structures (mise en forme, assemblage, fissuration,...), la plupart des grands codes de calcul d'éléments finis ont vu se développer des modules de traitement de ces conditions aux limites très particulières. La modélisation et un traitement numérique approprié du frottement sont des problèmes difficiles qui demeurent largement ouverts. Les développements dans ce domaine devront s'appuyer sur une forte concertation entre mécaniciens des structures, tribologues, mathématiciens, numériciens et physiciens du milieu industriel.

Toutefois, cette dernière décennie a vu se développer une formulation mathématique et des méthodes de calcul adaptées aux modèles de frottement de Coulomb avec contact unilatéral. Des modules de traitement numérique correspondants sont ou peuvent être implantés dans des grands codes de calcul et sont à comparer avec des méthodes d'inspiration plus physique comme les éléments finis de contact (troisième corps évanescent) ou certains processus itératifs sur des conditions aux limites.

La formulation en termes d'inéquations variationnelles se prête elle-même à différentes méthodes de résolution. Nous nous proposons ici de présenter l'une d'entre elles en soulignant ses avantages et ses inconvénients par rapport à d'autres approches : il s'agit d'un algorithme diagonal de point fixe s'appuyant sur une méthode de Gauss-Seidel surrelaxée projetée et de différentes améliorations conduisant à des gains de performances. Cette méthode est développée dans le code PROTIS du Laboratoire de Mécanique et d'Acoustique et implantée également dans différents codes de grande taille (MODULEF, EVPCYCL,...).

Les qualités d'une méthode de résolution numérique associée à une formulation donnée doivent être appréciées sous différents aspects. Il s'agira non seulement de la précision (dans la description du phénomène), du coût de calcul (facteur qui tend à devenir moins déterminant) mais également de la facilité à implanter le module de calcul correspondant dans un code existant. C'est ce que nous allons examiner pour la méthode proposée.

2. LE MODELE ET SA FORMULATION VARIATIONNELLE

Notre approche s'appuie sur la formulation proposée par Duvaut dans [1] sous la terminologie de frottement pour un problème statique. Nous reviendrons sur l'extension de ce modèle aux problèmes d'évolution quasi statiques à la fin de ce paragraphe.

Soit un solide occupant le domaine ouvert Ω de $\mathbb{R}^3$ de frontière suffisamment régulière $\partial\Omega = \Gamma_1 \cup \Gamma_2 \cup \Gamma_3$. Des conditions aux limites homogènes sont appliquées sur la partie Γ_1 de la frontière. Le solide est soumis à une densité de force volumique ϕ_1 dans Ω et une densité de force surfacique ϕ_2 sur Γ_2. Sur la partie Γ_3 de la frontière le solide est soumis à des conditions de contact unilatéral avec frottement avec un obstacle rigide.

Sur Γ_3, nous allons décomposer la trace $\gamma u_{|\Gamma_3}$ du déplacement u sur cette partie de la frontière et la densité F de la force de contact exercée par l'obstacle sur le solide, de la façon suivante :

(1) $$\gamma u_{|\Gamma_3} = u_N n + u_T \qquad \text{avec} \quad \begin{cases} u_N = \{\gamma u\}_i\, n_i \\ \{u_T\}_i = \{\gamma u\}_i - u_N n_i \end{cases}$$

(2) $$F = F_N n + F_T \qquad \text{avec} \quad F = \sigma n$$

$$\begin{cases} F_N = \sigma_{ij} n_j n_i \\ \{F_T\}_i = \sigma_{ij} n_j - F_N n_i \end{cases}$$

où n désigne le vecteur de composantes n_i, normal à la frontière Γ_3 en un point de celle-ci et σ le tenseur des contraintes de composantes σ_{ij}. Ainsi, les conditions de contact unilatéral s'écrivent :

(3) $$\begin{cases} u_N \leq 0 \\ F_N \leq 0 \\ u_N . F_N = 0 \end{cases}$$

On peut noter la faiblesse de la modélisation du contact unilatéral en petites déformations. La condition u_N négatif ou nul implique que la surface de contact après déformation sera au plus égale à la surface de contact initiale Γ_3 : on pourra avoir des décollements sur Γ_3 mais on ne pourra pas caractériser une extension de la zone de contact. Seule une formulation du problème en grandes déformations permettra de prendre en compte ce phénomène en introduisant une fonction à valeurs finies caractérisant la distance initiale du solide à l'obstacle. Toutefois nous avons effectué une approche numérique en écrivant que u_N doit être inférieur ou égal à une fonction finie donnée α de ce type et en vérifiant a posteriori que la solution calculée u est bien du même ordre que les valeurs de la fonction donnée α (cohérence du modèle en petites déformations).

Nous écrivons la loi de Coulomb sous la forme suivante (cas statique) associée aux relations (3) ($F_N \leq 0$).

$$(4)\quad \text{avec :}\quad \begin{cases} |F_T| \leq -\mu F_N \\ |F_T| < -\mu F_N \Rightarrow u_T = 0 \\ |F_T| = -\mu F_N \Rightarrow \begin{cases} u_T = -\lambda F_T \quad \lambda > 0 & \text{si } |F_T| > 0 \\ \text{pas de conditions sur } u_T & \text{si } |F_T| = 0 \end{cases} \end{cases}$$

La dernière éventualité correspond soit au cas limite de frottement nul ($\mu = 0$), soit au cas de non contact ($F_N = 0$).

La formulation variationnelle des équations d'équilibre conduit au problème P_1 en tenant compte des relations cinématiques (hypothèse des petites déformations) et des relations (3) et (4).

<u>Problème P_1</u> :

$$(5)\quad \left| \begin{array}{l} \text{Trouver } u \in \mathbf{K} = \{v \in (H^1(\Omega))^3 \text{ tel que } \gamma v_{|\Gamma_1} = 0 \text{ et } v_N \leq 0 \text{ sur } \Gamma_3\} \\ \text{tel que : } \forall v \in \mathbf{K} \\ a(u,v-u) - (f,v-u) + j(v,v) - j(u,v) \geq 0 \end{array} \right.$$

où - $a(u,v)$ est une forme bilinéaire, symétrique, définie positive et coercive :

$$(6)\qquad a(u,v) = \int_\Omega \text{grad}_s\, u\, K\, \text{grad}_s\, v\, dx.$$

K est la matrice d'élasticité, grad_s désigne l'opérateur gradient symétrique.

- (f,v) est une forme linéaire associée au chargement :

$$(7)\qquad (f,v) = \int_\Omega \phi_1\, v\, dx + \int_{\Gamma_2} \phi_2\, \gamma v_{|\Gamma_2} d\ell.$$

- $j(u,v)$ est une forme non différentiable

$$(8)\qquad j(u,v) = \int_{\Gamma_3} - \mu F_N(u)\, |v_T|\, d\ell.$$

Le problème P_1 est un problème difficile qui relève des inéquations quasivariationnelles (ceci apparaît plus directement dans la forme duale du problème écrite en contrainte). C'est pourquoi, nous préférons la formulation suivante :

Problème P_2 : Trouver la fonction $g \in (L^2(\Gamma_3))^3$, point fixe de l'application :

(9) $$g = -\mu F_N(S(g))$$

où S est l'application de $(L^2(\Gamma_3))^3$ dans $\mathbb{K}$

$S : g \longmapsto u$ défini par :

(10) $$\begin{cases} u \in \mathbb{K} \text{ tel que : } \forall v \in \mathbb{K} \\ a(u,v-u) - (f,v-u) + j(v) - j(u) \geq 0 \end{cases}$$

avec

(11) $$j(v) = \int_{\Gamma_2} g \mid v_T \mid d\ell.$$

Le problème s'appuie maintenant sur la résolution de l'inéquation variationnelle (10) (contenant des termes non différentiables j(v) (11)) couplée au procédé de point fixe sur la fonction g qui représente le seuil de glissement. Pour une valeur fixée de g, la relation (10) modélise un problème de frottement de Tresca. Le frottement de Coulomb est traité à l'aide de la méthode de point fixe (9).

Enfin, la forme bilinéaire a(u,v) étant symétrique, le problème (10) est équivalent au problème d'optimisation avec contraintes suivant (minimisation de l'énergie potentielle) :

Problème P_3 : Trouver $u \in \mathbb{K}$ tel que $\forall v \in \mathbb{K}$ $J(u) \leq J(v)$
avec :

(12) $$J(v) = \frac{1}{2} a(v,v) - (f,v) + j(v)$$

On établit l'équivalence des problèmes P_1 et P_2, la convergence du procédé (9), l'existence et l'unicité (pour un frottement faible) pour les solutions d'une forme régularisée de ces problèmes qui correspond à une loi de frottement non locale (voir [2], [3], [4], [5]). Nous n'aborderons pas ces questions ici, d'autant que les écritures adoptées dans ce texte sont restées assez formelles.

Du point de vue numérique, on notera que la convergence du procédé de point fixe (9) est assez rapide : 5 à 10 itérations dans les cas les plus défavorables. De toutes manières, nous verrons dans le paragraphe suivant un procédé d'accélération de convergence de la recherche du point fixe.

Commentaires :

Il faut tout d'abord apporter quelques précisions quant au domaine de validité et aux précautions d'utilisation de la loi de frottement (4). La formulation naturelle de la loi de Coulomb s'écrit en termes de conditions sur la vitesse de déplacement tangentiel alors que les relations (4) font intervenir seulement le déplacement tangentiel lui-même. Par ailleurs, la formulation naturelle des conditions unilatérales de non pénétration dans l'obstacle s'écrivent sur les déplacements (relations (3)). Le choix de l'inconnue pour le problème complet conduit à un compromis soit sur les conditions unilatérales dans le cas de formulation en vitesse, soit sur les conditions de frottement dans le cas de formulation en déplacement. Il s'agit de ce dernier choix dans cette présentation.

La validité du modèle précédent, problème d'élasticité statique en petites déformations, peut être étendue au cas de phénomènes évoluant de manière quasistatique moyennant certaines précautions. Les relations (4) seront alors considérées comme une formulation incrémentale de la loi de Coulomb et la résolution du problème P_2 se fera à chaque instant t sur la configuration actualisée. Ceci peut apparaître pénalisant pour les problèmes simples mais s'inscrit bien dans une optique de généralisation ultérieure aux problèmes de grands déplacements, grandes déformations. Les phénomènes quasi statiques évoqués relèvent aussi bien de problèmes évolutifs d'élasticité, que de problèmes de viscoélasticité [6] ou de viscoplasticité [9]. Il est clair que, vu la nature fortement non linéaire du frottement, la solution pour ces problèmes d'évolution, dépend du trajet de chargement. La résolution du problème P_2 sur la configuration actualisée permet le calcul de la solution pour un pas de temps fixé.

On notera également l'utilisation de ce type de modèle (sans frottement) dans l'étude du comportement de solides fissurés sous chargements complexes pour décrire correctement les phénomènes de contact sur les lèvres de la fissure (voir Raous [16]).

Une deuxième remarque concerne le choix présenté ici entre formulation en déplacement ou formulation duale en contrainte (voir [7], [8]). La formulation en déplacement nous a paru plus intéressante à cause de son côté relativement standard dans beaucoup de grands codes de calcul. Notre objectif est de présenter ici un procédé de calcul qui se prête bien à une intégration dans un code de calcul existant : nous verrons qu'il s'agira essentiellement pour un code "en déplacement" de remplacer le module de résolution du système linéaire par le module décrit au paragraphe 2. Par ailleurs, nous avons montré que cette formulation en déplacement se prête à une extension du modèle pour le traitement de problèmes de viscoélasticité [6] ou de viscoplasticité [9].

La formulation duale en contrainte présente toutefois certains avantages. En particulier, l'extension du modèle au cas du contact de deux solides déformables est plus simple. Ici, la généralisation de l'algorithme de projection décrit au paragraphe 2 pour ce cas, est plus délicate : en effet, le procédé de projection doit assurer également la continuité des forces sur la zone de contact des deux solides.

3. METHODE NUMERIQUE ET ALGORITHME

Le premier choix directement lié à la modélisation adoptée se situe au niveau de la classe de méthodes envisagées : méthodes variationnelles, éléments finis de contact, point fixe sur les conditions aux limites.

Parmi les approches de type éléments finis de contact il faut distinguer deux types de méthodes. Les unes effectuent une modélisation d'un troisième corps fictif, évanescent lors du contact : les caractéristiques de ce troisième corps permettent d'introduire le frottement et les conditions de non pénétration (exemple code CASTOR). D'autres introduisent ces relations entre noeuds en vis à vis sur la frontière susceptibles de contact sous la forme d'éléments "barres" ou "tiges" (exemple codes MARC,TITUS,MEF..).

Une autre classe de méthodes utilisées peut être désignée par le vocable de "point fixe sur les conditions aux limites". Ce sont des méthodes itératives où chaque itération est constituée d'une résolution complète d'un problème élastique : on ajoute une condition aux limites lorsqu'il y a pénétration, on en supprime une lorsque des forces de traction apparaissent sur le bord de contact. Le traitement du frottement est plus délicat. Malgré l'absence de résultats théoriques de convergence et les risques d'oscillations, il semble que ce procédé empirique soit assez efficace pour les problèmes de contact unilatéral sans frottement dans le cas de géométrie simple.

Notre approche s'inscrit dans le cadre des méthodes variationnelles. Il existe plusieurs types de méthodes de résolution d'inéquations variationnelles ou du problème de complémentarité posé initialement.

Les méthodes directes relevant de la programmation non linéaire permettent de traiter le problème de complémentarité (3) : la plus connue est due à Cottle-Dantzig et Lemke [10]. Elle est utilisée par Klarbring [11] par exemple pour traiter le problème avec frottement.

Les méthodes itératives relèvent des méthodes d'optimisation sous contraintes adaptées à la résolution du problème posé sous sa forme (12). Les plus utilisées sont les méthodes de pénalisation d'une part, et les méthodes de relaxation ponctuelle avec projection d'autre part. Les méthodes de pénalisation sont très utilisées aussi bien sur le problème primal que sur le problème dual. La difficulté réside dans la mise au point d'une stratégie convenable dans l'évolution du ou des coefficients de pénalisation au cours des calculs. Les conditions de non interpénétration ne sont en général pas vérifiées exactement. Ce sont des méthodes efficaces dont les performances dépendent de la qualité de la stratégie de la pénalisation (par exemple [4], [7], [8]).

Nous allons maintenant présenter et commenter la méthode de Gauss-Seidel surrelaxée avec projection (GSRP) que nous utilisons.

3.1 Algorithme utilisé dans le code PROTIS

Le code PROTIS est le code d'Eléments Finis développé au Laboratoire par J.C. LATIL [13]. Il s'agit d'un code de taille moyenne traitant de problèmes unilatéraux avec frottement en élasticité, viscoélasticité, viscoplasticité sous les hypothèses de contraintes planes, déformations planes ou symétrie de révolution.

Nous nous appuyons sur une méthode de relaxation ponctuelle dont les résultats de convergence sont établis dans [12]. On se ramène à une suite de problèmes de minimisation sous contraintes dans $\mathbb{R}$ ou $\mathbb{R}^-$: la minimisation se fait sur une composante après l'autre du vecteur constitué par les déplacements nodaux u_i ($i=1,n$; n est le nombre de degrés de liberté) et elle est donc élémentaire. Une étape $k+1$ du calcul consiste à résoudre le problème suivant pour $i = 1,n$:

<u>Problème P_4</u> : Trouver $u_i^{k+1} \in \mathbb{R}^-$ si u_i est une composante normale d'un noeud de contact

$u_i^{k+1} \in \mathbb{R}$ sinon

tel que $\forall v \in \mathbb{R}^-$ ou $v \in \mathbb{R}$, comme précédemment

$$J(u_1^{k+1},\dots,u_{i-1}^{k+1},u_i^{k+1},u_{i+1}^{k},\dots,u_n^{k}) \leq J(u_1^{k+1},\dots,u_{i-1}^{k+1},v,u_{i+1}^{k},\dots,u_n^{k}) \tag{13}$$

On effectue pour les noeuds de contact un changement de variables en passant en composantes normales et tangentielles par une rotation locale de coordonnées. La méthode de minimisation utilisée dans le contexte de la relaxation ponctuelle est une méthode de Gauss-Seidel surrelaxée avec projection modifiée pour tenir compte du terme non différentiable (11). Le procédé est le suivant :

- pour un noeud n'appartenant pas à la frontière de contact Γ_3 : traitement de Gauss-Seidel surrelaxée (GSR) classique :

$$(u_i^{k+1})_{GS} = -\frac{1}{a_{ii}}\left(\sum_{j=1}^{i-1} a_{ij}\,u_j^{k+1} + \sum_{j=i+1}^{n} a_{ij}\,u_j^{k} - b_i\right) \tag{14}$$

$$u_i^{k+1} = (1-\omega)\,u_i^{k} + \omega(u_i^{k+1})_{GS} \tag{15}$$

où : - u_i^k désigne la ième composante du vecteur des déplacements nodaux à l'itération k

- a_{ij} désigne le terme général de la matrice de rigidité globale

- b_i désigne la ième composante du vecteur second membre (charge et éventuellement contraintes thermiques, contraintes viscoélastiques ou viscoplastiques couplées à d'autres relations de comportement)

- ω est le coefficient de relaxation ; $0 < \omega < 2$

- pour la composante normale d'un noeud de la frontière de contact Γ_3 : traitement de Gauss-Seidel surrelaxée avec projection (GSRP). La projection est ici une opération simple puisqu'elle s'effectue sur le demi plan négatif. On applique (14) et (15).
  Si u_i^{k+1} est strictement positif, on pose $u_i^{k+1} = 0$;

- pour la composante tangentielle d'un noeud de la frontière de contact Γ_3, nous avons à résoudre un problème de minimisation dans $\mathbb{R}$ d'une fonctionnelle quadratique en u_i à laquelle on ajoute un terme en $|u_i|$. Ceci est simplement traité en examinant les deux éventualités de signe pour u_i : nous effectuons au plus deux calculs des étapes (14) et (15).

Ce procédé nous conduit à la solution u^ν du problème (12) ou (10) pour g_i^ν fixé. Les composantes du vecteur g^ν sont les seuils de glissement en chaque noeud de contact. L'itération de point fixe (9) s'écrit alors :

(16) $$g_i^{\nu+1} = - \mu F_{N_i}(u^\nu)$$

Notons ici un détail technique essentiel : les forces de contact F_{N_i} ne sont pas déduites du calcul des contraintes mais sont calculées directement à partir des résidus $R_i^\nu = \sum_{j=1}^{n} a_{ij} u_j^\nu - b_i$. Ce procédé à la fois précis et simple conduit à une bonne convergence de l'algorithme de point fixe.

Nous allons utiliser en fait un procédé diagonal décrit au paragraphe 3.3 : les mises à jour du seuil de glissement g^ν sont effectuées au cours des itérations de relaxation. On applique donc la méthode GSRP une seule fois avec modification de la fonction g au cours du calcul.

3.2 Commentaires

Il faut tout d'abord noter que l'aspect unilatéral et le frottement sont traités simultanément et qu'il n'y a pas d'itérations entre la résolution d'un problème unilatéral sans frottement et celle d'un problème de frottement avec contact imposé. Il n'y a pas de sous itérations : on utilise une seule fois la méthode GSRP modifiée pour tenir compte du frottement.

Suivant les critères posés dans l'introduction, on peut dire que la méthode est précise, aisée à mettre en oeuvre et relativement coûteuse sous la forme brute présentée ci-dessus.

La très bonne qualité de la précision sera appréciée sur les applications présentées dans le paragraphe 5. Les conditions (3) et (4) sont vérifiées avec une précision du millième sur l'égalité $|F_T| = -\mu F_N$.

La facilité de mise en oeuvre est extrêmement intéressante pour implanter cette méthode dans un code de calcul existant : il suffit de faire une rotation locale de coordonnées des noeuds de contact pour passer en composantes normales-tangentielles et de remplacer le "solveur" par le module approprié décrit ci-dessus.

Mais sous la forme simple présentée la méthode est relativement coûteuse. Les temps de calcul seront commentés par référence au temps de calcul nécessaire au traitement du même problème où les conditions unilatérales et de frottement sont remplacées par des conditions aux limites classiques : on l'appellera problème de référence (PR). Ceci permet une bonne estimation du coût de traitement du problème de contact pour n'importe quel code de calcul.

Les commentaires concernant les performances sont illustrés par des résultats relatifs à l'exemple présenté au paragraphe 5. La discussion menée de façon précise sur cet exemple donne la tendance générale du comportement de la méthode et des améliorations proposées, observées à travers le traitement d'un certain nombre d'exemples significatifs.

Dans le tableau 1, nous observons que la méthode sous sa forme simple de mise en oeuvre coûte environ 30 fois plus cher que le traitement du problème de référence (PR). Observant que les temps d'exécution du module de résolution seul sont dans un rapport de 130, on en déduit immédiatement que pour un problème élémentaire à un seul cas de charge la méthode est utilisable sous sa forme simple et que pour des cas conduisant à de nombreuses résolutions (charges évolutives, viscoélasticité, viscoplasticité...), il est nécessaire d'apporter à l'algorithme des améliorations présentées dans le paragraphe 3.3.

Ainsi en utilisant un stockage de type Morse pour la matrice, la résolution d'un problème de frottement coûte seulement 4 fois plus cher que celle d'un problème élastique classique (voir tableau 1). Le rapport entre les temps d'exécution des modules de résolution est de 8. Ce rapport devient de l'ordre de 3 si on utilise une méthode de condensation. Ceci signifie qu'après condensation le coût de la résolution du problème de contact unilatéral avec frottement est inférieur au triple de celui d'un problème d'élasticité classique.

Remarque : la question du coefficient de relaxation ω.

Pour une géométrie donnée, le seul coefficient à ajuster est le coefficient de relaxation ω. On notera que ce choix est plus simple que celui du ou des paramètres de pénalisation dont la stratégie d'ajustement est à la fois difficile et essentielle pour ces autres méthodes.

Les résultats théoriques concernant la détermination du ω optimal nécessite le calcul du rayon spectral de la matrice de Jacobi associée et ne concerne de toutes les façons que l'algorithme sans projection et sans traitement des termes non différentiables. Une exploration numérique est présentée sur la Fig. 1 qui donne le nombre d'itérations nécessaires en fonction de ω pour le problème sans conditions de contact, le problème avec contact unilatéral, et le problème avec contact unilatéral et frottement. Le test d'arrêt est un test de variation relative du vecteur solution (norme L^2) ; il est égal à .5 10^{-5}.

On note tout d'abord sur la Fig. 1, que l'algorithme de Gauss-Seidel surrelaxée est peu perturbé par l'opération de projection (Courbe (a) -GSR/ Courbe (b) -GSRP). Ceci provient du fait que le nombre de composantes soumises à la projection est faible devant le nombre total de variables (une sur dix environ).

La courbe (c) concerne le procédé diagonal avec modification du seuil de glissement g au cours des itérations de relaxation. La dépendance du nombre d'itérations en fonction de ω est évidemment plus erratique (car le procédé de résolution est plus complexe) et nous avons tracé une courbe moyenne donnant l'allure de cette évolution.

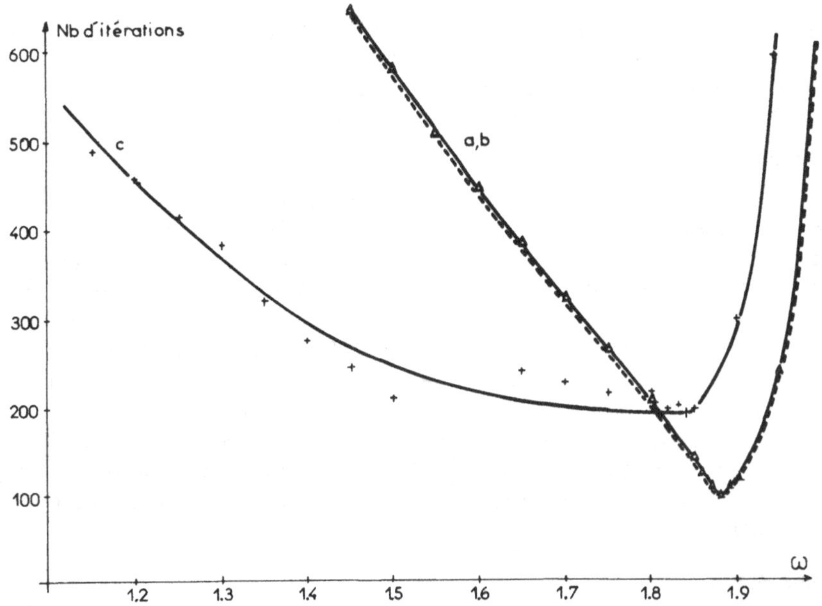

Figure 1 : Influence de ω sur le nombre d'itérations pour :
o--- Courbe (a). Gauss-Seidel surrelaxée
Δ—— Courbe (b). Gauss-Seidel surrelaxée projetée
+▬▬ Courbe (c). Gauss-Seidel surrelaxée projetée avec frottement

Il est très intéressant de noter que pour le traitement du frottement (procédé diagonal), le nombre d'itérations est moins sensible aux variations de ω que dans les deux autres cas. Ainsi pour le problème de frottement une détermination grossière du ω optimum suffit. Cette remarque est évidemment très importante. Par ailleurs, les courbes de la Fig. 1 montrent qu'il est préférable de sous-estimer ω que le surestimer. La détermination d'une valeur optimum du coefficient de relaxation ω est effectuée seulement dans le cas du traitement de plusieurs cas de charge pour un même problème.

4. AMELIORATIONS DE PERFORMANCES

Le procédé diagonal décrit est utilisé systématiquement. Les autres améliorations ou méthodes présentées ici nécessitent un travail supplémentaire qui peut être important suivant l'architecture du code de calcul où on veut les implanter ; elles seront donc justifiées si on doit effectuer un grand nombre de résolutions du problème de contact (évolution des chargements et/ou lois de comportement non élémentaires). Les temps de calcul correspondants sont présentés dans le tableau 1 et seront comparés aux temps de calcul du problème de référence (PR) (problème classique d'élasticité sur la même géométrie avec des conditions aux limites) traité à l'aide d'une méthode de Gauss LDL^T avec un stockage profil.

Les évaluations comparatives présentées concernent l'exemple du paragraphe 5. Elles ne sont qu'indicatives puisque les performances dépendent en particulier du rapport nombre de noeuds de contact/nombre de noeuds total, du rang de la matrice, de sa largeur de bande, ...

4.1 Procédé diagonal

Il consiste à effectuer les mises à jour du vecteur, dont les composantes g_i sont les seuils de glissement aux noeuds, au cours des itérations de relaxation. Il suffit d'attendre le début de stabilisation du procédé de relaxation. La date de mise à jour de g est contrôlée par la valeur du coefficient de convergence ε concernant le test de variations relatives. On commence avec une valeur élevée ($\varepsilon=10^{-1}$) et on le diminue au fur et à mesure des mises à jour jusqu'à la valeur finale $\varepsilon = .5\,10^{-5}$. Si le point fixe n'est pas atteint ($|g^{\nu-1}| > \ell$), on continue le processus avec la dernière valeur de ε. Cette méthode donne d'excellents résultats puisqu'elle nous fait passer de 506 itérations à 191. Le gain du temps de résolution est dans un rapport de 2,6 (voir Tableau 1).

4.2 Stockage Morse

Le stockage "profil" ou "skyline" convient très bien aux méthodes directes de type Choleski ou de "Gauss LDL^T". Toutefois pour les méthodes itératives, il conduit à un grand nombre d'opérations inutiles relatives aux zéros contenus dans le "profil". Le stockage "Morse" n'utilise que les éléments non nuls de la matrice en leur associant une matrice d'adresses. Il est performant pour les méthodes itératives puisqu'il conduit à gain de temps d'exécution du module de résolutions dans un rapport de 6 (voir tableau 1) pour un volume de stockage total équivalent ou plus faible (pour l'ensemble des 2 matrices).

4.3 Condensation

Il s'agit d'un travail préparatoire sur la matrice qui permet de poser le problème seulement sur les composantes de déplacement des noeuds de contact. On pourra trouver les détails de mise en oeuvre dans F. Taallah [14]. Il y a deux étapes :

. préparation du problème réduit : celà consiste à éliminer les variables ne concernant pas le contact. L'inverse d'une matrice de rang élevé extraite de la matrice de rigidité intervient dans ces calculs préparatoires. On doit effectuer m résolutions (où m est deux fois le nombre de noeuds de contact) d'un système linéaire de rang (n-m). On utilise à ce niveau une méthode directe de Gauss LDL^T où la matrice n'est factorisée qu'une fois ;

. résolution : elle se fait sur la matrice réduite. On applique la méthode GSRP au problème portant alors seulement sur les noeuds de contact. La matrice concernée est de petite taille (rang m) et pleine.

Dans le cas de l'utilisation de la méthode de condensation, on utilise un stockage de type profil car le stockage "morse" se prête mal au traitement par des méhtodes directes (étape de préparation), et n'offre aucun intérêt quand la matrice est pleine (matrice réduite de la deuxième étape).

La méthode de condensation conduit à un gain dans un rapport de 1 à 3 au niveau du module de résolution (Tableau 1). L'opération de condensation n'est pas spécialement coûteuse puisque le temps total reste du même ordre que celui concernant le traitement de la matrice complète (Morse). Toutefois, elle nécessite la programmation d'un module de calcul supplémentaire. Cette méthode sera donc utilisée dans le cas de nombreuses résolutions relatives à plusieurs cas de charges pour une géométrie et des caractéristiques fixées : elle est alors extrêmement efficace puisque une résolution du problème avec contact unilatéral et frottement coûte (seulement) de 2 à 3 fois plus cher qu'une résolution du problème de référence (PR).

On trouve dans le tableau suivant les temps d'exécution de l'ensemble du programme et de celui du module de résolution pour les calculs menés sur IBM 3081 K à l'aide du code PROTIS. Il concerne le test présenté au paragraphe 5. Pour le problème de référence (PR) correspondant (problème classique avec conditions aux limites) le temps total d'exécution est 1,84s et le temps d'exécution du module de résolution est de 0,39s (méthode LDL^T stockage profil).

Méthode	Stockage de la matrice	Temps total (en s.)	Temps d'exécution du "solveur"	Commentaires
Standard de base itérations de point fixe sur g^{ν} (voir (16))	Profil	53,88	52,37	Non utilisé
Algorithme diagonal (voir 4.1)	Profil	21,47	20,10	Bon et facilement implanté dans un code
Algorithme diagonal (voir 4.2)	Morse	7,26	3,25	Optimum dans le cas d'une seule résolution
Condensation	Profil	7,86	1,08	Optimum dans le cas de plusieurs résolutions

Tableau 1 : Temps de calcul pour le problème de contact unilatéral avec frottement décrit au paragraphe 5.

4.4 Méthode multigrille

Les méthodes multigrilles sont très séduisantes pour ce type de problèmes. En effet, nous utilisons un procédé itératif pour lequel un travail partiel sur grille grossière permettra d'accélérer le procédé. Par ailleurs, plus que l'alternance grille grossière/grille fine, c'est ici une alternance de type grille fine localement (contact)/grille fine globale que nous utiliserons.

Les premiers résultats pourront être trouvés dans F. LEBON [15].

4.5 Vectorisation

On sait maintenant vectoriser ce type d'algorithme [16]. Par conséquent, la méthode proposée s'intégrera bien dans un code d'éléments finis vectorisé. Un travail est en cours à ce sujet.

5. APPLICATIONS

Nous présentons ici les résultats concernant un exemple test que nous avons proposé au Groupe "Validation de codes de calcul" du GRECO "Grandes Déformations et Endommagement". Il a déjà été traité à l'aide de sept autres codes de calcul :

MARC utilisé par Boyer (INSA-Lyon)
TITUS utilisé par Level-Ravalart-Oudin (LGM-Valenciennes)
MEF module de frottement développé par Parenti (CSI-Compiègne)
CASTOR module de frottement développé par Mme Grolabe au CETIM (Senlis)
TACT développé par Curnier (Université de Lausanne)
Code de Surry-Jeannin à l'Ecole Centrale de Lyon
Code de Feijoo-Barbosa (L.C.C.- Rio de Janeiro).

5.1 Description de l'exemple

Il s'agit d'un exemple académique. On considère un barreau comprimé sur un plan sur lequel il glisse avec frottement et décollement éventuel (voir Fig.2).

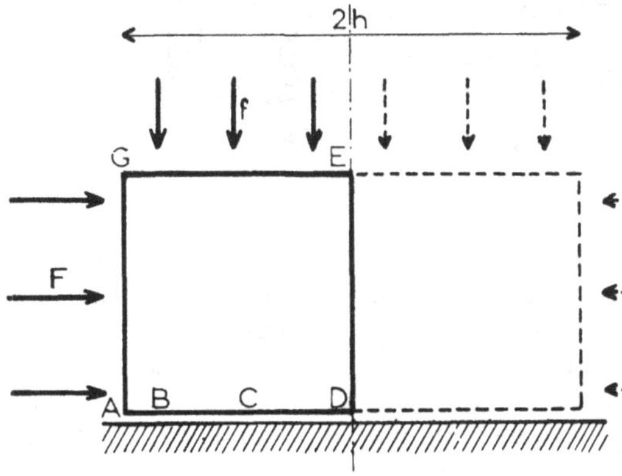

Figure 2 : Géométrie de l'exemple test (section du barreau).

Géométrie : h = 40 mm. Par raison de symétrie on n'étudie que le demi barreau.

Loi de comportement : Elasticité- Déformations planes.

$$\begin{cases} E = 13000 \text{ daN/mm}^2 \\ \nu = 0,2 \end{cases}$$

Conditions aux limites : On impose $u_x = 0$ sur le côté ED, et $u_x = u_y = 0$ au point D.

Conditions de contact unilatéral avec frottement : elles sont imposées sur le côté AD. Le coefficient de frottement μ sera pris égal à 1 ou 0,2 suivant les cas traités (Tableau 2).

Chargement : On applique :
. sur le côté GE une densité de force f
. sur le côté GA une densité de force F

Les valeurs de f et F sont précisées dans le tableau 2.

Discrétisation : Nous utilisons un maillage à 230 noeuds dont 33 noeuds de contact et à 394 éléments (triangles P_1) (voir Fig. 3).

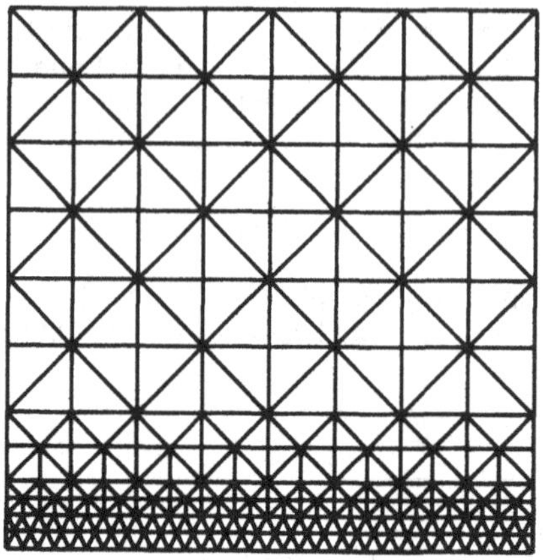

Figure 3 : Maillage.

5.2 Résultats

Nous présentons les résultats concernant cinq cas de calcul avec différentes combinaisons des trois grandeurs μ, f, F (frottement et chargement). Ces configurations de chargement conduisent à des états très différents de la cinématique du bord de contact. On présente dans le tableau 2 les nombres de noeuds du bord de contact qui sont en décollement, en glissement ou en blocage.

μ	F daN/mm²	f daN/mm²	longueur AB (décollement)	longueur BC (glissement)	longueur CD (blocage)
1	10	-5	3,75 (3 noeuds)	18,75 (15 noeuds)	17,5 (15 noeuds)
1	15	-5	3,75 (3 noeuds)	26,25 (21 noeuds)	10 (9 noeuds)
0,2	10	-5	0,	40 (32 noeuds)	0, (1 noeud)
0,2	10	-15	0,	23,75 (19 noeuds)	16,25 (14 noeuds)
0,2	10	-25	0,	3,75 (3 noeuds)	36,25 (30 noeuds)

Tableau 2 - Résultats : état du contact.

Il est intéressant de noter que sur un exemple aussi simple, la loi de Coulomb met en évidence un comportement complexe du bord de contact qu'un préjugement naîf ne permet pas de prévoir : on aurait pu penser par exemple que pour le premier cas de charge ($\mu = 1$) le rapport $|F| / |f|$ étant égal à 2, on allait avoir glissement de tous les noeuds de contact. Cette remarque veut souligner que la loi de Coulomb souvent qualifiée d'académique est susceptible de caractériser des états de contact complexes, lorsqu'on dispose d'un outil numérique convenable pour la traiter.

Nous donnons les résultats complets pour le premier cas traité ($\mu=1$, $f=-5$, $F=10$). On trouvera dans le tableau 3 les déplacements et les forces nodales (composantes normales et tangentielles) sur le bord de contact. Les figures 4, 5 et 6 présentent les courbes isocontraintes de σ_{xx}, σ_{yy} et σ_{xy}. On notera également que la partie bloquée de la frontière est clairement observée sur la Fig. 4 de par ses effets sur la composante σ_{xx} du tenseur des contraintes.

Le tableau 4 concerne un cas de traitement avec $\mu = 0,2$ ($f=-15$, $F=10$). On notera sur les tableaux 3 et 4 la précision avec laquelle sont vérifiées les conditions de contact (3) et de frottement (4) :

- lorsqu'il y a décollement ($u_N < 0$) les forces de contact sont nulles (de l'ordre du dix millième des autres forces de réaction) ;
- lorsqu'il y a contact avec glissement ($u_N = 0$, $|u_T| > 0$) on a bien $F_N < 0$ et l'égalité $|F_T| = -\mu F_N$ est vérifiée exactement (précision de l'ordre de 0,02 %) ;
- lorsqu'il y a contact avec blocage ($u_N = 0$, $u_T = 0$) on a $F_N < 0$ et $|F_T| < -\mu F_N$.

La précision sur l'équilibre de la structure dépend alors seulement de la discrétisation par éléments finis utilisée puisque l'état de contact étant maintenant connu, on obtient la même solution en résolvant un problème d'élasticité classique où les conditions unilatérales et de frottement sont remplacées, pour le premier exemple, par les conditions suivantes :

. bord libre pour la partie qui décolle (noeuds 1,2,3)
. bord encastré pour la partie bloquée (noeuds 19 à 33)
. déplacement en y nul et forces tangentielles données par le tableau 3, pour la partie glissante (noeuds 4 à 18).

Cette procédure constitue une évaluation simple de la qualité de la solution. Elle permet de s'assurer que les forces et les déplacements calculés sur le bord de contact par l'algorithme conduisent bien à l'équilibre de la structure. Une estimation de l'erreur est ainsi donnée par les résultats classiques concernant le type d'éléments utilisés.

* DEPLACEMENTS ET FORCES NODALES (COMPOSANTES NORMALE ET TANGENTIELLE)
SUR LE BORD DE CONTACT (CALCUL A PARTIR DES RESIDUS)
(PAS FIXE SUR LA FRONTIERE)

NOEUD NO	ANGLE.	DEPLACEMENT U NORMAL	DEPLACEMENT U TANGENT.	FORCE NORMALE	FORCE TANGENT.
1	-89.99994	-0.00057	0.01466	0.00045	-0.00050
2	-89.99994	-0.00027	0.01375	0.00022	-0.00081
3	-89.99994	-0.00005	0.01282	0.00006	-0.00029
4	-89.99994	0.00000	0.01179	-1.15652	-1.16000
5	-89.99994	0.00000	0.01073	-2.15722	-2.15707
6	-89.99994	0.00000	0.00968	-2.83476	-2.83372
7	-89.99994	0.00000	0.00865	-3.40153	-3.40005
8	-89.99994	0.00000	0.00765	-3.87173	-3.87015
9	-89.99994	0.00000	0.00668	-4.29144	-4.29044
10	-89.99994	0.00000	0.00576	-4.65716	-4.65636
11	-89.99994	0.00000	0.00487	-5.00823	-5.00775
12	-89.99994	0.00000	0.00403	-5.32867	-5.32829
13	-89.99994	0.00000	0.00324	-5.64373	-5.64381
14	-89.99994	0.00000	0.00250	-5.94789	-5.94853
15	-89.99994	0.00000	0.00181	-6.26836	-6.26953
16	-89.99994	0.00000	0.00120	-6.60693	-6.60803
17	-89.99994	0.00000	0.00067	-6.99333	-6.99474
18	-89.99994	0.00000	0.00024	-7.50336	-7.50571
19	-89.99994	0.00000	0.00000	-8.16213	-7.59714
20	-89.99994	0.00000	0.00000	-8.59421	-5.07222
21	-89.99994	0.00000	0.00000	-8.67540	-4.33170
22	-89.99994	0.00000	0.00000	-8.74927	-3.72717
23	-89.99994	0.00000	0.00000	-8.79855	-3.22507
24	-89.99994	0.00000	0.00000	-8.86453	-2.78680
25	-89.99994	0.00000	0.00000	-8.90199	-2.40985
26	-89.99994	0.00000	0.00000	-8.94228	-2.06015
27	-89.99994	0.00000	0.00000	-8.96011	-1.72634
28	-89.99994	0.00000	0.00000	-8.99416	-1.40853
29	-89.99994	0.00000	0.00000	-9.01044	-1.11337
30	-89.99994	0.00000	0.00000	-9.03463	-0.82676
31	-89.99994	0.00000	0.00000	-9.04294	-0.54228
32	-89.99994	0.00000	0.00000	-9.07299	-0.25986

Tableau 3 - Déplacements et forces aux noeuds de contact - cas $\mu = 1$, $f = -5$, $F = 10$

* DEPLACEMENTS ET FORCES NODALES (COMPOSANTES NORMALE ET TANGENTIELLE)
SUR LE BORD DE CONTACT (CALCUL A PARTIR DES RESIDUS)
(PAS FIXE SUR LA FRONTIERE)

NOEUD NO	ANGLE	DEPLACEMENT U NORMAL	DEPLACEMENT U TANGENT.	FORCE NORMALE	FORCE TANGENT.
1	-89.99994	0.00000	0.00762	-3.35073	-0.67084
2	-89.99994	0.00000	0.00707	-11.74459	-2.34853
3	-89.99994	0.00000	0.00641	-13.95058	-2.79039
4	-89.99994	0.00000	0.00380	-15.20170	-3.04091
5	-89.99994	0.00000	0.00522	-16.04510	-3.20920
6	-89.99994	0.00000	0.00468	-16.66504	-3.33357
7	-89.99994	0.00000	0.00416	-17.14960	-3.43053
8	-89.99994	0.00000	0.00367	-17.54442	-3.50909
9	-89.99994	0.00000	0.00320	-17.87204	-3.57529
10	-89.99994	0.00000	0.00276	-18.15105	-3.63036
11	-89.99994	0.00000	0.00234	-18.40039	-3.68025
12	-89.99994	0.00000	0.00194	-18.62941	-3.72610
13	-89.99994	0.00000	0.00157	-18.84053	-3.76857
14	-89.99994	0.00000	0.00123	-19.04073	-3.80850
15	-89.99994	0.00000	0.00092	-19.23584	-3.84738
16	-89.99994	0.00000	0.00063	-19.43391	-3.88702
17	-89.99994	0.00000	0.00039	-19.63220	-3.92649
18	-89.99994	0.00000	0.00019	-19.85364	-3.97043
19	-89.99994	0.00000	0.00005	-20.11948	-4.02414
20	-89.99994	0.00000	0.00000	-20.37927	-3.43888
21	-89.99994	0.00000	0.00000	-20.50497	-2.67520
22	-89.99994	0.00000	0.00000	-20.56311	-2.31541
23	-89.99994	0.00000	0.00000	-20.60228	-2.01358
24	-89.99994	0.00000	0.00000	-20.65053	-1.74785
25	-89.99994	0.00000	0.00000	-20.67986	-1.51782
26	-89.99994	0.00000	0.00000	-20.70970	-1.30232
27	-89.99994	0.00000	0.00000	-20.72472	-1.09461
28	-89.99994	0.00000	0.00000	-20.74921	-0.89516
29	-89.99994	0.00000	0.00000	-20.76195	-0.70878
30	-89.99994	0.00000	0.00000	-20.77946	-0.52688
31	-89.99994	0.00000	0.00000	-20.78633	-0.34579
32	-89.99994	0.00000	0.00000	-20.80659	-0.16576

Tableau 4 - Déplacements et forces aux noeuds de contact - cas $\mu = 0,2$, $f = -15$, $F = 10$

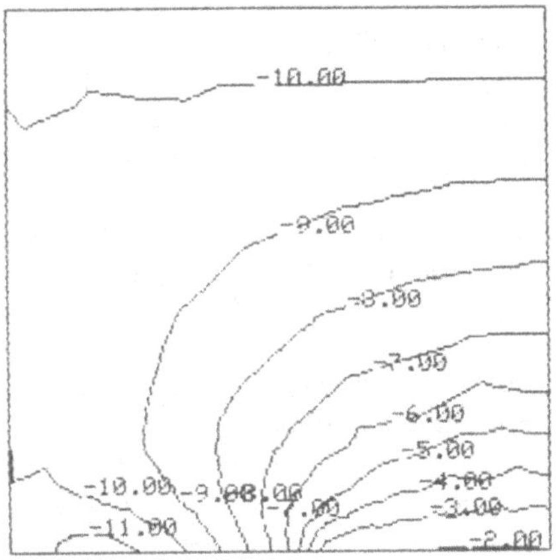

Figure 4 - Courbes d'isocontraintes σ_{xx}(daN/mm²)

$\mu = 1, \quad f = -5, \quad F = 10$

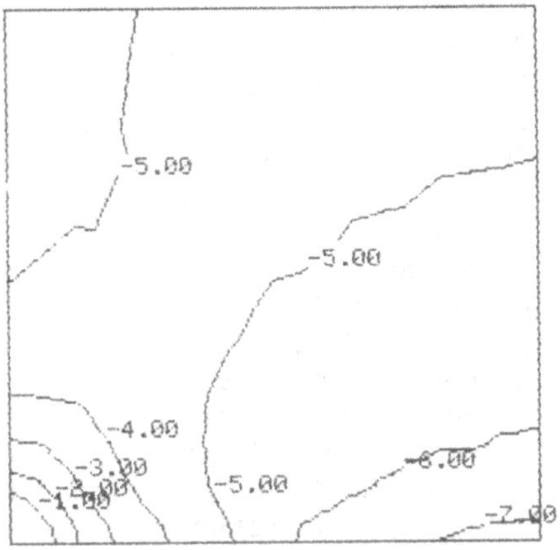

Figure 5 - Courbes d'isocontraintes σ_{yy} (daN/mm²)

$\mu = 1, \quad f = -5, \quad F = 10$

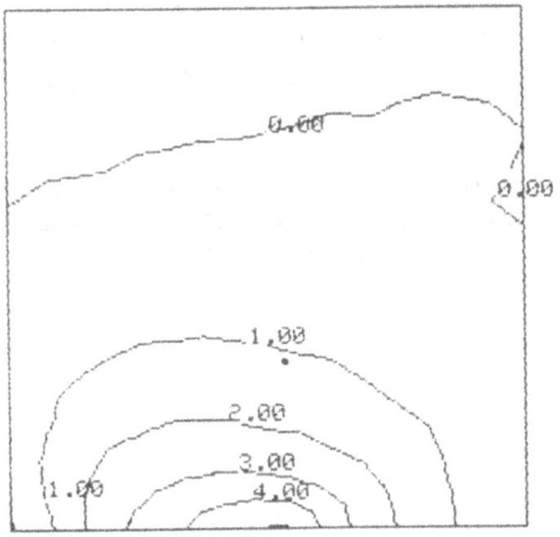

Figure 6 - Courbes d'isocontraintes σ_{xy} (daN/mm²)

$\mu = 1, \quad f = -5, \quad F = 10$

REFERENCES :

1. Duvaut, G., Lions, J.L. : Les inéquations en Mécanique et en Physique, Dunod, Paris, 1972.
2. Duvaut, G. : Equilibre d'un solide élastique avec contact unilatéral et frottement de Coulomb, C.R. Acad. Sci., Paris, Série A, t. 290, 263, 1980.
3. Cocu, M. : Existence of solutions of Signorini problem with friction, Int. J. Engn. Sci., 22, n°5, 567, 1984.
4. Campos, L.T., Oden, J.T., Kikuchi, N. : A numerical Analysis of a class of contact problems with friction in elastostatics, Computer Meth. in Applied Mech. and Eng., 34, 821, 1982.
5. Panagiotopoulos, P.D. : Inequality Problems in Mechanics and Applications, Birkhäuser, Boston-Basel, Stuttgart, 1985.
6. Raous, M. : Contacts unilatéraux avec frottement en viscoélasticité, in "Unilateral Problems in Structural Analysis", Ed. G. Del Piero, F. Maceri, CISM Publisher, Springer-Verlag, Vienne, 1985.
7. Kalker, J.J. : On the Contact Problem in Elastostatics, in "Unilateral Problems in Structural Analysis", Ed. G. Del Piero, F. Maceri, CISM Publishers, Springer-Verlag, Vienne, 1985.
8. Curnier, A. : A theory of friction, Int. J. of Solids & Structures, 20, 7, 637, 1984.
9. Raous, M. : Comportement d'un solide fissuré sous charges alternatives en viscoplasticité, Journal de Mécanique Théorique et Appliquée, n° Spécial "4e Colloque Franco-Polonais", 1982.
10. Bazaraa, M.S., Shetty, C.M., Nonlinear Programming, John Wiley & Sons, New York, 1975/
11. Klarbring, A.: Contact problems in linear elasticity, Thesis, Dpt of Mech. Engn., Linkoping University, Sweden, 1985.
12. Glowinski, R., Lions, J.L., Tremolieres, R., Analyse Numérique des Inéquations Variationnelles, Dunod, Paris, 1976.
13. Latil, J.C., Méthode d'Eléments Finis et Inéquations. Notice du Code PROTIS, Note Interne L.M.A. n° 2188, 1986.
14. Taallah, F. : Méthode de condensation pour un algorithme de Gauss-Seidel relaxé avec projection, Mémoire de DEA-Math. Appliquées, Université de Provence, 1985.
15. Lebon F. : Résolution d'inéquations variationnelles par méthodes multigrilles, Mémoire de DEA-Math. Appliquées, Université de Provence, 1985.
16. Raous, M. : On Two Variational Inequalities arising from a Periodic Viscoelastic Unilateral Problem, in "Variational Inequalities and Complementarity Problems", Ed. Gianessi-Cottle-Lions, J.Wiley, 1979.

CONVEX PROBLEMS IN STRUCTURAL MECHANICS

G. Romano*, E. Sacco**
*Istituto di Scienza delle Costruzioni, University of Naples
**Dipartimento di Ingegneria Civile, 2nd University of Rome

ABSTRACT

The object of the paper is a presentation of the basic theory of those problems in structural mechanics in which the relations between the internal (stress-strain) and the external (displacement-reaction) fields are governed by general convex (non differentiable) potentials.
The analysis is developed in the formal framework of Convex Analysis and is based on the Fenchel duality theory.
Admissibility conditions on the data and their variational formulations are discussed. Necessary and sufficient conditions defining limit loads and distorsions are given.
Existence under admissible (but not limit) data is proved, variational and extremum principles are developed.

1. INTRODUCTION

In the recent times an increasing interest has been devoted by researchers in structural analysis to problems where the constraint conditions between the external and the internal state variables are of an unilateral or of a (more general) convex type.

Starting from the leading contributions of J.J. Moreau [1], the analysis of structural problems governed by convex potentials has greatly been developed both in the theoretical and the computational aspects. A comprehensive exposition of the main results in this field and extensive references can be found in the recent book by P.D. Panagiotopoulos [2].

The mathematical theory of convex problems was laid down in its modern form by Fenchel [3] in the early fifties and subsequently developed and extended by Moreau [4] and Rockafellar [5].

There is no doubt that the concepts and the methods of convex analysis are the most suitable tools to carry out a general theory of structural models where the constraint conditions can be expressed in terms of convex potentials.

Such class of problems includes a wide number of interesting models for structural engineering, including elastic problems with non linear external and internal constitutive relations, perfect plasticity, limit analysis and viscous flow problems.

As often occurs a general abstract theory yields a clear conceptual picture of the basic features of the class of problems under investigation, while more special and apparently simpler treatments may happen to hide the real significance of some fundamental aspects of the problem at hand.

In this line of reasoning, the present paper develops a general theoretical approach to convex problems in structural mechanics.

The formulation of the governing field equations and constraint conditions is carried out in an abstract formal context which emphasizes the mechanical meaning more that the technical mathematical aspects.

In the sequel the convex or concave potentials and the convex sets involved in the analysis will be assumed to be regular enough to ensure the validity of the invoked mathematical results.

Admissibility conditions on the data and related variational formulations are discussed in detail. Existence of a solution is proved for safe values of the data, that is for data which are admissible but not limit. General extremum and stationarity principles are developed.

The presentation makes explicit reference to elastic problems defined by convex potentials, but viscous flow and perfectly plastic problems are easily covered by the theory by considering velocities and strain rates as kinematic state variables.

2. STATEMENT OF THE PROBLEM

We shall deal with structural models in which non linear geometric effects are not taken into account, so that displacement and velocity spaces can be identified.

The state variables involved in the analysis will then belong to the following linear spaces:

- V displacements,
- D deformations,
- F external forces,
- S internal forces.

The pairs of spaces F, V and S, D are assumed to be set in separating duality by means of two bilinear forms which have the mechanical meaning of external and internal virtual work. The same symbol $\langle .,. \rangle$ will denote both virtual works.

The linear operator:

$T : V \to D$ deformation operator

associates to each displacement field in V the corresponding deformation field in D.

The equilibrium operator:

$T' : S \to F$

is defined to be the dual of T by:

$$\langle s, Tv \rangle = \langle T's, v \rangle \qquad \forall s \in S, v \in V .$$

It associates to each internal force s in S the corresponding external force in F.

We shall consider the problem defined by the linear field equations:

$$Tu = \varepsilon \quad \text{kinematic compatibility,} \tag{2.1}$$

$$T's = f \quad \text{static equilibrium.} \tag{2.2}$$

The external force f is assumed to be the sum of an assigned load ℓ and of an unknown reaction force r.

The deformation field ε is assumed to be the sum of an imposed distorsion δ and an unknown elastic strain e.

The statement of the problem is completed by the definition of the external and the internal constraint conditions, which relate the dual pairs of unknown fields u, r and s, e, and will be introduced in the next paragraph.

3. CONSTRAINT CONDITIONS

We shall consider constraint relations defined by (possibly non--differentiable) maximally ciclically monotone laws as defined in [5], pag. 238.

The internal constraint condition (constitutive relation) is suitably defined by a non-decreasing monotone law, and the external one by a non-increasing monotone law.

This is equivalent to consider two functionals:
the internal elastic strain energy, which is a convex potential ϕ defined on D, and
the external constraint potential, which is a concave potential g defined on V.

As usual in convex analysis, we shall consider the convex (concave) functionals to be extended to the whole space by setting their values to $+\infty(-\infty)$ outside of the domain of definition [5] pag. 23.

The elastic constitutive relations are:

$$s \in \partial \phi(e) \quad \text{internal constraint}, \tag{3.1}$$

and

$$r \in \partial g(u) \quad \text{external constraint}. \tag{3.2}$$

The symbol ∂ denotes the operator which associates, to each argument in the domain of the functional, the convex set of its slopes at that point (fig. 1-2).

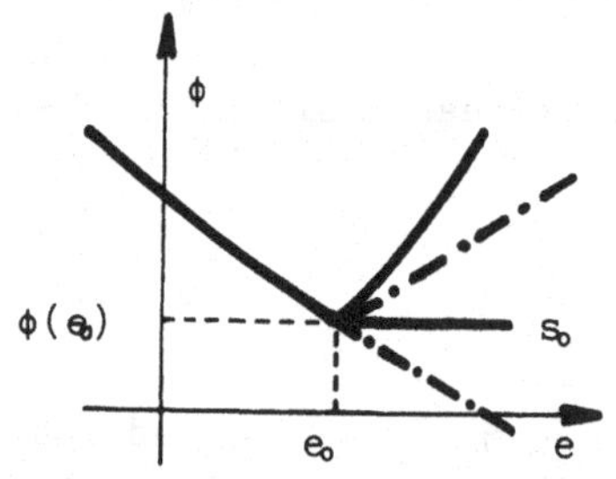

Fig.1a) Convex potential

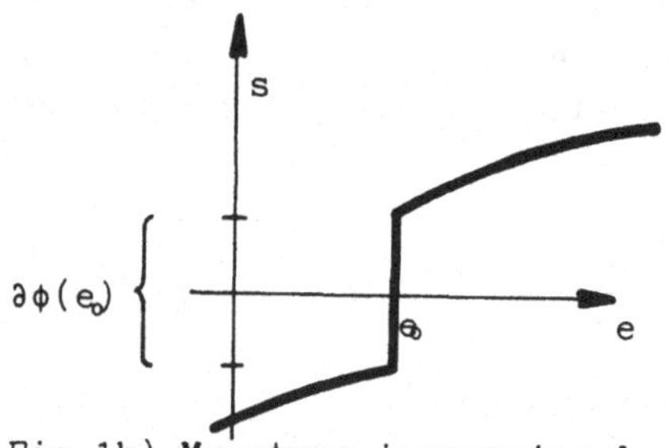

Fig.1b) Monotone increasing law

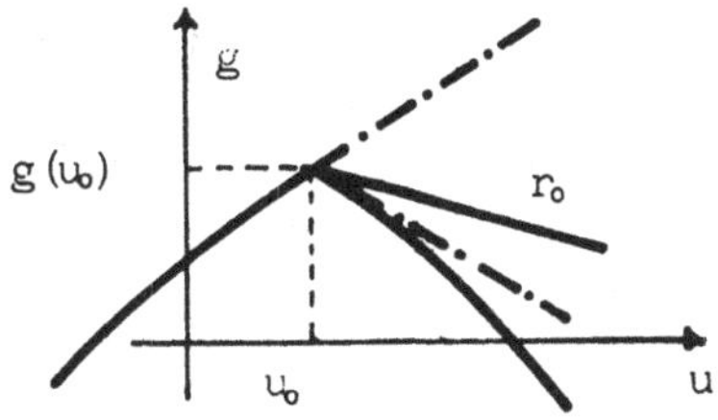

Fig.2a) Concave potential

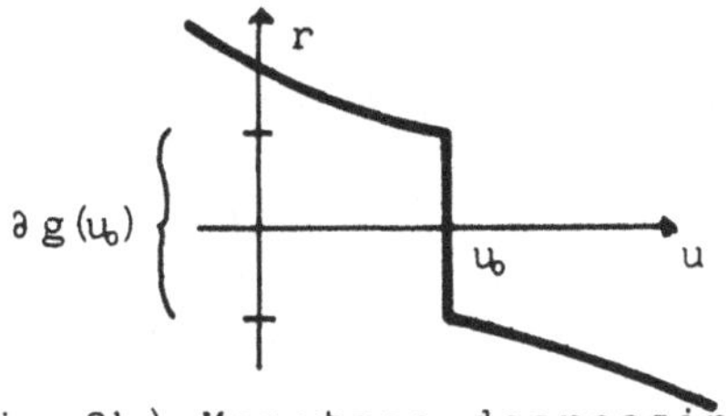

Fig.2b) Monotone decreasing law

In the literature ∂ is called the subdifferential (superdifferential) operator when acting on convex (concave) functionals [5]. The slopes are referred to as the sub (super) gradients of the functional.

We prefer the names "slope operator" and "slope" which are more natural in a geometric interpretation of the involved operations both in the convex and in the concave case.

The inclusions (3.1) and (3.2) are equivalent to the variational inequalities:

$$\phi(\eta) - \phi(e) \geq \langle s, \eta - e \rangle \qquad \forall\, \eta \in D, \tag{3.3}$$

$$g(v) - g(u) \leq \langle r, v - u \rangle \qquad \forall\, v \in V. \tag{3.4}$$

An equivalent (inverse) formulation of the relations (3.1) and (3.2) is:

$$e \in \partial\, \phi^*(s), \tag{3.5}$$

$$u \in \partial\, g^*(r), \tag{3.6}$$

where ϕ^* and g^* are the Fenchel conjugates of ϕ and g, defined by:

$$\phi^*(s) = \sup_{\eta} \{ \langle s, \eta \rangle - \phi(\eta) \} \tag{3.7}$$

$$g^*(r) = \inf_{v} \{ \langle r, v \rangle - g(v) \}, \tag{3.8}$$

which represent the complementary elastic strain energy and the complementary external constraint potential, respectively.

The supremum in (3.7) is attained at $\eta = e$ satisfying the inclusion

(3.5). Analogously the infimum in (3.8) is attained at $v=u$ satisfying the inclusion (3.6).

A further statement equivalent to either (3.1),(3.2) or (3.5),(3.6) is given by the relations:

$$\phi(e) + \phi^*(s) = \langle s, e \rangle ,$$

$$g(u) + g^*(r) = \langle r, u \rangle ,$$

known as the Fenchel transforms.

The domains of the potentials ϕ and g and their conjugates ϕ^* and g^* will be denoted by:

$C = \text{dom } g$	set of admissible	displacements,
$R = \text{dom } g^*$	" " "	reactions,
$E = \text{dom } \phi$	" " "	strains,
$Q = \text{dom } \phi^*$	" " "	stresses.

All the sets C, R, E, Q are convex sets.

A simple example of external constraint relations defined by a concave potential is given by the unilateral support condition that the set of admissible displacements is the conical variety:

$$C = w + C_0 ,$$

where w is an assigned constraint displacement and C_0 is the convex cone of the additional admissible displacements.

The concave potential g is then given by the concave indicator function of the convex conical variety C:

$$g(u) = \hat{\chi}_C(u) = \begin{cases} -\infty & \text{if } u \notin C \\ 0 & \text{if } u \in C, \end{cases}$$

and the conjugate potential g^* by:

$$g^*(r) = \hat{\chi}_C^*(r),$$

where $\hat{\chi}_C^*$ is the concave support function of C:

$$\hat{\chi}_C^*(r) = \inf\{ \langle r, v\rangle / v \in C\} = \hat{\chi}_{C_0^+}(r) + \langle r, w \rangle ,$$

and C_0^+ is the positive polar cone of C_0:

$$C_0^+ = \{f \in F: \langle f, v \rangle \geq 0 \quad \forall\, v \in C_0\} \;.$$

In the one dimensional case, the graph of the monotone non increasing displacement-reaction relation and of the related concave potentials g and g^*, is given in fig. 3.

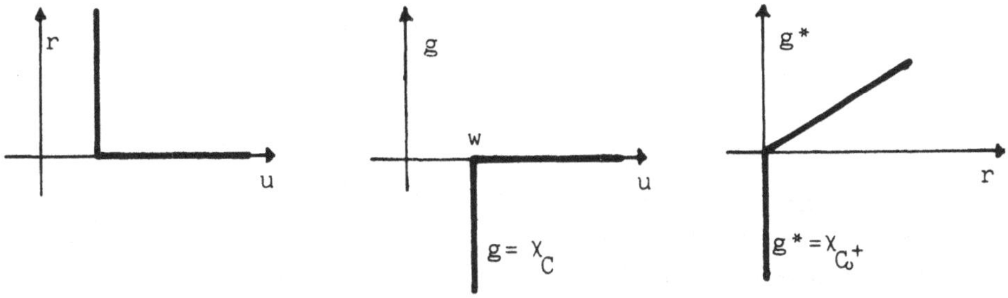

Fig.3) Unilateral external constraints

An interesting special example of internal constraint defined by a convex potential is provided by the constitutive relation of an elastic material without tensile strenght.

Such a material is defined by the characteristic property that the admissible stresses belong to a convex cone Q and the total strain is the sum of an elastic part and an anelastic crack satisfying the normality rule:

$$e = S^{-1}\, s + \delta \qquad \text{(S elastic stiffness operator)},$$

$$\delta \in \partial\, \chi_Q(s) \qquad \text{(convex set of outward normals to Q at s)},$$

where χ_Q is the convex indicator function of Q.

The convex complementary elastic strain energy potential is given by [6]:

$$\phi^*(s) = \frac{1}{2}\, \|\, s\, \|^2_{S^{-1}} + \chi_Q(s)\;,$$

and the conjugate elastic strain energy potential by:

$$\phi(e) = \frac{1}{2}\, \|\, e - \Pi\,(e)\, \|^2_S \;,$$

where Π is the orthogonal projector on Q^- (the negative polar of Q) in the energy of S [7][8].

In the one dimensional case, the graph of the monotone non decreasing strain-stress relation and the related convex potential ϕ^* and ϕ, is given in fig. 4.

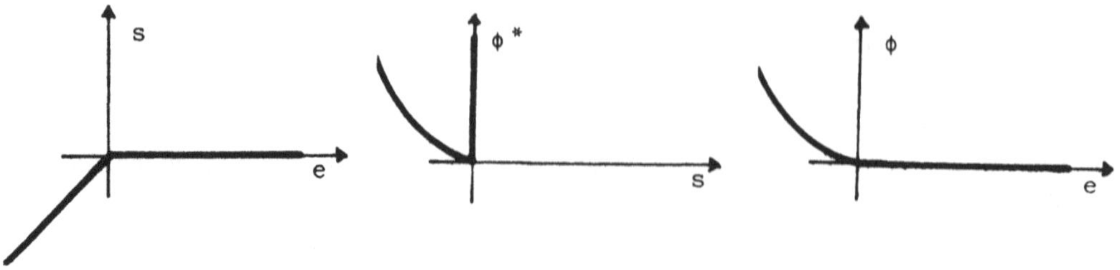

Fig.4) No-tension elastic material

4. THE CONVEX ELASTOSTATIC PROBLEM

The elastostatic problem for the structural model described in par. 2 and subject to the convex constraint conditions discussed in par. 3, can be formulated as follows:

Given a load $\ell \in F$ and a distorsion $\delta \in D$

Find an admissible displacement $u_0 \in C$
" " elastic strain $e_0 \in E$
" " reaction $r_0 \in R$
" " stress $s_0 \in Q$

satisfying the field equations:

$$Tu_0 = e_0 + \delta \quad \text{kinematic compatibility} \tag{4.1}$$

$$T's_0 = r_0 + \ell \quad \text{static equilibrium ,} \tag{4.2}$$

and the constraint conditions:

$$s_0 \in \partial\, \phi(e_0) \;;\; e_0 \in \partial\, \phi^*(s_0) \quad \text{internal constraint} \tag{4.3}$$

$$r_0 \in \partial\, g(u_0) \;;\; u_0 \in \partial\, g^*(r_0) \quad \text{external constraint .} \tag{4.4}$$

The admissibility conditions on the data, their variational formulation and the proof of the existence of a solution are developed in the next paragraphs.

5. ADMISSIBILITY CONDITIONS ON THE DATA

The applied load ℓ and distorsion δ will be said to be admissible if:

$$\ell \in \Lambda \doteq T'(Q) - R \,, \tag{5.1}$$

$$\delta \in \Delta \doteq T(C) - E \,. \tag{5.2}$$

Defining the convex set of stress fields equilibrated with the load ℓ :

$$\Sigma = \{s \in S : T's \in \ell + R\} ,$$

and the convex set of displacement fields compatible with the distorsion δ :

$$U = \{u \in V : Tu \in \delta + E\} ,$$

the conditions (5.1) and (5.2) can be rewritten in the equivalent form:

$$\Sigma \cap Q \quad \text{not empty,} \tag{5.3}$$

$$U \cap C \quad \text{not empty.} \tag{5.4}$$

The admissible data will be called limit or safe respectively if they belong to the boundary or to the interior of the admissible sets Λ and Δ.

If the linear envelope F_Λ and D_Δ of Λ and Δ are properly included in the ambient spaces F and D, the boundary and the interior must be considered in the relative sense, that is by assuming the linear envelopes F_Λ and D_Δ as ambient space [5] pag. 44.

6. VARIATIONAL CONDITIONS OF ADMISSIBILITY

Both from the theoretical and the applied point of view it is interesting to give a variational formulation to the admissibility conditions on the data.

To this end it is sufficient to observe that:

$$\ell \in \Lambda \Leftrightarrow \chi_\Lambda(\ell) = 0 , \tag{6.1}$$

where χ_Λ is the convex indicator function of Λ:

$$\chi_\Lambda(\ell) = \begin{cases} 0 & \text{if } \ell \in \Lambda \\ +\infty & \text{if } \ell \notin \Lambda . \end{cases}$$

Under the assumption that Λ is a closed convex set we have:

$$\chi_\Lambda^{**}(\ell) = \chi_\Lambda(\ell) , \tag{6.2}$$

where χ^{**} is the iterated conjugate functional of χ, defined by:

$$\chi_{\Lambda}^{**}(\ell) = \sup_{v}\{<\ell, v> - \chi_{\Lambda}^{*}(v)\} , \tag{6.3}$$

that is the conjugate of the support functional:

$$\chi_{\Lambda}^{*}(v) = \sup_{\lambda \in \Lambda} <\lambda, v>. \tag{6.4}$$

Hence (6.1) by virtue of (6.2), (6.3) and (6.4) can be written:

$$<\ell, v> \leqq \sup_{\lambda \in \Lambda} <\lambda, v> \tag{6.5}$$

Since $\Lambda = T'(Q) - R$ we have:

$$\sup_{\lambda \in \Lambda} <\lambda, v> = \sup_{\substack{s \in Q \\ r \in R}} <T's - r, v> = \sup_{s \in Q} <s, Tv> - \inf_{r \in R} <r, v> \tag{6.6}$$

Thus $\ell \in \Lambda$ is equivalent to:

$$<\ell, v> + \inf_{r \in R} <r, v> \leqq \sup_{s \in Q} <s, Tv> \qquad \forall\, v \in V , \tag{6.7}$$

that is:
the minimal external virtual work must be not greater than the corresponding maximal internal virtual work, for any displacement field.

The significant test fields v in the inequality (5.2) are those for which the first member is not $-\infty$ and/or the second is not $+\infty$. Then we may restrict ourselves to consider the test fields such that:

$$v \in N_R^{in} \qquad Tv \in N_Q^{out} , \tag{6.8}$$

where N_R^{in} is the inward normal cone to the convex R,
and N_Q^{out} is the outward normal cone to the convex Q.

It is interesting to remark that the variational formulation (6.7) of the admissibility conditions on the applied load is in fact the general statement of the fundamental theorem of the limit analysis:
"If the stress fields and the constraint reactions are imposed to belong to the convex sets Q and R, it is possible to satisfy the equilibrium equation if the applied load verifies the variational inequality (6.7)". It can be shown that, by virtue of this general result, the kinematic theorem of plastic collapse can be extended to non-standard plastic

materials [10] [11].

An identical procedure allows to prove that the admissibility condition on the imposed distorsion δ is equivalent to the variational inequality:

$$< s, \delta > + \sup_{\eta \in E} < s, \eta > \geq \inf_{v \in C} < T's, v > \qquad \forall\, s \in S \,. \qquad (6.9)$$

The significant test fields are obtained by choosing

$$s \in N_E^{out} \cap N_{TC}^{in} \,. \qquad (6.10)$$

The convex cone N_{TC}^{in} is in fact the cone of selfequilibrated stresses corresponding to selfequilibrated locking reactions:

$$N_{TC}^{in} = \Sigma_0 = \{ s \in S \; : \; T's \in R_0 \} \,,$$

where $R_0 = N_C^{in}$ is the convex cone of constraint reactions associated to limit displacements (that is: displacements on the boundary of the admissible set C, see fig. 5).

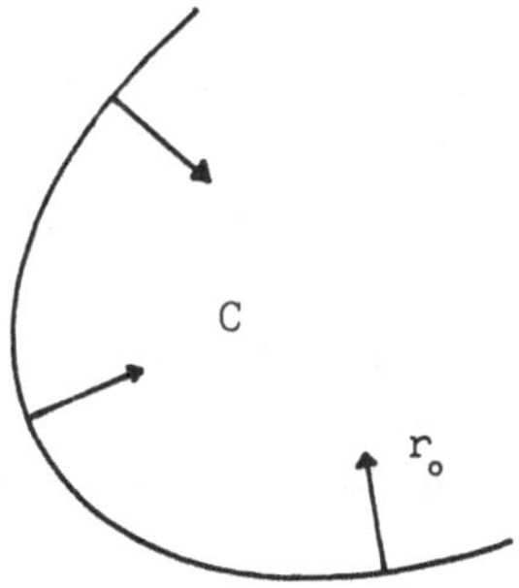

Fig.5) Locking reactions

It is worth observing that the variational inequalities (6.7) and (6.9) can be equivalently written as:

$$< \ell, v > + \sup_{r \in R} < r, v > \geq \inf_{s \in Q} < s, Tv > \qquad \forall\, v \in V \,, \qquad (6.11)$$

$$< s, \delta > + \inf_{\eta \in E} < s, \eta > \leq \sup_{v \in C} < T's, v > \qquad \forall\, s \in S \,. \qquad (6.12)$$

The significant test fields in (6.11) and (6.12) will then be the opposites of the fields satisfying (6.8) and (6.10).

A variational condition, which is necessary and sufficient in order that a load ℓ be a limit load, is provided by the following theorem:

an assigned load $\ell \in \Lambda$ is a limit load iff:

$$\exists\, v_0 \notin F_\Lambda^\perp : \langle \ell, v_0 \rangle = \chi_\Lambda^*(v_0) \tag{6.13}$$

or more explicitly:

$$\exists\, v_0 \notin F_\Lambda^\perp : \langle \ell, v_0 \rangle + \inf_{r \in R} \langle r, v_0 \rangle = \sup_{s \in Q} \langle s, Tv_0 \rangle \,. \tag{6.14}$$

This condition amounts in requiring that the variational inequality (6.7) be satisfied as an equality for at least a test field v_0 not orthogonal to F_Λ.

Let us first prove the sufficiency:
since $\ell \in \Lambda$ we have $\chi_\Lambda(\ell)=0$; the condition (6.13) can then be equivalently written:

$$\langle \ell, v_0 \rangle = \chi_\Lambda^*(v_0) + \chi_\Lambda(\ell) \,,$$

which is in turn equivalent to [5] th. 23.5:

$$v_0 \in \partial\, \chi_\Lambda(\ell) = N_\Lambda^{out}(\ell) \,, \tag{6.14}$$

where $N_\Lambda^{out}(\ell)$ is the convex cone of the outward normal to Λ at the point ℓ.

Now if the admissible load ℓ would be safe, the normal cone $N_\Lambda^{out}(\ell)$ should be exactly $F_\Lambda^\perp$, and this is in contradiction with the assumption that $v_0 \notin F_\Lambda^\perp$. Hence ℓ must belong to the relative boundary of Λ.

The proof of the necessity is easily carried out by following the previous steps in the inverse order.

An analogous reasoning yields the necessary and sufficient condition ensuring that an imposed distorsion δ is a limit distorsion.

Denoting by D_Δ the linear envelope of Δ in the strain space D, we have that: an imposed distorsion $\delta \in \Delta$ is a limit distorsion iff:

$$\exists\, \sigma_0 \notin D_\Delta^\perp : \langle \sigma_0, \delta \rangle + \sup_{\eta \in E} \langle \sigma_0, \eta \rangle = \inf_{v \in C} \langle T'\sigma_0, v \rangle \,. \tag{6.15}$$

In fact (6.15) can be written as:

$$<\sigma_0, \delta> = \chi^*_\Delta(\sigma_0) + \chi_\Delta(\delta) \ ,$$

which is equivalent to:

$$\sigma_0 \in \partial\, \chi_\Delta(\delta) = N^{out}(\delta) \ .$$

Since by assumption $\sigma_0 \notin D_\Delta^\perp$, the admissible distorsion δ cannot belong to the relative interior of Δ, and hence will be a limit distorsion.

A simple graphic sketch of the limit distorsion is given in fig. 6b.

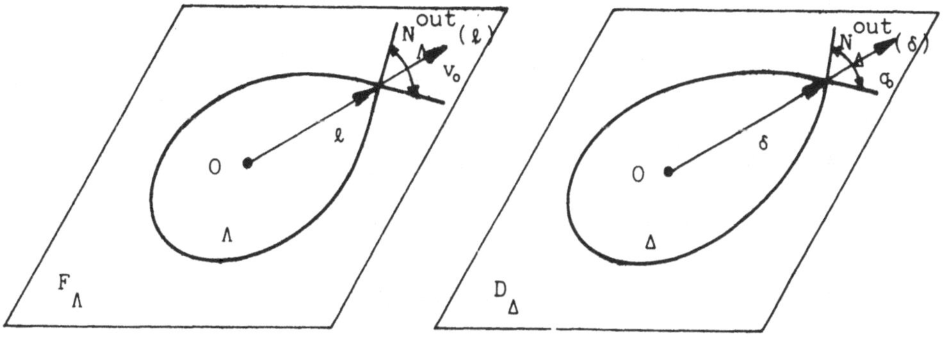

Fig.6a) Limit load Fig.6b) Limit distorsion

7. EXISTENCE OF A SOLUTION

The admissibility conditions (5.1) and (5.2) are obviously necessary for the existence of a solution of the convex elastostatic problem.

A sufficient condition is proved by the stronger assumption that the data are safe.

The proof can be alternatively developed in terms of displacement or of stress fields.

Let us first choose the former alternative.

Let us denote by:

$$\Phi(u) = \phi(Tu - \delta) \quad \text{with} \quad u \in U \tag{7.1}$$

the convex functional of the elastic strain energy expressed in terms of displacements. The conjugate functional Φ^* is given by:

$$\Phi^*(f) = \inf_\tau \{\phi^*(\tau) + <\tau, \delta> / T'\tau = f\} \ . \tag{7.2}$$

By virtue of (5.4) the domain of definition of Φ is not empty and then it can be proved that the conjugate functional Φ^* is proper, i.e., that the infimum in (7.2) is not $-\infty$.

The expression (7.2) for $\Phi^*(f)$ can be verified by proving that its conjugate is equal to Φ:

$$\begin{aligned}
\Phi^{**}(u) &= \sup_f \{ \langle f, u \rangle - \Phi^*(f) \} = \\
&= \sup_f \{ \langle f, u \rangle - \inf_\tau \{ \phi^*(\tau) + \langle \tau, \delta \rangle / T'\tau = f \} \} \\
&= \sup_f \sup_\tau \{ \langle f, u \rangle - \langle \tau, \delta \rangle - \phi^*(\tau) / T'\tau = f \} \\
&= \sup_\tau \{ \langle \tau, Tu - \delta \rangle - \phi^*(\tau) \} = \phi(Tu - \delta) = \Phi(u) \,. \qquad (7.3)
\end{aligned}$$

From (7.2) we get:

$$\text{dom } \Phi^* = T'(\text{dom } \phi^*) = T'(Q) \,,$$

and hence the inclusion (5.1) can be written as:

$$\ell \in \text{dom } \Phi^* - \text{dom } g^* \,. \qquad (7.4)$$

The regularity of the convex (concave) functionals involved in the analysis is assumed to be such that their slopes can be evaluated at any point in their domain. Hence we have that:

$$\text{dom } \Phi^* = \text{dom } \partial \Phi^* \,,$$

$$\text{dom } g^* = \text{dom } \partial g^* \,.$$

Now from the equivalence:

$$u \in \partial \Phi^*(f) \Leftrightarrow f \in \partial \Phi(u) \,,$$

it follows that: $\text{dom } \partial \Phi^* = \text{im } \partial \Phi = \partial \Phi(U)$,

and likewise : $\text{dom } \partial g^* = \text{im } \partial g = \partial g(C)$.

The admissibility condition (5.2) in the equivalent form (5.4) allows then to rewrite (7.4) as follows (see [5], th. 9.3):

$$\ell \in \partial \Phi(U) - \partial g(C) = \partial(\Phi - g)(U \cap C) \,. \qquad (7.5)$$

From (7.5) we infer that it exists at least a displacement field $u_0 \in U \cap C$ such that:

$$\ell \in \partial(\Phi - g)(u_0) .$$

By the assumption that ℓ is a safe load ([5], th. 23.8), we have:

$$\ell \in \partial \Phi(u_0) - \partial g(u_0) . \tag{7.6}$$

The generalized chain rule [9]:

$$\partial \Phi(u_0) = T' \partial \phi(Tu_0 - \delta) \tag{7.7}$$

allows then to write:

$$\ell \in T' \partial \phi(Tu_0 - \delta) - \partial g(u_0) , \tag{7.8}$$

and setting

$$e_0 = Tu_0 - \delta$$

$$s_0 \in \partial \phi(e_0)$$

$$r_0 \in \partial g(u_0) ,$$

we get the conclusion that u_0, s_0, e_0, r_0 is a solution of the convex elastostatic problem.

An analogous development can be carried out in terms of stress fields by considering the concave functional:

$$\psi^*(s) = g^*(T's - \ell) \quad \text{with} \quad s \in \Sigma , \tag{7.9}$$

and the conjugate functional:

$$\psi(e) = \sup_v \{ g(v) - \langle \ell, v \rangle / Tv = e \} . \tag{7.10}$$

Noting that:

$$\text{dom } \psi = T\, C$$

from (5.2) we get

$$\delta \in \operatorname{dom} \psi - \operatorname{dom} \phi = \partial \psi^*(\Sigma) - \partial \phi^*(Q)$$

and by the intersection property (5.3):

$$\delta \in \partial(\psi^* - \phi^*)\ (\Sigma \cap Q)\ . \tag{7.11}$$

Then there exists $s_0 \in \Sigma \cap Q$ such that:

$$\delta \in \partial\, \psi^*(s_0) - \partial\, \phi^*(s_0)\ . \tag{7.12}$$

Since by the chain rule:

$$\partial\, \psi^*(s_0) = T\ \partial\, g^*(T' s_0 - \ell) \tag{7.13}$$

setting:

$$r_0 = T'\ s_0 - \ell$$

$$u_0 \in \partial\ g^*(r_0)$$

$$\ell_0 \in \partial\ \phi^*(s_0)\ ,$$

we get the result.

8. VARIATIONAL AND EXTREMUM PRINCIPLES

An equivalent formulation of the convex elastostatic problem can be obtained by substituting the field equations (4.1) and (4.2) into the constraint conditions (4.3) and (4.4) to get:

$$Tu_0 - \delta \in \partial\, \phi^*(s_0)\ , \tag{8.1}$$

$$T' s_0 - \ell \in \partial\ g(u_0)\ . \tag{8.2}$$

By the defining property of Fenchel's transform we get:

$$\begin{aligned} &\phi(Tu - \delta) + \phi^*(s) \geq \langle s, Tu - \delta \rangle \qquad \forall\ s \in \Sigma\ ,\quad u \in U\ ,\\ &g(u) + g^*(T's - \ell) \leq \langle T's - \ell, u \rangle \qquad \forall\ \ell \in \Lambda\ ,\quad \delta \in \Delta\ , \end{aligned} \tag{8.3}$$

where the equality holds if u_0, s_0 satisfy (8.1) and (8.2).

Defining the convex functionals:

$$F(u) = \phi(Tu - \delta) - g(u) - \langle \ell, u \rangle \qquad \text{potential energy} , \tag{8.4}$$

$$G(s) = \phi^*(s) - g^*(T's - \ell) + \langle s, \delta \rangle \quad \text{complementary energy} , \tag{8.5}$$

from (8.3) we get:

$$F(u) + G(s) \geq 0 \qquad \forall\, u \in U, \quad s \in \Sigma ,$$

and also:

$$F(u_0) + G(s_0) = 0 , \tag{8.6}$$

and

$$F(u_0) = \min_u F(u) \tag{8.7}$$

and

$$G(s_0) = \min_s G(s) , \tag{8.8}$$

that are the extensions, to the convex elastostatic problem, of the well known principles of the minimum potential and complementary energy.

A general free variational principle, generalizing the Hu-Washizu principle of linear elasticity, is obtained by considering the functional:

$$H(u, s, r, e) = \phi(e) + g^*(r) + \langle s, Tu \rangle - \langle s, e + \delta \rangle + \\ - \langle \ell + r, u \rangle \tag{8.9}$$

In fact we have:

$$\begin{array}{l} H \text{ stationary} \Leftrightarrow \\ \text{at } x_0 = (u_0, s_0, r_0, e_0) \end{array} \left\{ \begin{array}{l} 0 \in \partial_u H(x_0) \Leftrightarrow \ell + r_0 = T's_0 \\ 0 \in \partial_s H(x_0) \Leftrightarrow \delta + e_0 = Tu_0 \\ 0 \in \partial_e H(x_0) \Leftrightarrow s_0 \in \partial \phi(e_0) \\ 0 \in \partial_r H(x_0) \Leftrightarrow u_0 \in \partial g^*(r_0) . \end{array} \right. \tag{8.10}$$

Note that H is linear in s, u, convex in e and concave in r, and at a solution we have:

$$H(u_0, s_0, r_0, e_0) = F(u_0) = - G(s_0) .$$

A minimax principle can be formulated by generalizing the Reissner functional to:

$$R(u, s) = - \phi^*(s) - g(u) + \langle s, Tu - \delta \rangle - \langle \ell, u \rangle . \qquad (8.11)$$

Then:

$$R \text{ stationary at } x_0=(u_0, s_0) \Leftrightarrow \begin{cases} 0 \in \partial_u R(x_0) \Leftrightarrow T's_0 - \ell \in \partial g(u_0) \\ 0 \in \partial_s R(x_0) \Leftrightarrow Tu_0 - \delta \in \partial \phi^*(s_0) . \end{cases} \qquad (8.12)$$

Note that R(u, s) is convex in u and concave in s so that:

$$R \text{ stationary} \Leftrightarrow \min_u \max_s R(u, s) = R(u_0, s_0) .$$

Furthermore

$$\min_u R(u, s) = - G(s) ,$$

$$\max_s R(u, s) = F(u) ,$$

so that:

$$\min_u \max_s R(u, s) = R(u_0, s_0) = F(u_0) = - G(s_0) .$$

The generalized Hu-Washizu functional, Reissner functional, the potential energy and the opposite of the complementary energy have then a common value at any solution of the Convex Elastostatic Problem.

REFERENCES

1. Moreau J.J.: La notion de sur-potential et les liaisons unilatérales en élastostatique. C.R. Acad. Sc. Paris, 267 (1968), 954-957.
2. Panagiotopoulos P.D.: Inequality Problems in Mechanics and Applications. Birkhauser, Boston-Basel-Stuttgart 1985.

3. Fenchel W.: On conjugate convex functions. Canad. J. Math., 1 (1949), 73-77.
4. Moreau J.J.: On unilateral constraints, friction and plasticity, in: New Variational Techniques in Mathematical Physics (Ed. Cremonese Roma 1974), C.I.M.E. Bressanone 17-26 giugno 1973, 171-322.
5. Rockafellar R.T.: Convex Analysis. Princeton University Press, Princeton 1970.
6. Romano G., Romano M.: Elastostatics of structures with unilateral conditions on stress and displacement fields, in: Unilateral Problems in Structural Analysis, Springer-Verlag Wien-New York (1985), CISM Courses and Lectures n. 288, Ravello September 22-24, 1983, 315-338.
7. Romano G., Sacco E.: Sul calcolo di strutture non resistenti a trazione. AIMETA VII Congresso Nazionale, Trieste 2-5 ottobre 1984, Sez. V, 217-226.
8. Romano G., Sacco E.: Materiali non resistenti a trazione: equazioni costitutive e metodi di calcolo. Atti Ist. Scienza delle Costruz., Facoltà di Ingegneria, Università di Napoli (1984).
9. Rockafellar R.T.: Directionally lipshitzian functions and subdifferential calculus. Proc. London Math. Soc. 39 (1979), 331-355.
10. Salençon J.: Calcul à la rupture et analyse limite. Presse de l'école National de Ponts et Chaussées, Paris (1983).
11. Romano G., Sacco E.: Analisi limite e collasso plastico per materiali non standard ben posti. Riv. Ital. Geotecnica, 1 (1985), 37-41.

UNILATERAL CONTACT BETWEEN A GEOMETRICALLY NON-LINEAR PLATE AND ELASTIC FOUNDATION. A MODEL OF DELAMINATION PHENOMENA IN A TWO-LAYERED COMPOSITE

D. Shilkrut

Department of Mechanical Engineering, Ben-Gurion University, Beer Sheva

ABSTRACT

The unilateral problem of a circular geometrically non-linear plate which rests on an elastic foundation is investigated. Between the foundation and the plate there exists a connection with a given bounded ultimate (failure) tensile stress. Naturally, a rupture of the above-mentioned connection may occur beginning from some value of external forces and then in some region or regions the contact is destroyed and delamination phenomena (cracks) occur. In the regions of contact the reaction of the elastic foundation is determined by means of Winkler's hypothesis. In the delamination zones no interaction between the plate and the foundation exists. This model of delamination (crack forming) process in a two-layered composite is investigated and some numerical results are given.

GOVERNING EQUATIONS. CONTACT CONDITIONS. THE STATE OF THE ART OF THE PROBLEM.

The differential equations in dimensionless variables and parameters of the Karman-Marguerre type which govern the axisymmetrical thermoelastic deformation (pre-buckling and post-buckling) of a geometrically non-linear, rotationally orthotropic, circular plate may be written as follows [1,2,3 4 and others]:

$$L_k(\omega) = k^2\left(-\frac{\theta^2}{2\rho} + n_T\right); \quad L_k(\theta) = -\frac{m_a}{\rho}\left[\int_0^\rho (q+q_R)\rho\, d\rho + \omega\theta\right] + m_T. \qquad (1)$$

$$(0 \leqslant \rho \leqslant 1).$$

where

$$L_k(\ ) = (\ )'' + (\ )'1/\rho - k^2(\ )/\rho^2; \quad k^2 = E_\phi/E_r;$$

$$\psi^2 = \alpha_\phi/\alpha_r; \quad m_a = 12(1-\mu_r\mu_\phi) = 12(1-k^2\mu_r^{\ 2}),$$

$$E_r\mu_\phi = E_\phi\mu_r; \quad \mu_r+\mu_\phi \leqslant 1. \qquad (2)$$

$$n_T(\rho) = (1-\psi^2)N_T(\rho)/\rho - \psi^2 N_T'(\rho); \quad m_T(\rho)/12 =$$

$$= [1-k^2\mu_r) - k^2\psi^2(1-\mu_r)]M_T(\rho)/\rho + (1+k^2\mu_r\psi^2)M_T'(\rho);$$

$$N_T = \alpha_r a^2\int_{-\frac{1}{2}}^{\frac{1}{2}} T(\rho,\zeta)\,d\zeta/h^2; \quad M_T = \alpha_r a^2\int_{-\frac{1}{2}}^{\frac{1}{2}} \zeta T(\rho,\zeta)\,d\zeta/h^2. \qquad (3)$$

The basic unknowns are the membrane stress function $\omega(\rho)$ and the angle of rotation $\theta(\rho)$ of the normal to the plate's midplane due to the deformation. By these functions all the parameters of the problem are expressed [2,3]. The argument ρ is the dimensionless radial coordinate of the points of the plate. The constant parameters $E_{r,\phi}$, $\mu_{r,\phi}$, and $\alpha_{r,\phi}$ are Young's moduli, Poisson's ratios, and the thermal coefficients in radial and circumferential directions, respectively, when the centers of orthotropy and of the middle plane coincide. These parameters are taken to be independent of the stationary temperature $T(\rho,\zeta)$, which can be a function of ζ and ρ. ζ is the dimensionless coordinate marked off from the middle plane normal to one. a and h are the plate's radius and constant thickness, respectively. $q(\rho)$ and $q_R(\rho)$ are the intensities of the distributed transverse forces and reactions of the foundation, respectively. The usual boundary conditions are the linear forms (4):

$$\omega(0) = \theta(0) = 0; \quad \alpha_1\omega'(1) + \beta_1\omega(1) = \gamma_1;$$

$$\alpha_2\theta'(1) + \beta_2\theta(1) = \gamma_2. \qquad (4)$$

The conditions at $\rho = 0$ are a consequence of the problem's symmetry and the boundness of the basic unknowns in the neighborhood of the center of the plate. The constants α_i and β_i depend on the type of the plate's edge supports. γ_1 define the given edge membrane force in the case of a movable (in the plate's plane) support. The constant γ_2 defines the given edge bending moment acted on hinged supports. These conditions are explained in detail in [2].

The downward deflections $W(\rho)$ are given by the formula:

$$W(\rho) = \xi - \int_0^{\rho} \theta(\rho)d\rho; \; \xi = W(0). \tag{5}$$

The state of the art of the problem presented here is the following. A plate, the deformation of which is described by equations (1) and (2), rests on an elastic foundation. The connection between the plate and its foundation has a given bounded ultimate (failure) tensile stress. Then, when the reaction of the foundation is greater than the above-mentioned ultimate stress of the connection, a rupture takes place and as a result a crack occurs. Thus, the process of delamination begins. In the crack region the reaction of the foundation is assumed to be zero. In regions where the plate-foundation connection is not destroyed (the contact between is completely intact), the reaction of the foundation is of the classic Winkler type. Thus, the contact conditions corresponding to the above-described type of contact can be expressed by the following relation for the foundation's reaction $q_R(\rho)$:

$$q_R(\rho) = \begin{cases} -\lambda W(\rho) \text{ if } W > -\frac{\chi}{\lambda} \\ 0 \text{ if } W < -\frac{\chi}{\lambda}; \; \chi \geq 0. \end{cases} \tag{6}$$

where λ is the coefficient of the rigidity of the foundation. $\chi \geq 0$ is the ultimate (failure) tensile stress of the connection. The first line of (6) corresponds to the regions without cracks; the second one takes place in the crack regions. (Tangential interaction between the plate and the foundation is not taken into account in the approach used here.) Thus, corresponding to the above-mentioned state of the art, there exists a complete connection of the Winkler type or there is no interaction between the plate and the foundation (crack regions). Something intermediary does not exist.

If $\chi = 0$, we have the case when no adhesion between the plate and the foundation exists, i.e., the plate simply rests on the foundation. The problem in question has two non-linearities of different types. One of these is the geometric non-linearity of the plate itself; the second one is due to the phenomenon of separation of the plate from the foundation where the allocation of the regions of separation is unknown. The first results of the investigations of this problem are described in [4]. A similar approach to the problem of delamination of a two-layered composite is considered in [5].

ALGORITHM AND SOME NUMERICAL RESULTS

It is convenient for the numerical solution to present the right hand side of the second equation (1) into another form not containing $W(\rho)$ explicitly. ($W(\rho)$ is introduced into the second equation (1) by means of q_R.) In order to eliminate W from the above-mentioned equation, we can use the following transformation of the integral (7).

$$\int_{\rho_i}^{\rho} \rho q_R d\rho = -\lambda\int_{\rho_i}^{\rho} \rho W d\rho = -\lambda\xi\int_{\rho_i}^{\rho} \rho d\rho + \lambda\int_{\rho_i}^{\rho} \rho\int_{0}^{\rho} \theta ds d\rho =$$

$$= -\frac{\lambda\xi}{2}(\rho^2-\rho_i^2) + \frac{\lambda}{2}[\rho^2\int_0^{\rho}\theta d\rho - \rho_i^2\int_0^{\rho_i}\theta d\rho - \int_{\rho_i}^{\rho}\rho^2\theta d\rho]. \tag{7}$$

where ρ_i is the coordinate determining the beginning of the considered contact region. Expression (7) was obtained by integration by parts of the double integral (7). By means of this expression the right hand side $Q(\rho)$ of the second equation (1), which has different expressions for different distributions of the contact and crack regions, can be written as follows. For example, let there be a full contact in the interval ($0\leq\rho\leq\rho_1<1$); in the region ($\rho_1\leq\rho\leq\rho_2<1$) there is a crack and the interval ($\rho_2\leq\rho\leq 1$) is again a contact region. Then we have (see second equation (1)):

$$Q(\rho) = \frac{m_a}{\rho}\{-\int_0^{\rho}\rho q d\rho + \frac{\lambda}{2}[\xi\rho^2 - \rho^2\int_0^{\rho}\theta d\rho + \int_0^{\rho}\rho^2\theta d\rho] + \omega\theta\} + m_T;$$

$$(0\leq\rho\leq\rho_1<1). \tag{8}$$

In the region of the crack where $q_R \equiv 0$: $(\rho_1\leq\rho\leq\rho_2<1)$

$$Q(\rho) = \frac{m_a}{\rho}\{-\int_0^{\rho}\rho q d\rho + \frac{\lambda}{2}[\xi\rho_1^2 - \rho_1^2\int_0^{\rho_1}\theta d\rho + \int_0^{\rho_1}\rho^2\theta d\rho] + \omega\theta\} + m_T. \tag{9}$$

In the last region, which is a contact one:

$$Q(\rho) = \frac{m_a}{\rho}\{-\int_0^{\rho}\rho q d\rho + \frac{\lambda}{2}[\xi\rho_1^2 - \rho_1^2\int_0^{\rho_1}\theta d\rho + \int_0^{\rho_1}\rho^2\theta d\rho + \xi(\rho^2-\rho_2^2) - \rho^2\int_0^{\rho}\theta d\rho +$$

$$+ \rho_2^2\int_0^{\rho_2}\theta d\rho + \int_{\rho_2}^{\rho}\rho^2\theta d\rho] + \omega\theta\} + m_T; \quad (\rho_2\leq\rho\leq 1) \tag{10}$$

Thus, the governing equations (1) have acquired an integro-differential form instead of the usual differential one. Aside from this, they contain explicitly the additional unknown parameter ξ (see (8)-(10)) which is absent in the usual case (no foundation). The presence of parameter ξ in the governing equation demands a supplementary boundary condition in addition to (4). Such a condition can always be written by taking into account the constraints on W determined by the edge supports. For example, if the plate's edge is supported by an immovable in the Z direction support, then the additional boundary condition is W(1) = 0 and ξ is determined by (5). If any constraints on the motion of the plate's edge in the Z direction do not exist, then the additional boundary condition takes the following form:

$$Q(1) = P \tag{11}$$

where P is the external shear force per unit length of the plate contour. The other two external forces (membrane force and bending moment) which might act on the same contour are taken into account by the right hand sides of the last two conditions (4).

The algorithm of the numerical solutions is based on the deformation map technique described for the classical case (when no foundation exists) in [3,6]. The foundation does not introduce principal changes in this technique. However, it is necessary to take into account the existence of the additional parameter ξ and to build maps for each of its values. On Fig. 1 am example of such a deformation map is given. This is built for the following conditions. $\xi = 1$; $q = 0$; the plate is elastic and thermoisotropic ($k = 1$, $\psi = 1$, see (2)); the plate rests without additional supports on the foundation ($\alpha_1 = 0$, $\beta_1 = 1$, $\alpha_2 = 1$, $\beta_2 = \mu_r$; (see (4)); $\lambda = 20$ and $\chi = \infty$ (the connection between the plate and the foundation is indestructible and we have the classical Winkler's case); $\mu_r = 0.3$; $T \equiv 0$ (the thermal stresses are absent); γ_1 is the dimensionless radial contour membrane force; γ_2 is the dimensionless boundary bending moment.

Conforming to the deformation map technique [3,6], each point of the map is a result of a numerical solution of a Cauchy problem (initial value problem) for the system of equations (1) when the values of the initial parameters $\omega'(\Delta\rho)$ and $\theta'(\Delta\rho)$ are given. The other two initial conditions are determined by (12), which results from the general properties of the multitude of the solutions of (1) (see [6]).

$$\omega(\Delta\rho) = \omega'(\Delta\rho)\Delta\rho/k; \quad \theta(\Delta\rho) = \theta'(\Delta\rho)\Delta\rho/k; \quad (0<\Delta\rho<<1) \tag{12}$$

The initial point is taken $\rho = \Delta\rho$ instead of $\rho = 0$ because some coefficients of equations (1) are irregular when $\rho = 0$. The numerical solution of Cauchy's problem can be realized by the Runge-Kutta method and other step methods. On each step it is necessary to check the value of W at the end point of this step in order to know how to calculate the reaction q_R (by the first or second line of expression (6)) on the next step.

By choosing some values of initial parameters $\omega'(\Delta\rho)$ and $\theta'(\Delta\rho)$ and calculating the values of $\omega(\Delta\rho)$ and $\theta(\rho)$ by means of formulas (12), we can use the Runge-Kutta method to obtain the corresponding values of γ_1

and γ_2 . These values of γ_1 and γ_2 determine a point on the deformation map. By changing the values of $\omega'(\Delta\rho)$ and $\theta'(\Delta\rho)$ we determine another point on the map and so on. This is drawn on the map in Fig. 1 and the maps used by us [2,3,4,6] represent a single parameter family of curves, the parameter of which is $\omega'(\Delta\rho)$. Each curve on the map is a locus of points $\omega'(\Delta\rho)$ = const. where the values of $\theta'(\Delta\rho)$ differ. (It is possible to build different versions of deformation maps, but the principle remains the same [3,6]). The map contains in a condensed form all the qualitative and quantitative information regarding the behavior of the investigated plate loaded by boundary forces γ_1 and γ_2 for the given values of ξ, the characteristics of the foundation, the value of the parameter of the transverse forces, the thermal field and the kind of supports. In short, the map contains the entire volume of information thay may be obtained from the considered boundary value problem of the system (1). The intersection of the curves of the deformation map with different lines determine the behavior of the plate for different types of loading. For example, the intersection of the map with straight lines parallel to the axis $\gamma_2(\gamma_1$ = const.) determines the behavior (axisymmetrical deformation or loss of stability) of the plate under variable (boundary bending moment) when the boundary membrane force is fixed. The intersection with an inclined straight line which passes through the initial point of the map determines the solution of the case when the boundary moment γ_2 depends linearly on boundary radial force γ_1, i.e., γ_1 is applied to the contour of the plate excentrically and not in the middle plane, and so on. It is useful to remember here that the point of the above-mentioned intersection gives the initial conditions for the corresponding initial value problem, the solution of which determines all the parameters of the investigated state of the plate.

Let us consider the case when the isotropic plate connected to the elastic foundation is deformed by boundary bending moments only, as is shown on Fig. 2. Other external forces (except for the reaction q_R in the regions where the connection plate-foundation is not destroyed), and thermal stresses are absent. In this case k=1; q=0; $n_T = m_T = 0$ (see (1), (2)); $\alpha_1=0$; $\beta_1=1$; $\gamma_1=0$; $\alpha_2=1$; $\beta_2 = \mu_r = \mu_r = \mu_\phi = 0.3$; $\gamma_2=M_r(1)$ (see 4)); P=0 (see (11)). Thus the plate does not have any supports except for the foundation itself. Take, for example, $\lambda=20$ and $\chi=40$. In this case according to (6) cracks occur in the regions where $W(\rho) < -2$. (Note that $W(\rho)$ is positive downward.) Based on the corresponding deformation maps built for various values of ξ, auxiliary graphs of γ_2 and Q(1) versus W(1) for different values of ξ can be obtained. Two examples of similar graphs are given in Figs. 3 and 4. In the case $\xi = 1$, the graphs Q(1) versus W(1) does not intersect the horizontal axis. This means that there does not exist any solution for the considered problem when $\xi = 1$ since the additional boundary condition (11) Q(1) = 0 is not satisfied in this case. Contrary to this, when $\xi = 0.5$ the graph Q(1) versus W(1) intersects the axis W(1) twice. Then two solutions exist -- one when $W(1) > -2$ and another when $W(1) < -2$. In the latter case a crack region in the vicinity of the edge exists; in the former one there is no crack on the periphery of the plate and the computations show that there are no cracks in general. For the specific case considered here only one crack region can exist if $\xi > 0$ which is situated in the neighborhood of the plate's contour. The results are described by the graph drawn for $\gamma_2 > 0$ in Fig. 5. This graph consists, in fact, of two distinct parts. The first part (where γ_2 is small) links the origin of the coordinate system and the bent point of the curve where the derivative $d\gamma_2/d\xi = \infty$. For all the states of the plate which correspond to this region of the curve there is a complete connection between the plate and the foundation since $W(\rho) > -2$ everywhere. The second region of the curve drawn on Fig. 5 starts from the above-mentioned bent point (where $\gamma_2/m = 4.5$). For $\gamma_2/m > 4.5$ each equilibrium state contains a crack situated in a belt adjoined to the contour $\rho = 1$ where $W(1) < -2$. (A number of such equilibrium states are shown in Fig. 6.) The dependence of the length ℓ of the crack γ_2 has a facing character (see Fig. 7 where ℓ is given relative to the plate's radius). Consequently, it is impossible to completely tear off the plate from the foundation by means of contour bending moment only.

Another numerical investigation is concerned with the bending of the same plate when the rigidity of the elastic foundation is $\lambda = 1$, and $\chi = 1$. The results are given in Fig. 8. In this figure by line 1 the graph γ_2/m versus ξ for positive and negative values of γ_2 when $\xi > -1$ is drawn. For $\gamma_2/m > 0.82$ (approximately) a crack occurs near the plate's edge. The graph of the length of this crack versus ξ is given by line 2. A crack region in the central part of the plate when $\xi < -1$ (because $\lambda = \chi = 1$ (see (6)) is possible. The graphs of γ_2/m and ℓ versus ξ for $\xi < -1$ are presented in Fig. 8 and denoted by 3 and 4, respectively.

We shall end the paper with the following remark. The bending process by contour moments is equivalent to a thermoelastic problem when the temperature field T is an odd function of Z and does not depend on ρ and when $\psi = k = 1$. Then $N_T \equiv n_T \equiv m_T \equiv 0$ (see (1) and (3)), but $M_T = \text{const.} \neq 0$. In this case all the thermal terms in system (1) are equal to zero and the influence of the temperature field is transferred by means of thermal edge moment M (see [2,3,6]).

REFERENCES

1. Stavsky, Y.: Non-Linear Axisymmetric Deformations of Heterogeneous Shells of Revolution, in: Contribution to Mechanics (Ed. D. Abir), Pergammon Press, Oxford 1970, 181-194.
2. Shilkrut, D.: Stability and vibrations of geometrically nonlinear cylindrically orthotropic circular plates, J. Appl. Mech., 5 (1984), 354-360.
3. Shilkrut, D.: The deformation map as a means for investigating the behaviour of deformable non-linear bodies, Mechanics Research Communications, 9 (1982), 77-84.

4. Shilkrut, D.: Axisymmetrical deformation of geometrically nonlinear cylindrically orthotropic circular plates on elastic foundation, Israel J. of Technology, 21 (1983), 85-90.
5. Ascione, L. and Bruno, D.: On the Delamination Problem of Two-Layer Plate, in: Unilateral Problems in Structural Analysis (Ed. G. Del-Piero and F. Maceri), Springer Verlag, Wien, N.Y., CISM, Udine Courses and Lectures, No. 388 (1985), 1-15.
6. Shilkrut, D.: Investigations of axisymmetric deformation of geometrically non-linear rotationally orthotropic circular plate, Int. J. Non-Linear Mech., 18(2) (1983), 95-118.

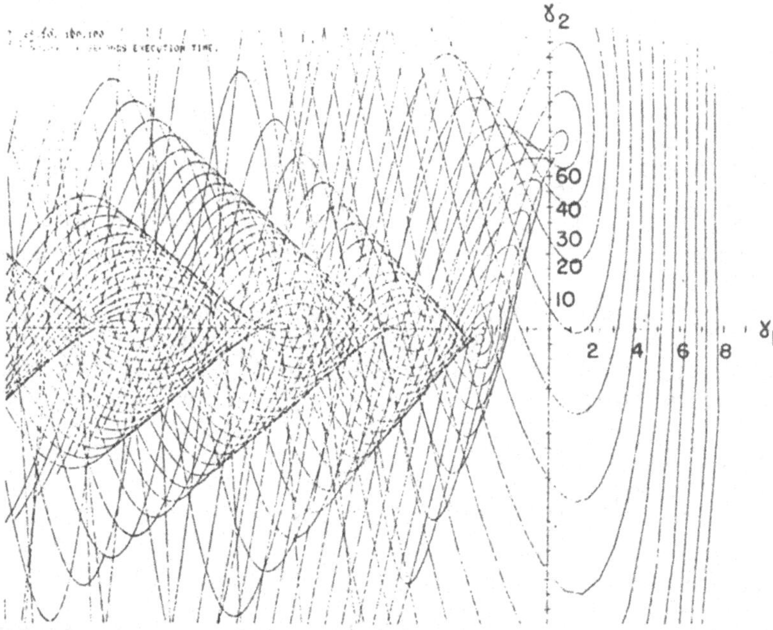

Fig. 1

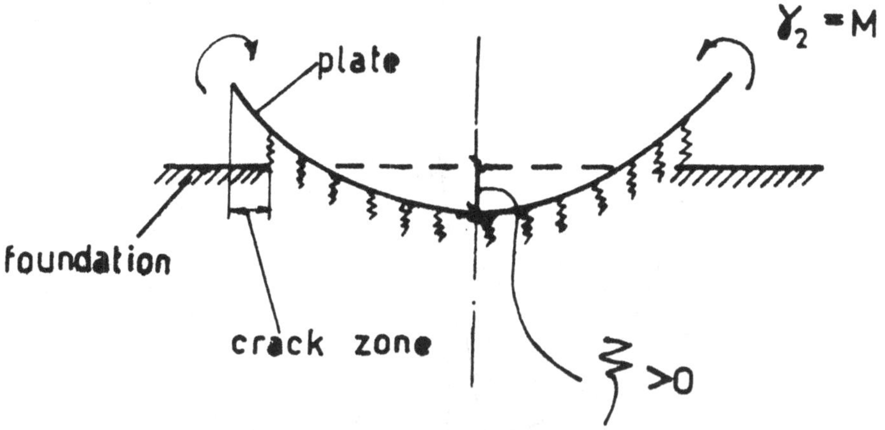

Fig. 2

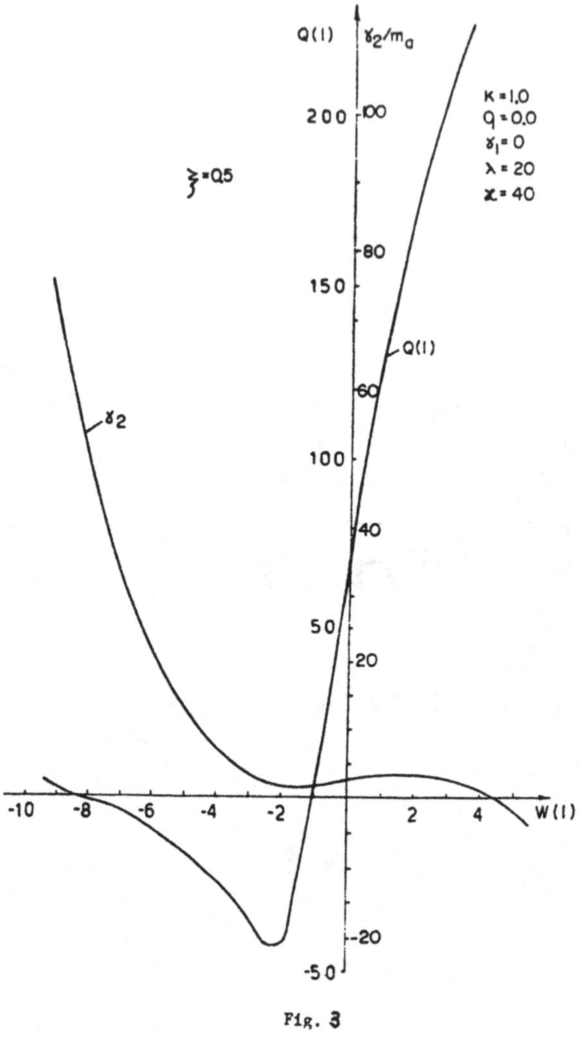
Q(1)
γ2/ma
κ=1.0
q=0.0
γ1=0
λ=20
χ=40
ξ=0.5
200
150
100
50
-50
100
80
60
40
20
-20
Q(1)
γ2
-10
-8
-6
-4
-2
2
4
W(1)

Fig. 3

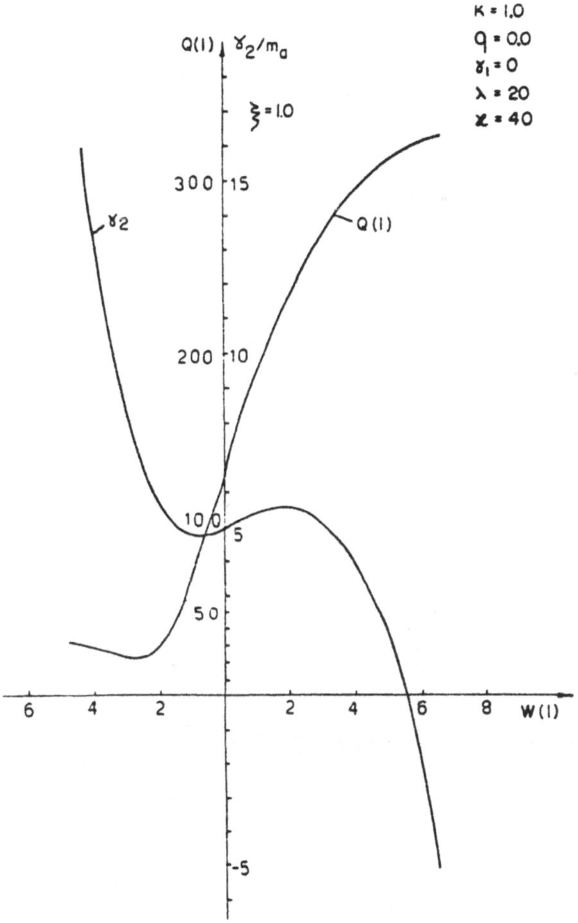

κ = 1.0
q = 0.0
γ₁ = 0
λ = 20
ϰ = 40
Q(1)
γ2/ma
ξ = 1.0
300
15
200
10
100
5
50
-5
γ2
Q(1)
6
4
2
2
4
6
8
W(1)

Fig. 4

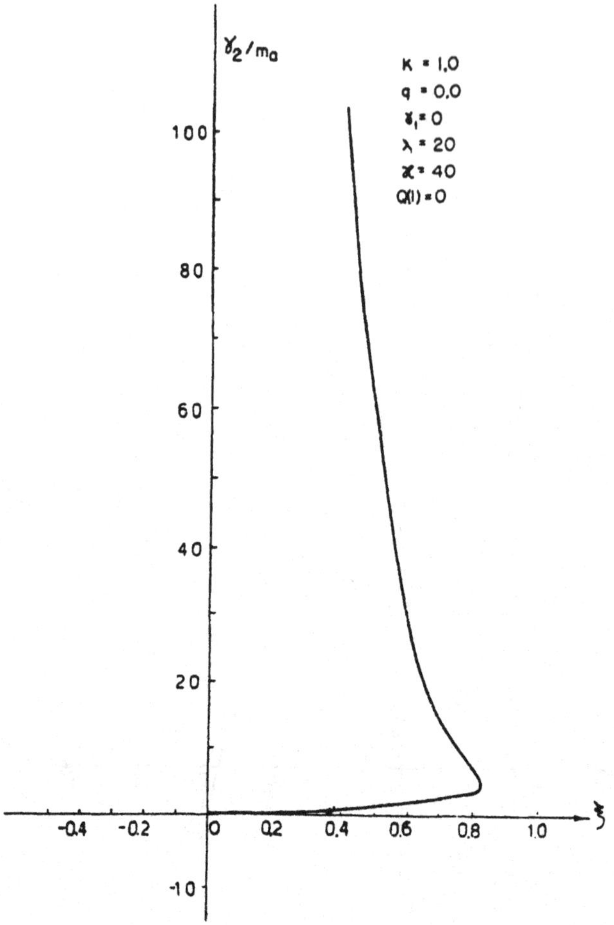

γ_2/m_a
κ = 1,0
q = 0,0
γ_1 = 0
λ = 20
χ = 40
Q(1) = 0
100
80
60
40
20
-10
-0.4
-0.2
0
0.2
0.4
0.6
0.8
1.0
ξ

Fig. 5

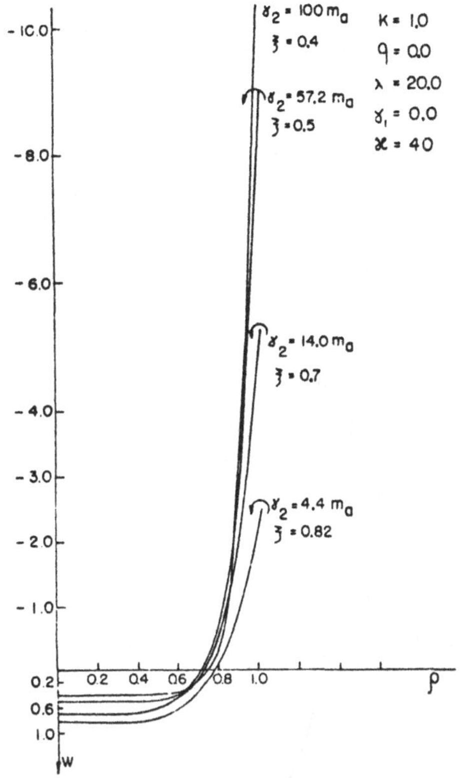
$\gamma_2 = 100\, m_a$
$\xi = 0.4$
$K = 1.0$
$q = 0.0$
$\lambda = 20.0$
$\gamma_1 = 0.0$
$\varkappa = 40$
$\gamma_2 = 57.2\, m_a$
$\xi = 0.5$
$\gamma_2 = 14.0\, m_a$
$\xi = 0.7$
$\gamma_2 = 4.4\, m_a$
$\xi = 0.82$
-10.0
-8.0
-6.0
-4.0
-3.0
-2.0
-1.0
0.2
0.6
1.0
w
0.2 0.4 0.6 0.8 1.0
ρ

Fig. 6

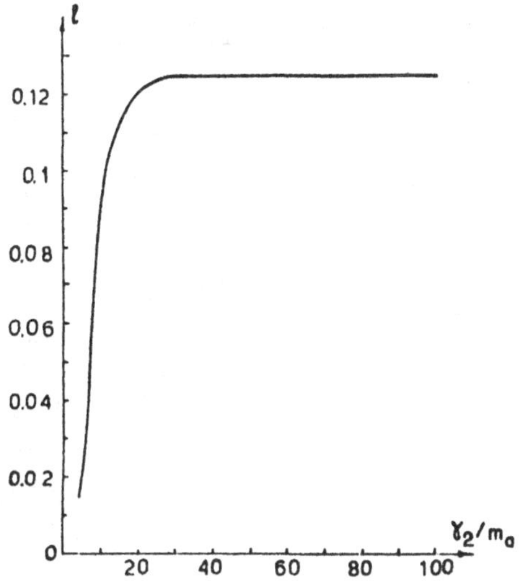
l
0.12
0.1
0.08
0.06
0.04
0.02
0
20 40 60 80 100
γ_2/m_a

Fig. 7

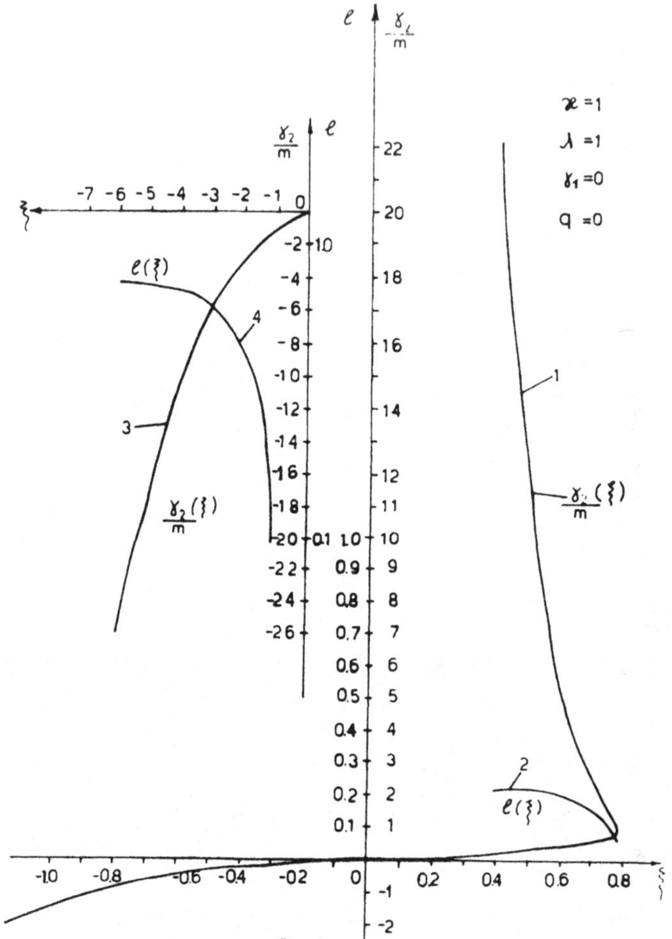

Fig 8

LIST OF PARTICIPANTS

Alessandro BARATTA — Istituto di Costruzioni — Facoltà di Architettura — Via Monteoliveto, 3 — 80134 NAPOLI — ITALY.

Luigi BIOLZI — Istituto di Meccanica Teorica ed Applicata — viale Ungheria, 43 — 33100 UDINE — ITALY.

D. BISCHOFF — Institut für Baumechanik und Numerische Mechanik — Callinstrasse 32 D-3000 HANNOVER — WEST GERMANY.

Silvia BRICCIOLI BATI — Dipartimento di Costruzioni — Facoltà di Architettura — Piazza Brunelleschi, 6 — 50121 FIRENZE — ITALY.

Domenico BRUNO — Dipartimento di Strutture — Università della Calabria — 87036 COSENZA — ITALY.

Elio CABIB — Istituto di Meccanica Teorica ed Applicata — viale Ungheria, 43 — 33100 UDINE — ITALY.

Mario COMO — Istituto di Tecnica delle Costruzioni — Facoltà di Ingegneria — via Claudio, 21 — 80125 NAPOLI — ITALY.

Alain CURNIER — Départment de Mécanique — Ecole Polytechnique Fédérale de Lausanne — Ecublens ME — CH — 1015 LAUSANNE — SWITZERLAND.

Cesare DAVINI — Istituto di Meccanica Teorica ed Applicata — viale Ungheria, 43 — 33100 UDINE — ITALY.

Gianpietro DEL PIERO — Istituto di Meccanica Teorica ed Applicata — viale Ungheria, 43 33100 UDINE — ITALY.

Salvatore DI PASQUALE — Dipartimento di Costruzioni — Facoltà di Architettura — Piazza Brunelleschi, 6 — 50121 FIRENZE — ITALY.

Michel FREMOND — Laboratoire Central des Ponts et Chaussées — 58, Boulevard Lefebvre 75732 PARIS — FRANCE

Fabio GASTALDI — Laboratorio di Analisi Numerica del CNR — 27100 PAVIA — ITALY.

Michel JEAN — Institut de Mathématiques — Université des Sciences et Techniques du Languedoc — Place Eugène Bataillon — 34060 MONTEPELLIER — FRANCE.

Anders KLARBRING — Institute of Technology — Department of Mechanical Engineering S-581 83 LINKOPING — SWEDEN.

Franco MACERI — Dipartimento di Ingegneria Civile Edile — II Università di Roma — Via Orazio Raimondo — 00173 ROMA — ITALY.

Helge-Otmar MAY — Institute for Technical Mechanics — RWTH Aachen — 51 AACHEN 51 AACHEN — WEST GERMANY.

Luciano NUNZIANTE — Istituto di Scienza delle Costruzioni — Facoltà di Ingegneria Piazzale Tecchio — 80125 NAPOLI — ITALY.

P.D. PANAGIOTOPOULOS — School of Technology — Aristotelian University — THESSALONIKI — GREECE.

Michele PARADISO — Dipartimento di Costruzioni — Facoltà di Architettura — Piazza Brunelleschi, 6 — 50121 FIRENZE — ITALY.

Michel RAOUS — Laboratoire de Mécanique et d'Acoustique — CNRS — 31 ch. Joseph-Aiguier — BP71 — 13277 — MARSEILLE — FRANCE.

Giovanni ROMANO — Istituto di Scienza delle Costruzioni — Facoltà di Ingegneria Piazzale Tecchio — 80125 NAPOLI — ITALY.

Elio SACCO — Istituto di Scienza delle Costruzioni — Facoltà di Ingegneria — Piazzale Tecchio — 80125 NAPOLI — ITALY.

Dov SHILKRUT — Mechanical Engineering Department — Ben-Gurion University of the Negev — BEER SHEVA 84 105 — ISRAEL.

Giacomo TEMPESTA — Dipartimento di Costruzioni — Facoltà di Architettura — Piazza Brunelleschi, 6 — 50121 FIRENZE — ITALY.

Thomas WICKS — Institute of Applied Mathematics — University of Missouri — ROLLA, Missouri 65401 — U.S.A.

CISM VOLUME No. 304

UNILATERAL PROBLEMS IN STRUCTURAL ANALYSIS - 2

ERRATA CORRIGE

p. 47 The title should be:
. . . TO INELASTIC CONTACT PROBLEMS

p. 259 The title should be:
. . . EN TERMES D'INEQUATIONS VARIATIONNELLES

Printed in the United States
By Bookmasters